Trace Elements in Coal

To Wyn, Patricia and Philip

Trace Elements in Coal

Dalway J. Swaine
CSIRO, Division of Coal Technology
North Ryde, NSW, Australia

Butterworths
London Boston Singapore Sydney Toronto Wellington

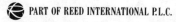 PART OF REED INTERNATIONAL P.L.C.

All rights reserved. No part of this publication may be reproduced in any material form (including photocopying or storing it in any medium by electronic means and whether or not transiently or incidentally to some other use of this publication) without the written permission of the copyright owner except in accordance with the provisions of the Copyright, Designs and Patents Act 1988 or under the terms of a licence issued by the Copyright Licensing Agency Ltd, 33–34 Alfred Place, London, England WC1E 7DP. Applications for the copyright owner's written permission to reproduce any part of this publication should be addressed to the Publishers.

Warning: The doing of an unauthorised act in relation to a copyright work may result in both a civil claim for damages and criminal prosecution.

This book is sold subject to the Standard Conditions of Sale of Net Books and may not be re-sold in the UK below the net price given by the Publishers in their current price list.

First published 1990

© Butterworth & Co. (Publishers) Ltd, 1990

British Library Cataloguing in Publication Data

Swaine, Dalway J.
   Trace elements in coal
   1. Coal. Structure & chemical properties
   I. Title
   553.2′4

ISBN 0-408-03309-6

Library of Congress Cataloging-in-Publication Data

Swaine, Dalway J.
   Trace elements in coal / Dalway J. Swaine.
     p.   cm.
   Includes bibliographical references.
   ISBN 0-408-03309-6
   1. Coal--Analysis.   2. Trace elements--Analysis.
   I. Title.
TP326.A8S85   1990   89–71215
662.6′22994--dc20   CIP

Photoset by Scribe Design, Gillingham, Kent
Printed and bound in Great Britain by Courier International Ltd, Tiptree, Essex

Preface

> That which I have neither seen, nor carefully considered after reading or hearing of, I have not written about.
>
> Agricola (1556)

This book encompasses all aspects of trace elements in coal and is based on a consideration of a vast array of publications. As well as direct references to trace elements in coal, relevant references from allied fields are given where necessary for explanation or comparison. Much of the information is from the past decade, in keeping with the upsurge of interest in most countries where coal is produced or used. However, pertinent earlier work, especially that of the pioneer V.M. Goldschmidt (Swaine, 1988), has not been neglected. Perhaps the main impetus for knowledge of trace elements in coal is now the environmental aspects arising from the mining, treatment and usage of coal. This book is intended to be a concise, yet complete, reference work for the specialist and for the tyro, and will surely lessen what Professor J. Lukasiewicz has termed 'the degree of ignorance' of published information. It is a *vade mecum* for coal technologists and for coal scientists.

The coverage of the origin and mode of occurrence of trace elements in coal and of some aspects of their relevance is necessarily limited. Indeed, each topic merits a separate book! However, sufficient information is given to provide a proper background for the chapters dealing in detail with trace elements in coal.

The preparation of this work required the cooperation of many individuals and libraries. It is a pleasure to thank all those who were helpful, in particular various colleagues in CSIRO. The high quality of our results depended on the skill of Miss M.C. Clark, W.C. Godbeer, N.C. Morgan and K.W. Riley of CSIRO, North Ryde, and J.J. Fardy and the late R.E. Porritt of CSIRO, Lucas Heights. I am grateful for the assistance of the CSIRO librarians, Miss D.K. Myers, Miss A.L. Stevenson and Mrs R. Thieben. An impetus to trace-element research in Australia was provided by grants from the National Energy Research, Development and Demonstration Program to CSIRO, SECV and BHP; in particular, I appreciate the encouragement of Dr K.D. Lyall. Financial support towards the preparation of the manuscript was gratefully received from the National Energy Research, Development and Demonstration Program and from the Australian Coal Association. The help of Butterworths is acknowledged, especially the ready cooperation of Patricia Horwood and Karen Panaghiston. Finally, I am most grateful to my wife for her patient support and help with proof reading.

Dalway J. Swaine
CSIRO Division of Coal Technology
North Ryde, NSW, Australia
Professorial Fellow
Department of Inorganic Chemistry
The University of Sydney
NSW, Australia

Contents

Preface	v
List of abbreviations	xi
1 Introduction	**1**
1.1 Relevant books and reviews	4
1.2 Background information on coal areas	5
1.3 Geochemical cycling	5
2 Origin of trace elements in coal	**8**
2.1 The formation of peat	9
2.2 Geological and geochemical aspects of coal seams	15
2.3 Geology of Australian coals	17
2.4 Coal mining in Australia	18
2.5 The constitution of coal	19
2.6 The history of trace elements in coal	22
3 Mode of occurrence of trace elements in coal	**27**
3.1 Trace element–organic matter associations	28
3.2 Methods for determining the total amount of mineral matter	37
3.3 The identification of minerals in coal	38
3.4 Trace element–mineral matter associations	39
3.5 Silicon-rich minerals	40
3.6 Carbonate minerals	41
3.7 Sulfide minerals	43
3.8 Oxides	46
3.9 Sulfates	47
3.10 Phosphates	47
3.11 Lignites and brown coals	47
4 Methods of analysis	**50**
4.1 Bulk sampling	51
4.2 Preparation of sample for analysis	52
4.3 Contamination	54
4.4 Reference materials	55
4.5 Chemical methods	56

	4.6	Atomic emission spectroscopy	58
	4.7	Inductively coupled plasma atomic emission spectrometry	61
	4.8	Atomic absorption spectrometry	63
	4.9	Neutron activation analysis	66
	4.10	X-ray fluorescence spectrometry	68
	4.11	Spark source mass spectrometry	70
	4.12	Other methods	71
	4.13	Consideration of results	72

5 Contents of trace elements in coals — 77

5.1	General comments	77
5.2	Detailed information for individual trace elements	78
	5.2.1 Antimony	79
	5.2.2 Arsenic	79
	5.2.3 Barium	83
	5.2.4 Beryllium	86
	5.2.5 Bismuth	89
	5.2.6 Boron	89
	5.2.7 Bromine	92
	5.2.8 Cadmium	96
	5.2.9 Caesium	99
	5.2.10 Chlorine	99
	5.2.11 Chromium	101
	5.2.12 Cobalt	103
	5.2.13 Copper	106
	5.2.14 Fluorine	109
	5.2.15 Gallium	113
	5.2.16 Germanium	115
	5.2.17 Gold	119
	5.2.18 Hafnium	120
	5.2.19 Indium	122
	5.2.20 Iodine	122
	5.2.21 Lead	123
	5.2.22 Lithium	124
	5.2.23 Manganese	128
	5.2.24 Mercury	128
	5.2.25 Molybdenum	134
	5.2.26 Nickel	136
	5.2.27 Niobium	137
	5.2.28 Phosphorus	140
	5.2.29 Platinum group elements	140
	5.2.30 Radium	142
	5.2.31 Rare earth elements	143
	5.2.32 Rhenium	146
	5.2.33 Rubidium	146
	5.2.34 Scandium	146
	5.2.35 Selenium	148
	5.2.36 Silver	154
	5.2.37 Strontium	154

5.2.38 Tantalum	156
5.2.39 Tellurium	156
5.2.40 Thallium	160
5.2.41 Thorium	160
5.2.42 Tin	164
5.2.43 Titanium	164
5.2.44 Tungsten	166
5.2.45 Uranium	170
5.2.46 Vanadium	171
5.2.47 Yttrium	174
5.2.48 Zinc	174
5.2.49 Zirconium	178

6 Comparisons of coal with shale and soil — 184

   6.1 Comparisons of coal with soil — 184
   6.2 Comparisons of coal with shale — 184

7 Variations within seams — 187

   7.1 Geographical variations — 187
   7.2 Stratigraphic variations — 188
   7.3 Vertical variations — 188
   7.4 Lateral variations — 189

8 Radioactivity and coal — 192

9 Relevance of trace elements in coal — 196

   9.1 Effects of beneficiation — 196
   9.2 Seam correlation and discrimination — 198
   9.3 Degree of marine influence — 200
   9.4 Environmental aspects — 201
      9.4.1 Mining and related operations — 202
      9.4.2 Rehabilitation after mining — 202
      9.4.3 Combustion for power production — 203
      9.4.4 Deposition of trace elements from the atmosphere — 206
   9.5 Health aspects — 210
   9.6 Coal as a possible source of metals and nonmetals — 212

10 Concluding remarks — 215

11 References — 218

Abbreviations

AAS	Atomic Absorption Spectrometry
AC	Alternating Current
AES	Atomic Emission Spectroscopy
AFS	Atomic Fluorescence Spectrometry
AS	Australian Standard
ASTM	American Society for Testing and Materials
BHP	The Broken Hill Proprietary Company Limited, Australia
BS	British Standard
CSIRO	Commonwealth Scientific and Industrial Research Organization
DC	Direct Current
DCP	Direct Current Plasma Spectrometry
EAAS	Electrothermal Atomic Absorption Spectrometry
EC	European Community
EDX	Energy Dispersive X-ray
EMPA	Electron Microprobe Analysis
EPR	Electron Paramagnetic Resonance Spectroscopy
ESR	Electron Spin Resonance Spectroscopy
et al.	And others
FAAS	Flame Atomic Absorption Spectrometry
FTIR	Fourier Transform Infrared Spectroscopy
ICPAES	Inductively Coupled Plasma Atomic Emission Spectrometry
ICPMS	Inductively Coupled Plasma Mass Spectrometry
IDSSMS	Isotope Dilution Spark Source Mass Spectrometry
IEA	International Energy Agency
INAA	Instrumental Neutron Activation Analysis
IR	Infrared Spectroscopy
KHM	Kol-Hälsa-Miljö, Swedish Coal, Health and Environment Project
LTA	Low Temperature Ashing using an oxygen plasma

NAA	Neutron Activation Analysis
NBS	National Bureau of Standards, USA. The NBS has been renamed The National Institute of Standards and Technology (NIST))
NCB	National Coal Board, UK
NERDDP	National Energy Research, Development and Demonstration Program, Australia
NMR	Nuclear Magnetic Resonance Spectroscopy
NSW	New South Wales, Australia
NZ	New Zealand
OFS	Orange Free State, South Africa
PGE	Platinum Group Elements (Ru,Rh,Pd,Os,Ir,Pt)
PIGE	Proton Induced Gamma-ray Emission
PIXE	Proton Induced X-ray Emission Spectroscopy
ppb	Part per billion (1 in 10^9; $ng\,g^{-1}$)
ppm	Part per million (1 in 10^6; $\mu g\,g^{-1}$; $mg\,kg^{-1}$)
ppt	Part per trillion (1 in 10^{12}; $pg\,g^{-1}$)
PTFE	Polytetrafluoroethylene
Qnsld	Queensland, Australia
REE	Rare Earth Elements (La–Lu)
RNAA	Radiochemical Neutron Activation Analysis
SA	South Australia
SABS	South African Bureau of Standards
SECV	State Electricity Commission of Victoria, Australia
SEM	Scanning Electron Microscopy
SIMS	Secondary Ion Mass Spectroscopy
SRM	Standard Reference Material, NBS
SSMS	Spark Source Mass Spectrometry
STEM	Scanning Transmission Electron Microscopy
TEM	Transmission Electron Microscopy
TEMS	Thermal Emission Mass Spectrometry
UK	United Kingdom
USA	United States of America
USGS	United States Geological Survey
USSR	Union of Soviet Socialist Republics
Vic.	Victoria, Australia
WA	Western Australia
XPS	X-ray Photoelectron Spectroscopy
XRD	X-ray Diffraction
XRFS	X-ray Fluorescence Spectrometry

| per
% per cent
~ approximately
< less than
> greater than

Chapter 1

Introduction

Although there has been some interest in trace elements in coal for about 100 years, intensive work started about 50 years ago with investigations by V.M. Goldschmidt and collaborators at the University of Göttingen (Goldschmidt, 1935, 1937). The earliest references to trace elements in Australian coals were the detection of several elements in ashes from three Victorian brown coals (Sinnatt and Baragwanath, 1938) and the determination of gallium and germanium in flue dusts from several coals (Cooke, 1938). Among early workers on relatively large numbers of coal samples were Zubovic and coworkers in the United States, Clark and Swaine in Australia, Yudovich and others in the Soviet Union and Gluskoter and coworkers in Illinois, USA. The main centres of continuing investigations during the past 30 years have been Reston and Denver (US Geological Survey), Champaign-Urbana (Illinois State Geological Survey) and North Ryde, Australia (CSIRO). Of course, good work, albeit on a lesser scale, has been carried out elsewhere, for example in the United Kingdom, Germany, the Soviet Union, Canada, South Africa and Belgium.

My aim is to assess relevant data for trace elements in all coals, most of the emphasis being on separate sections for the elements arranged alphabetically. The term 'trace' is preferred to 'uncommon' or 'rare' and the distinction between 'minor' and 'trace' will not be made. Trace elements are defined as those elements with concentrations in most coals in the range up to about 1000 ppm in dry coal. 'As expected for a naturally occurring material formed under varying conditions over a long period, coal contains most of the elements in the Periodic Classification.' (Swaine, 1985a.) Medvedev and Batrakova (1959) regard 62 of the elements as trace or rare, while Swaine (1981a) referred to CSIRO data for 66 trace elements in coals from various parts of Australia. Although He, Ar and Ne occur in coals (Krejci-Graf, 1983), these gases will not be referred to again. Although there are published data for trace elements in coal ash, it is considered that proper comparisons should be made on a coal basis. Hence, the stress is on data in coal. There are still statements inferring that ash is an inherent part of coal, when mineral matter is what is meant. In an attempt to overcome this error, one should always refer to mineral matter in coal and to ash yield, which is derived from heating coal. However, passing references will be made to data for coal ash. In some cases, unusual coals have been studied and their high trace-element contents have not been delineated from values commonly found in other coals. Hence, an attempt will be made to ascertain the ranges of values commonly found in coals, avoiding the concept of a normal coal, which is meaningless. Results will be given as parts per million (ppm) which is the unit usually preferred by geochemists, although analytical chemists now use $\mu g\,g^{-1}$. The term 'heavy metal' is often used in environmental science, but it is most unsatisfactory because some important elements are not metals and are not heavy!

Although the current interest in environmental aspects of trace elements, for example their redistribution during coal cleaning and during combustion for power production, is the probable reason for continuing work on trace elements in coal, there are also several other reasons. During the mining of coal, particularly surface mining, some trace elements may be mobilized, especially under oxidizing conditions which affect pyrite, thereby producing acidic conditions (Swaine, 1978a). The resulting redistribution of trace elements may produce changes in the concentrations of some elements in nearby surface and underground waters and in the overburden. These changes should be considered when rehabilitation of an area is being undertaken, especially in the case of pastures and grazing animals, where deficiencies or excesses of some elements may be in question. Trace elements in coal have been relevant to several practical situations. An example is the requirement of less than 2 ppm boron in carbon used as a moderator in nuclear reactors. This means that the coal used must be even lower in boron (Hutcheon, 1953). According to Khan and Sen (1959), low boron coke (0.3 ppm B) suitable for making reactor graphite can be prepared by treating coke by chlorination, hydrofluorination and freonation. The interest in boron in certain steels is due to its effects on mechanical properties, and the possibility of stabilizing the nitrogen content (Borrowdale, Jenkins and Shanahan, 1959). Some trace elements, for example arsenic and phosphorus, may cause embrittlement in steel (Kurmanov et al., 1957; Steven and Balajiva, 1959) and relatively low concentrations of phosphorus are often demanded in metallurgical cokes. Traces of gallium and germanium in steel derived from coke were considered to be unimportant (Josefsson, 1954). There have been reports of arsenic in beer due to the hops having been dried in coal-fired kilns (Duck and Himus, 1951), but other reports have stated that the amounts of arsenic were too low to be dangerous (Moon, 1901; Jones and Dawson, 1945; Johnson, 1947). The main source of arsenic in beer was traced to glucose and invert sugars derived from sulfuric acid (Manchester Brewers' Association Commission, 1901). Chicory dried by coke firing may contain some arsenic (Atkinson, Dickinson and Harris, 1950). There should be no meaningful contamination of hops or chicory dried in coal- or coke-fired kilns if suitable low-arsenic coals are used. It is important to realize that the arsenic content of coals is variable, a fact that was not appreciated in a famous law case in the United Kingdom, where a person was hanged on evidence regarding the arsenic content of ash from a domestic fire (Casswell, 1961). Arsenic is said to have poisoned some pastures in a region of heavy coal burning (Haught, 1954) and bees were affected by arsenic from an electricity-generating plant in Czechoslovakia (Svoboda, 1958). In these two instances, it is probable that particle-attenuation devices (for example electrostatic precipitators) were not used.

Coal burnt in spreader stoker-fired boilers and in chain-grate stoker-fired boilers may give rise to phosphatic deposits, high in boron, arsenic and other trace elements (Brown and Swaine, 1964). Among compounds identified on the water tubes were boron phosphate and boron arsenate (Swaine and Taylor, 1970). Some high-chlorine coals may give rise to troublesome boiler deposits under certain conditions of firing. It can be assumed that coals with less than 3000 ppm Cl are unlikely to cause such problems (Wandless, 1958) but higher concentrations of chlorine must always be considered in relation to possible corrosion problems during coal usage (Meadowcroft and Manning, 1983).

Fluorine in coal may cause some etching in direct coal-fired glass annealing furnaces (Horton, 1950), and bricks and mortar may be affected by fluorine-containing gases in annular kilns (Keil and Gille, 1950). Malt dried in a coal-fired kiln may pick up some fluorine, but most is found in the spent grains (Webber and Taylor, 1952). Much attention has been given to coal and the carbonization and combustion products of coal as possible sources of several elements, especially germanium, which Goldschmidt (1935) had suggested could be produced commercially from coal. Perhaps the best coal-derived source of germanium is flue dust (Thompson and Musgrave, 1953). Some germanium has been produced commercially in the United Kingdom, the Soviet Union and Japan from coal flue dusts. Gallium was extracted from coal flue dusts in the United Kingdom (Morgan and Davies, 1937), and Inagaki (1956) extracted gallium from coal flue dusts and from gas liquor. Goldschmidt (1954) suggested that a 'study of coal ashes for uranium might also be worthwhile'. Berkovitch (1956) considered that coal might also be a possible source of selenium, and Kuhl (1957) has suggested that Polish coal ash could be a possible source of cobalt. It has been stated that 'economic factors must determine whether coal will ever be used as a commercial source of a particular trace element' (Swaine, 1962a).

Flyash is a major by-product of coal combustion and may be defined as 'the solid material extracted by electrical or mechanical means from the flue gases of a boiler fired with pulverized coal' (AS, 1971). Perhaps the main use of flyash is as a partial replacement of cement in concrete, but there are also many other uses (Reid, 1981; Swaine, 1981b).

Flyash is a potential source of some trace elements for plants and hence flyash has a role in soil fertilization. Under some conditions manganese could be harmful (Rees and Sidrak, 1956) and boron could affect plant growth (Holliday et al., 1958), but such effects decrease as weathering proceeds (Jones and Lewis, 1960). Before adding flyash to soils, relevant experiments should be carried out to ensure that harmful amounts of trace elements would not be available to plants or to grazing animals (Page, Elseewi and Straughan, 1979). Flyash from brown coal is suitable as a liming agent for calcium-deficient soils (Endell, 1958).

The relative importance of trace elements in coal in connection with coal usage is not easily established and depends on several variables including concentration in coal and biological availability. The results of two assessments are shown in Table 1.1. PECH (1980) groups the elements

4 Introduction

Table 1.1 Assessments of environmental significance of trace elements in coal

PECH (1980)					Swaine (1982)	
(a)	(b)	(c)	(d)	(e)	(i)	(ii)
As	Cr	Ba	Po	Ag	As	B Sb
B	Cu	Br	Ra	Be	Cd	Be Sn
Cd	F	Cl	Rn	Sn	Cr	Cl Th
Hg	Ni	Co	Th	Tl	F	Co Tl
Mo	V	Ge	U		Hg	Cu U
Pb	Zn	Li			Ni	Mn V
Se		Mn			Pb	Mo Zn
		Sr			Se	

Note: see text for categories (a)–(e), (i), (ii).

into six categories, namely, (a) elements of greatest concern; (b) those of moderate concern; (c) those of minor concern; (d) radioactive elements 'generally considered to be of minor concern, but adequate information is lacking for a proper assessment'; (e) those of concern but 'with negligible concentrations in coal and coal residues'; and (f) elements of 'no immediate concern'. Swaine (1982) has grouped the elements into (i) those of 'prime environmental interest'; and (ii) those 'that could be of environmental interest'. In general, there is a measure of agreement in these assessments, which is perhaps surprising as they are based on somewhat difficult criteria.

Most of the reported possible adverse effects of trace elements in coal are unlikely to occur with modern methods of coal utilization. The following statement by McCarroll (1980) is appropriate: 'It can be concluded that with presently available control technology, no detectable adverse health effects will be observed associated with increased coal combustion.'

1.1 Relevant books and reviews

There are many books on the formation and general properties of coal, for example Raistrick and Marshall (1939), Moore (1947), Francis (1961), van Krevelen (1961). Schobert (1987) gives a general introduction to coal and its uses, while Berkowitz (1979) gives an introduction to coal technology. Coal utilization is described in detail by Lowry (1945, 1963), later expanded by Elliott (1981) into a comprehensive survey of all aspects of coal, including useful references to trace elements and mineral matter. Fundamental aspects of coal science, including coal origins, physical and chemical structure, reactivity and reactions, and coal as an organic rock, are covered well by Gould (1966). Other useful reference books are on coal petrology (Stach *et al.*, 1982), on the geochemistry of coal (Bouska, 1981), on the chemistry of low-rank coals (Schobert, 1984) and on the geology of Gulf Coast lignites (Finkelman and Casagrande, 1986). Information on trace elements in coals, especially those from eastern

Europe, has been compiled by Valkovic (1983) and there is information on Soviet coals in Yudovich (1978) and in Yudovich, Ketris and Merts (1985).

Environmental aspects of coal are dealt with by PECH (1980) and KHM (1983) and trace elements in nutrition are treated in detail by Underwood (1977).

Given (1984) has dealt with the organic geochemistry of coal, especially the relevance of recent studies to its structure. The prime reference to the mode of occurrence of trace elements in coal is the report on the thorough investigation by Finkelman (1981a). There are several reviews on trace elements in coal, including those by Briggs (1934), Gibson and Selvig (1944), Clark and Swaine (1962a), Swaine (1962a), Bethell (1963), Abernethy and Gibson (1963), Abernethy, Peterson and Gibson (1969), Durie and Swaine (1971), Yudovich *et al.* (1972), Swaine (1975), Gluskoter, Shimp and Ruch (1981) and Krejci-Graf (1983). References to US literature have been compiled by Averitt *et al.* (1972, 1976) and there is a more comprehensive bibliography covering the period to 1980 (Valkovic, 1981).

1.2 Background information on coal areas

Based on data for 1987, world production of coal was around 4625 Mt, slightly less than half being from the West (MM, 1988). Production figures for hard coal and for lignite/brown coal have been calculated for 1750 mines, representing about 80% of the output for the West (MM, 1987). General information on the Soviet coal industry, including data for the geographic distribution of coal production, is available (Shabad, 1986). There is also information on the coalfields (location, accessible coal, brief description of area) of the world (IEA, 1983). In order to compare data for coal resources properly, 'a standardized, definitive, broadly applicable classification system' has been developed (Wood *et al.*, 1983).

Coal and coal-bearing strata are dealt with by Murchison and Westoll (1968), while coal geology and coal technology are covered by Ward (1985). Information on Australian coals, especially their geology, is given by Traves and King (1975). Other relevant matters include all aspects of the Australian coal industry (ACYB, 1987), the occurrence, mining and use of Australian black coal (Cook, 1975), mining methods in Australia (WC, 1980) and Latrobe Valley brown coal, 'probably the largest single deposit of brown coal in the world' (ACM, 1985).

1.3 Geochemical cycling

There has long been recognition of the relevance of chemistry to earth science. In 1821 Berzelius referred to mineralogy as 'the chemistry of the earth's crust' (Wedepohl, 1969). However, the first appreciation of geochemistry as such originates from the use of a term *Geochemie* by Schönbein (1938). Goldschmidt (1954) regarded geochemistry as the study of the composition of earth materials and the atmosphere and of the migration and distribution of elements. He also realized the importance

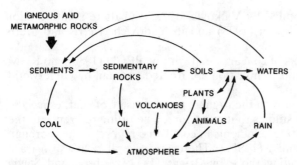

Figure 1.1 A simplified version of a geochemical cycle (Trudinger, Swaine and Skyring, 1979; reprinted with permission from Elsevier)

of the concept of the cycling of elements which depends on chemical and biological factors. There has been increasing emphasis on biogeochemical cycling of elements (Trudinger and Swaine, 1979). 'A realistic appraisal of the role of an element and of the relevance of its place in a particular part of the geochemical cycle depends on the fact that the system is *dynamic*, not static.' (Trudinger, Swaine and Skyring, 1979.) A simplified version of a geochemical cycle is shown in Figure 1.1. Although most of this book is concerned with coal *per se*, sediments are relevant as a source of trace elements and trace elements reach the atmosphere during coal usage, thereby being actively involved in part of the geochemical cycle. The mobilization of trace elements in soils and natural waters is important in the proper assessment of the environmental role of trace elements (Fishbein, 1981). Information on the amounts of elements mobilized by the combustion of fossil fuels and by weathering has been compiled by Bertine and Goldberg (1971). Although such calculated estimates are useful general guides, lack of data for some trace elements is a limitation. Global trace metal cycles have been dealt with in detail by Mackenzie, Lantzy and Paterson (1979), and atmospheric inputs have been assessed by Lantzy and Mackenzie (1979) and by Nriagu (1979). Studies of trace elements in ice from glaciers have indicated that natural processes were the dominant source of the eight elements investigated (Jaworowski, Bysiek and Kownacka, 1981). There is no doubt that the combustion of coal mobilizes trace elements into the atmosphere, the amounts varying greatly for different coals, that is, for different concentrations, and for different operating conditions. It is not easy to assess the significance of these additions of trace elements to the atmosphere and to geochemical cycles. In terms of amounts it has been estimated that natural sources account for 4–5 times more than anthropogenic sources (Meszaros, 1981). Results of a study of the deposition of trace elements in the environs of an Australian power station, together with a review of published results, are given by Swaine, Godbeer and Morgan (1984a).

Among the natural sources of trace elements of the atmosphere are volcanoes, forest fires, the weathering of rocks, thermal springs, erosion of metal-rich surface deposits, reactions at water surfaces and plant growth (Swaine, 1984a). Investigations of particulates from smoke clouds from eruptions of the Icelandic volcano, Syrtlingur, showed a great variety of particle compositions (McClaine, Allen and McConnell, 1968). It was suggested that a process of selective vaporization from the magma was followed by condensation from the vapour phase. Studies of high-temperature gases from the Merapi volcano in Indonesia indicated that most of the elements in the gas arose from incomplete degassing, although there were also some particles and effects from wall rock reactions (Symonds *et al.*, 1987). There are data for most trace elements in volcanic emissions, especially volatile elements, for example from Heimaey (Mroz and Zoller, 1975), from Augustine volcano (Lepel, Stefansson and Zoller, 1978), from Kilauea (Greenland and Aruscavage, 1986) and from Mt St Helens (Varekamp *et al.*, 1986). It has been calculated that chlorine and fluorine are emitted by volcanoes mainly as HCl and HF and that 'volcanoes should be regarded as a significant natural source of tropospheric and stratospheric HF' (Symonds, Rose and Reed, 1988). The weathering of rocks is the precursor of soils and of much of the dust in the atmosphere. While the main source of trace elements in coal is associated with what occurs in the swamp during coalification, there could have been some inputs from the atmosphere.

Chapter 2

Origin of trace elements in coal

The origin of trace elements in coal is fundamentally concerned with the history of the coal swamp, the geological changes that have occurred and atmospheric inputs which would have taken place from time to time. Some reference will be made to the early peat stages, to geological effects and to some aspects of mining, all of this being regarded as a framework for the consideration of processes that could have affected the loss and retention of trace elements. Large amounts of vegetation were required for the formation of peat and later of coal seams. It has been estimated that 12–20 m of vegetation were needed to form 1 m of coal (Berkowitz, 1985). However, Velikovsky (1956) assumed that 10 m of plant remains formed 1 m of peat and that 12 m of peat were needed to form a 1 m thick layer of coal. Various estimates suggest that the ratios of peat thickness to thickness of bituminous coal range from around 2:1 to 30:1 (Ryer and Langer, 1980). Although it is difficult to know much about the plant precursors of many coals, scientific studies of Victorian brown coals have provided information that has enabled an artist, C. Woodford, to depict a primeval forest in the Latrobe Valley (Figure 2.1). Among the trees shown are brown pine, banksia, she-oak, celery top pine, King William pine and kauri.

Figure 2.1 An artist's depiction of a primeval forest prior to the formation of brown coal in the Latrobe Valley, Victoria (copy of painting by C. Woodford, provided by G.E. Baragwanath and reproduced with permission from the SECV)

Some of the problems of coal geochemistry, as at 1949, were pointed out by Miller (1949). A selective review of the geochemistry of coal, covering references mainly from 1949 to 1957, summarized views on the origin of coal with some reference to certain trace elements (Breger, 1958). It was stressed that more work should be done on peat and lignite because of the prospect of eliciting 'geochemical knowledge regarding the formation of coal'. Breger (1958) also suggested that more work was needed 'to determine the affinity of humic substances for various elements'. Given (1984) has discussed in depth the organic geochemistry of coal.

Clymo (1987b) has made a challenging statement that

> coals are often assumed to have passed through a stage when they would have been called peat. But the botanical remains in the wide variety of present-day peats are very different from those in coals. This observation prompts the question: to what extent does the organic chemistry of coals depend on that of the plants from which they formed, and to what extent is it determined by the <u>processes</u> of peat formation and independent of the original botanical composition?

Despite this question, Clymo (1987b) and Moore (1987) regard the study of modern peats as helpful to an understanding of coal formation. Various aspects of ancient and modern peat systems have been discussed by several authors (Lyons and Alpern, 1989).

2.1 The formation of peat

Peat is formed by bacteria and fungi and the chemical products produced by them in the partial decomposition and disintegration of plant and animal material under wet anaerobic conditions. According to Clymo (1987b) 'as much as 99% of the original plant materials may have disappeared during peat formation and losses are selective. It is this highly modified residue which may be the material in which coal formation proper begins.' Peat is formed in bogs where *Sphagnum* mosses predominate and in marshes where sedges predominate. Some aspects of palaeobotany relevant to the origin of coal have been summarized by Given *et al.* (1980). Some peats have been subjected to silting and drainage from mineral soils (minerotrophic peats), while others (ombotrophic peats) receive their mineral inputs only from atmospheric precipitation (Gorham, 1967). At some stage the situation would have arisen where the plant growth within a peat swamp could not have competed with the ingress of sediments, thereby allowing sedimentation to occur; that is, interseam sediments would have been laid down, ultimately becoming dirt-bands in coal. Mattson and Koutler-Andersson (1955) cited several sources of dry deposition, namely, biogenic (for example pollen, spores and seeds), pedogenic (soil particles), oceanogenic (salt spray), pyrogenic (from fires), plutogenic (from volcanoes), cosmogenic (cosmic and meteoritic dust) and industrial dust. Meteorite ablation products have been identified in coal (Finkelman and Stanton, 1978).

Volcanic ash layers in coal have been investigated by Triplehorn and Bohor (1984) who suggested that the 'environments of coal deposition provide one of the best places for preservation of air-fall volcanic materials'. Altered volcanic ash partings were used by Ryer et al. (1980) to reconstruct the depositional history of a coal bed in central Utah. Snelling and Mackay (1985) proposed that volcanism was instrumental in the rapid deposition of seam materials in the Walloon Coal Measures of New South Wales and Queensland. The accumulation of various trace elements in brown coals from Ilnitsk, Western Carpathia, was favoured in areas of high volcanic and post-volcanic activity (Smirnov, 1969a). Gorham (1967) has reviewed the chemical properties of peat that differentiate minerotrophic fens from ombotrophic bogs in the United Kingdom. The ecology of peatlands has been reviewed by Clymo (1987a) who cites four main ways in which plant matter can become available for peat formation: (i) large amounts of *Sphagnum* are common in ombotrophic bogs, the plants behaving so that the upper branches shade those below which eventually die; (ii) plants rooted in peat yield their leaves and branches to the surface; (iii) underground stems formed just below the surface ultimately die; and (iv) roots growing in older peat eventually die. The first three kinds of plant material can be attacked by bacteria and fungi, possibly aided by invertebrate animals, in oxygenated water. Clymo (1987a) has summarized the sequence of breakdown processes rather neatly, namely aerobic decay → structural collapse → much lower hydraulic conductivity, then this plus excess precipitation → waterlogging → lower rate of transport of oxygen, then this plus microorganisms → anoxic conditions → much slower anaerobic decay. The contents of sulfur and ash in coals probably depend on the 'pH-controlled levels of bacterial activity in ancestral peat swamps' (Affolter and Hatch, 1984). Bog development and peat accumulation in the northern sector of the United Kingdom and the effects of major ecological factors on bog development generally were reviewed by Gorham (1957).

Several aspects of peat formation are discussed in Given and Cohen (1977), including palaeobotany of petrified peat, peat stratigraphy and microbial activities in the decay of leaves. The effects of changes in palaeoclimate on peat formation in the Central Appalachian Basin, USA, were studied by Cecil et al. (1985), who suggested that geochemical conditions were probably controlled mainly by the palaeoclimate. Other aspects of palaeoclimate are referred to by Phillips and Cecil (1985) in an overview of papers dealing with recent studies of the Pennsylvanian System of North America. The relevance of palaeoclimate to the occurrence and quality of coal are discussed by Cecil et al. (1983). The origin of peat and its composition and nature are referred to in a review of the use and production of peat by Spedding (1988). Detailed information on peat deposits in the Okefenokee Swamp, southwest Georgia, especially in relation to their relevance to coal formation, is given by Cohen et al. (1984b), with specific chemical and organic geochemical aspects being dealt with by Cohen et al. (1984a) and Casagrande (1984) respectively. Organic substances and their formation in peat, including the complex material known as humic acid, have been dealt with by Manskaya and Drozdova (1968). There is an extensive bibliography on the ecology, hydrology and biogeochemistry of *Sphagnum* bogs (Gorham, Santellmann and McAllister, 1984). From a detailed study of a *Sphagnum* moss-dominated raised bog in Maine, USA, Raymond, Cameron and Cohen (1987) showed that 'the geometry of the depositional basin, the level of the water table, the proximity to marine waters, and the influx of inorganic material into the swamp/bog complex have had major effects on the botanical composition and inorganic content of the resulting peat deposit'. It follows that these factors would also affect the trace-element status of peat.

Baker (1973, 1978) found that humic acids were able to mobilize metals, including gold, in soils under certain conditions. However, other conditions could favour the accumulation of trace elements probably by mobilization prior to fixation. Examples of the role of humic acids are given by Szalay (1957, 1969) and Kulinenko (1972). Bonnett and Cousins (1987) carried out experiments in which humic acids derived from peat were shown to be capable of taking up various trace elements, some more readily than others. Kawamura and Kaplan (1987) suggested that metal complexes formed with low molecular weight dicarboxylic acids may take part in the mobilization, transport and accumulation of trace elements in some geochemical situations. Humin, which is the insoluble fraction of sedimentary humic material has been investigated by nuclear magnetic resonance (NMR) which suggested that it is not a clay–humic acid complex, but that it is 'composed of a significant fraction of paraffinic carbons derived most likely from algal or microbial sources' (Hatcher *et al.*, 1985). In developing a geochemical coal model, Cecil *et al.* (1985) related basic compositional parameters of coal to the chemistry of the coal swamp. The origin of facies in the Upper Freeport coal bed, west central Pennsylvania, has been considered in terms of a geochemical model which involves peat formation 'resulting from the interaction of climate, plant types, rainfall, groundwater geochemistry, nutrient supply, and sedimentation' (Stanton *et al.*, 1986). Similar factors are cited by Hunt (1982) for 'the extent of alteration of the peat-forming material' and he adds 'and most importantly the level and composition of swamp water and fluctuations of the groundwater table'. In view of the statement by Donaldson, Renton and Presley (1985) that 'very little of the original biomass of swamp vegetation apparently survives to become coal even under most favourable conditions', it would seem that special conditions of geology, biochemistry and chemistry are needed for coal formation. Decomposition of humin under anaerobic conditions can yield coal directly (Hatcher *et al.*, 1985), some trace elements being available for biological use, while at the same time there could be some losses and retention of some elements. Various aspects of the occurrence and distribution of mineral matter in peat have been dealt with by Raymond and Andrejko (1983) and by Shotyk (1988), and there are some data for trace-element contents in peat, for example Mitchell (1954), Swaine (1957), Kochenov and Kreshtapova (1967), Cameron (1975), Casagrande and Erchull (1977), Cameron and Schruben (1983), Dissanayake (1984), Carlson (1985) and Bonnett and Cousins (1987). The peat probably acquired trace elements from the original plant material and by adsorption and cation exchange from swamp waters. It is known that *Sphagnum* species remove cations by ion exchange (Bosserman, Auble and Hamilton, 1984). A study of the Okefenokee peat-forming environment led Casagrande and Erchull (1977) to suggest that 'the variations in metal distributions in peat-forming systems may contribute to the often wide variations in metal distributions in horizontal and vertical directions in a coal seam'. An interesting application of trace elements in peat was their use in prospecting for ore bodies, assumed to be the source of relatively high concentrations of several metals (Salmi, 1955). Szadeczky-Kardoss and Vogl (1955) studied several coals and concluded that trace-element enrichment took place during the formation of peat, as well as during subsequent coalification processes. There is an informative review of the inorganic chemistry of peat including the occurrence and distribution of minerals and trace elements (Shotyk, 1988).

Stressing the fact that coal is a sedimentary rock, McCabe (1984, 1987) discussed various depositional environments of coal and coal-bearing strata, starting with the cyclothem era and finishing with the post-cyclothem era. The cyclothem era involved the recognition of repeated sequences of rocks in coal-bearing strata, interpreting the sequences as products of depositional (and erosional) events', while the post-cyclothem era related the formation of peats to 'modern clastic depositional environments' (McCabe, 1984). A historical review of the sedimentology of coal and coal-bearing sequences of the United States (Rahmani and Flores, 1984a) precedes an update of the subject covering a wide range of relevant topics (Rahmani and Flores, 1984b).

Although a few coals may be allochthonous, i.e. formed from plant material that was transported from outside the coal basin, most coals are regarded as autochthonous (Stach *et al.*, 1982), i.e. formed from plant material *in situ*. The origin of trace elements in coal is surely at least partly attributable to the stage of peat formation, namely the uptake of elements by plants, the removal of elements by ion exchange from swamp waters, the direct addition of mineral matter from incoming waters and from the atmosphere, and minerals formed *in situ*. Additions of minerals depend on the types of nearby rocks, weathering and the means of transport into the peat area (Cohen, Spackman and Raymond, 1987), these factors determining the relative importance of detrital inputs. Changes in pH may enhance the retention of trace elements or their removal. For example, sulfate-reducing bacteria do not favour low pH (Baas Becking, Kaplan and Moore, 1960), say below about 4, while low pH conditions favour leaching and formation of soluble compounds, which could be lost from the swamp. The importance of pH during peat formation to the mineral matter content of coal has been stressed by Cecil *et al.* (1982). It seems that the hummocky topography found in many peats, notably highmoors, could limit the effects of additions of cations and of fine particles of mineral matter, simply because the tops of the hummocks had been below the water surface. Raised swamps are probably precursors of low-ash coal because of the relatively low input of rock particles (Cairncross and Cadle, 1988). Microorganisms take part in the decomposition of plant material, initially under aerobic conditions and latterly under anaerobic conditions, to give a range of organic compounds (Manskaya and Drozdova, 1968; Flaig, 1968). Studies of microbial activity in some peat swamps in Florida, USA, showed a variety of effects, 'but for some reason the level of mineralizing activity is strictly limited' (Given *et al.*, 1983b). An investigation of Australian coals by transmission electron microscopy (TEM) has shown that biodegraded material and the remains of microorganisms were present from alteration 'at a very early stage of deposition' (Taylor and Liu, 1987). It would seem reasonable to surmise that during their growth these organisms took up some trace elements, which remained fixed in the dead material, thereby becoming a source of trace elements in peat and even coal.

Casagrande and Given (1980) proposed that amino acids of microbial origin 'appear to be quite significant participants in the input to coalification'. Various trace elements could be associated with some amino acids, at least at the peat stage. A study of fossil leaves from brown coal in the Latrobe Valley, Victoria, showed that 'selective resistance to fungal attack' had led to preservation of structures in the coal due to selective loss of carbohydrates (Wilson *et al.*, 1987). Admakin (1974) found that germanium and several other trace elements occurred in the highest concentrations in coal formed from plants that had undergone the highest degree of transformation.

Larsen *et al.* (1984) summarized the two views of the chemical changes during coalification. 'One pictures the coalification process as a condensation polymerization of a variety of small molecules formed by the degradation of biological macromolecules', whereas the other view envisages 'alterations and reactions of the biological macromolecules themselves.' In other words, is coalification a polymerization process or does it involve changing existing macromolecules? 'It seems likely that each process may occur during different stages of the coalification process.' (Larsen *et al.*, 1984.) NMR techniques are being used to provide useful information on peat and coal (Wilson, 1987). For example, a study of several coals has led to the conclusion that 'coalification is believed to occur primarily through alteration of the lignin molecule' at least for the brown coal to high-rank bituminous coal stage (Wilson and Pugmire, 1984). After studying several coals and macerals by cross-polarization magic angle spinning carbon-13 NMR spectrometric techniques, Wilson *et al.* (1984) concluded that 'as coal rank increases, the aromatic rings are defunctionalized, and the functional groups are replaced by hydrogen faster than cross-linking reactions occur'.

Despite the vast amount of work that has been done on peat and coal formation, the situation has been assessed recently by the down-to-earth statement that 'our understanding of the role of cellulose and lignin in contributing to coal formation is still deficient' (Spackman *et al.*, 1988). In an attempt to resolve this matter, careful detailed experiments have been carried out on selected samples of peat, which were fractionated mechanically (Spackman *et al.*, 1988). Early results indicated that the fractionation did 'achieve a separation having both botanic and chemical significance' and that plant tissues and organs separated from the peats were not preserved to the same extent (Spackman *et al.*, 1988). Earlier work showed the need for improved fractionation procedures (Lucas, Given and Spackman, 1988) to produce samples which could be defined botanically, thereby improving the interpretation of chemical results (Given *et al.*, 1983a). Barnes, Barnes and Bustin (1984) have discussed the chemical changes of organic matter during diagenesis. The chemistry of the transformation of plants to coal has been reviewed succinctly by Neavel (1981).

Because of its environmental importance, the origin of sulfur in peat has been studied in detail. Also, several trace elements are associated with sulfur in coal. It is very likely that at least some sulfur is introduced into organic matter through biosynthesis (Francois, 1987). Casagrande *et al.* (1977) studied peats from the Okefenokee swamp, Georgia, and the Everglades swamp, Florida, and their results indicated that virtually all the sulfur in coal could be incorporated during peat formation, that hydrogen sulfide was commonly found and that 'plants appear to fix and, subsequently, contribute much of the sulfur found in peat'.

Based on isotopic studies of Australian coals, Smith, Gould and Rigby (1982) are of the opinion that 'the bulk of organically combined sulfur in coal is generally introduced during plant growth and/or by reaction of sulfur species with plant debris during the early accumulation of the latter', and that 'freshwater conditions presumably prevailed' during the early formation of low-sulfur coals, say less than 1% total S, as noted by Smith and Batts (1974). Later work by Casagrande and Siefert (1977) suggested that organic sulfur occurs in two forms in coal, namely associated with carbon–sulfur bonds (for example cysteine and thiophenes) and carbon–oxygen–sulfur bonds (for example thiol esters). Casagrande, Gronli and Sutton (1980) found that the organic matter was dominantly humin and that ester sulfate was a major part of the total sulfur content and was found mainly in the fulvic acid fraction. The extensive studies of sulfur in coal by Casagrande and co-workers have been summarized well by Casagrande (1987). At least in low-sulfur coals, the organic sulfur seems to have originated from sulfur in the original plants (Westgate and Anderson, 1984). Altschuler *et al.* (1983) studied sulfur in Everglades peat and postulated that 'pyrite formation in organic-rich swamps depends on the use of organic oxysulfur compounds in dissimilatory respiration by sulfur-reducing bacteria'. Perhaps such a process would allow the removal of some trace elements by association with the pyrite. The implications from the above studies on sulfur in peat are that there could have been fixation of some trace elements by hydrogen sulfide produced during peat formation and by some organic sulfur groups, for example mercapto (−SH). Studies of organic matter–metal interactions in sediments suggest that under certain conditions the competition between metal ions, humic acids and bisulfide ions may yield metal–organic complexes rather than sulfides. During the early stages of coalification sulfate-containing waters entering the peat swamp would have provided reduced sulfur species by the action of sulfate-reducing bacteria. There is an interesting suggestion by Affolter and Stricker (1989) that the organic sulfur content of US coals may be limited by 'the latitude at which the original swamp developed', the lower temperatures of the higher latitudes affecting the 'rate of microbiological activity' and *ipso facto* inhibiting sulfate reduction, which 'could suppress the amount of reduced sulfur available for incorporation in the peat as organic sulfur'. Such conditions may well also limit the formation of pyrite and other sulfide minerals. The comment by Given and Miller (1985) that 'there are still a number of significant unsolved problems related to the fixation of sulfur in peats and coals' surely applies to trace elements. It is likely that the conditions of restricted drainage that would occur in peat swamps would enhance the uptake of several trace elements by some plants because of the increased mobilization of the trace elements, as was found for Scottish soils (Swaine, 1951; Swaine and Mitchell, 1960). Perhaps some of the content of trace elements in plants could be released during the decomposition of the plants and then react with hydrogen sulfide or with plant sulfur to form insoluble sulfides. At some stages metal–organic complexes could have formed, thereby either mobilizing or immobilizing trace elements, depending on several factors, *inter alia* pH and Eh. Saprykin and Kulachkova (1975) felt that the migration of trace elements depended on the structure and molecular weight of active groups, for example, −COOH and −OH, and on the ionic radius and charge of the trace elements, as well as on the geochemical environment.

2.2 Geological and geochemical aspects of coal seams

The ages and nature of the main coal deposits of the world have been dealt with by Francis (1961). In terms of total world reserves, the age distribution of coal has been estimated (Bouska, 1981) as Carboniferous (24%), Permian (31%), Triassic (1%), Jurassic (16%), Cretaceous (13%) and Tertiary (13%). Australian bituminous coals are predominantly Permian and the Victorian brown coals are Tertiary. The reasons for the age distribution of coals are discussed by Butler, Marsh and Goodarzi (1988) who regard climate as the dominant factor and cite the relatively high mean global precipitation and temperatures as the reason for the high rate of coal formation in the Carboniferous and Permian. It has long been considered that most coal seams were formed *in situ*, that is, from vegetation that grew in the place where the coal is found, or from material that was transported by water to the place where coalification occurred, the so-called Drift Theory. During coalification there were geological events, mainly subsidence and elevation, which were necessary for the attainment of suitable conditions required for coalification from the peat stage. Coalification (or Inkohlung) involves the changes in a series comprising peat, brown coal, lignite, subbituminous coal, bituminous coal and anthracite. Trace-element data will be given for these coals. The early stage of coalification is sometimes called the biochemical stage, while the stage from about lignite is called the geochemical stage. The above series represents increases in maturation, i.e. the extent of coalification, usually known as rank. General background information on the formation of coal seams is given by Mott (1943), Moore (1947), Bennett (1964), Murchison and Westoll (1968), Zubovic (1976), Stach *et al.* (1982), Ward (1985) and Scott (1987). Aspects of the sedimentology of coal, namely, the stratigraphy, sedimentation and depositional environments have been discussed by Rahmani and Flores (1984b). Suitable conditions for coal formation may be found in alluvial, deltaic and coastal environments.

'The accumulation of any thickness of coal is also dependent upon the basin within which the peat develops undergoing subsidence.' (Moore, 1987), based on Ingram, 1982.) The relative significance of tectonics, sea level fluctuations, and palaeoclimate in relation to Cretaceous coals in the United States have been dealt with in detail by Beeson (1984), while Butler, Marsh and Goodarzi (1988) discuss the genesis of coalfields in relation to plate tectonics which influenced coal formation 'because there must be a gently subsiding basin or a prograding continental margin on which the plant debris accumulates and without which no deposit would be preserved'. It has been realized for some time that movements of the earth's crust have caused tilting and folding of coal beds (Stutzer, 1940). Tectonic effects occurred not only during the early stages of coalification in the Sydney Basin, but 'other tectonic movements, probably related more to the general pattern of the relief of stress in the coal measure rocks, have occurred up to recent geological times' (Taylor and Warne, 1960).

During coalification, geological changes would have affected the distribution of trace elements laterally and vertically to different extents. 'A coal seam as a whole is a product of particular genetic conditions but the share and effect of each of these conditions were not the same in all parts of the coal-forming basin.' (Bouska, 1981.) Although Nicholls and Loring (1962) suggested that 'the only elements likely to have been accumulated by the coal-forming plants are Co, Ni and Mo' in some British coals, several other trace elements would have been taken up by various plants during growth prior to and during the peat stage. It is not possible to know how much, indeed, if any, of these trace elements remained with the organic matter as coalification proceeded. It has been traditional to regard those elements assimilated by plants as part of the inherent mineral matter in coal (Sprunk and O'Donnell, 1942), but this term is also used for those elements associated with the organic matter. It is likely that trace elements assimilated by plants may be lost, at least in part, during the decomposition of the plants, but on the other hand, the partly decomposed plant material and peat may remove and fix some elements from circulating solutions. The nature and concentrations of trace elements associated with this organic matter may be dictated by 'changes in the ability of the organic matter to fix metals' and there could perhaps be 'a preferential chelation effect for a particular humic substance' as postulated by Nissenbaum and Swaine (1976) for marine humic substances.

Interactions between geological, physico-chemical and biochemical factors are important determinants of the concentration and distribution of trace elements in coal seams. Among the geological factors are the rate of subsidence in the coal basin, the rate of uplift of the drainage area, the nature of the source rocks in the areas adjacent to the coal basin and the relationship between the drainage area and the area of the coal basin (Zubovic, 1976). On the basis of studies of coals from the Interior coal province, USA, Zubovic (1966a) postulated that 'metals forming smaller more highly charged ions, which generally form more stable bonds, are concentrated in near-source areas of a coal basin and are depleted elsewhere'. The effective accumulation of organic matter for coal formation requires 'an optimum balance of tectonic and climatic conditions' (Zubovic, 1976). According to Horne et al. (1978) 'rapid subsidence during sedimentation generally results in abrupt variations in coal seams' with resultant lower trace-element contents compared with those where slower subsidence occurs. The general nature and occurrence of igneous intrusions associated with coal seams have been reviewed by Gurevich and Shishlov (1987) who refer to instances in the Sydney Basin, New South Wales, and in Queensland. In a study of some Permian coals in the southern part of the Sydney Basin, Facer, Cook and Beck (1980) found that thermal effects were limited to less than 100 m above the base of the coal measures and that there were minimal effects on the rank of coals above the volcanic rocks. It would seem reasonable to expect some effect of local thermal activity on volatile trace elements in coal seams near such activity, but this effect may be limited to a change of location of such elements within the seam, rather than to losses from the seam.

Although Leutwein (1956) regards marine influence as of secondary importance for trace-element contents, Renton (1978) found that the variability of mineral matter in some US coals reflected marine or freshwater influences on the depositional environment. Swaine (1962b) considers that the boron content of the bituminous coals of New South Wales is a measure of the extent of marine influence during the early stages of coalification. Zodrow (1986) found that the concentration of elements in coals from the Sydney Coalfield, Nova Scotia, is higher in thinner seams than in thicker seams, probably because of different conditions at the peat stage, as proposed for differences in the spatial distribution of ash yield and sulfur content for the Upper Hance coal seam in southeastern Kentucky (Esterle and Ferm, 1986). The results of an investigation of the Lower Kittanning seam, western Pennsylvania, by Rimmer and Davis (1984a) indicated the importance of subsidence rates on the thickness and mineralogy of the coal deposits. Variations in trace-element contents in coals may decrease towards the centre of a basin, probably because the amounts left in solution decreased after passing the peripheral areas where there would have been some removal of trace elements by adsorption and reactions with coaly matter. This 'enrichment at the margins of the seam is almost certainly a result of groundwater transport' (Drever, Murphy and Surdam, 1977).

The processes of coal formation have been dealt with by Schopf (1948), Gould (1966) and Bouska (1981), while much relevant information on low-rank coals has been compiled by Schobert (1982). From results presented in this latter compilation, it was considered that more work is needed on clay minerals, on distinguishing detrital and authigenic minerals, on the associations of trace elements with macerals and on the geochemistry of trace elements during coalification. Various aspects of coal geology in the United States in relation to future research have been summarized by Meissner, Cecil and Stricker (1977) and details of coal provinces and coal basins in the United States are summarized by Nelson (1987). Goodarzi (1987a) examined the effects of weathering on the concentrations of some trace elements in selected coals from British Columbia, Canada, and found slight changes which were different for organically bound and inorganically bound elements.

2.3 Geology of Australian coals

The specific geology of world coals is outside the scope of this book, but some reference will be made to the geology of Australian coals. Major sources of geological information on Australian coal occurrences are in Traves and King (1975) and Mallett (1982).

Hobday (1987) has discussed the tectonic setting and depositional systems of Gondwana coal basins in Australia and South Africa, where he delineated 'palaeotopography and depositional environment' as 'the most important factors in determining coal seam geometry' in most cases. He regarded 'rapid tectonic subsidence, changes in groundwater regime or gradual marine transgression' as major controls of some seams and as influences on their inorganic composition.

Among other relevant topics are the geology of the Bowen Basin (Davis, 1971), the formation and nature of Victorian brown coal (Baragwanath, 1962), cyclic sedimentation in coals of the Sydney Basin (Duff, 1967), geological interpretation of ply structure of the Bulli seam (Shibaoka and Bennett, 1975), structure and sedimentation in the Western coalfield of New South Wales (Branagan, 1960, 1961), stratigraphy and sedimentation in the Western coalfield of New South Wales (Bembrick, 1983), environmental interpretations of Gondwana coal measure sequences in the Sydney Basin (Britten et al., 1975), the relevance of the isotopic composition of total and organically combined sulfur in low-sulfur coals to deposition in fluvio-deltaic environments (Hunt and Smith, 1985), and the recognition of intervals of peat accumulation by detailed facies analysis of samples from the Callide seam, Queensland (Flood, 1985). Petrographic data have been used by Leblang, Rayment and Smyth (1981) to reconstruct the conditions of deposition in the Austinvale coal deposit, Queensland, and Smyth (1970) compiled type seam sequences, based on petrographic studies, which were interpreted as coming from peat-forming areas which had become progressively drier. The petrographic composition of coal seams in coal basins in eastern Australia has been related to the rate of subsidence, for example seams deposited under stable conditions had few complete cycles and were relatively poor in vitrinite, while seams deposited under unstable conditions had incomplete cycles which were rich in vitrinite (Shibaoka and Smyth, 1975).

Australian coals are of various ages, for example Permian (Sydney Basin and Bowen Basin), Triassic (Ipswich area and Leigh Creek), Jurassic (Walloon) and Tertiary (Victorian brown coals).

A study of coal seams from the East Whitbank coal area in South Africa led Cairncross and Cadle (1988) to suggest that differences in coal characteristics between Carboniferous and Gondwana coals 'are probably related to different palaeoclimatic conditions and basic tectonics' at the peat stage. This would seem to apply to Australian coals also.

2.4 Coal mining in Australia

As elsewhere, coal is mined in Australia by underground and surface techniques, using the most modern approaches and equipment. Large-scale surface (open-cut) mining of Australian coal commenced in the mid-1960s (Isles, 1986) and is now widely used. In 1987, 62% of the active bituminous coal mines were underground and 38% on the surface. The importance of surface mining is shown by the fact that 23 of the 25 mines with the largest production (about 2 million tonnes or more each) were surface mines, 14 being in Queensland and nine in New South Wales. In 1987, two-thirds of the production of Australian bituminous coal was from surface mines. Detailed information on the production of bituminous coal, measured and indicated reserves, locations of collieries, coal use in Australia and exports is given by JCB (1988). Laverick and Grice (1980) have given a general account of Australian coals used for coke making and power production, with special reference to the export market. As well as bituminous coal with an annual production of about 130 Mt, there are extensive deposits of brown coal in Victoria, especially in the Latrobe Valley, where the annual production is about 45 Mt from seams up to 140 m thick. Subbituminous coal is mined in the Collie Basin, Western Australia, where the annual production is about 4 Mt (ACM, 1987) and hard brown coal is mined at Leigh Creek, South Australia (about 2.7 Mt). A relatively small amount of bituminous coal (0.5 Mt) is mined in north-eastern Tasmania (Bacon, 1986).

Information on environmental controls for coal mining is given by Hannan (1980), and Swaine (1980a) has discussed trace-element aspects of coal mining. Trace elements in coal in relation to rehabilitation after mining are considered by Swaine (1978a) as part of a broader account of the management of lands affected by mining (Rummery and Howes, 1978). Socio-economic aspects of coal-mining developments in Australia, including resource management, are found in Hannan (1982).

2.5 The constitution of coal

The organic-rich components in coal can be identified microscopically by the techniques of coal petrology. These petrographic components are termed 'macerals' and can be regarded as the organic counterparts of the mineral matter in coal. Macerals, for example vitrinite, exinite, fusinite, are derived from certain plants or parts of plants which have been degraded under rather critical conditions. Details of coal petrology, including the methods of examination and the origin of the petrographic constituents, have been covered fully by Stach et al. (1982). The terminology of coal petrography has undergone much change in the past 30 years and will probably continue to do so.

There has been a great deal of interest and research on the constitution or structure of coal, but this matter is far from settled (Wender et al., 1981; Meyers, 1982) and modern techniques, for example NMR, are being used to shed more light on this important aspect of coal science. Horton (1958) reviewed the situation and it was clear then that hydroxyl groups were present in coals with up to about 89% carbon, and that carboxylic acid groups were a feature of low-rank coals. Yohe (1958) gave a good definition:

> coal is a complex mixture of high molecular weight substances. Aromatic (benzene) ring systems and condensed aromatic ring systems are present, and these condensed nuclei increase in size and form an increasingly greater part of the total substance as the rank of the coal

increases. Aliphatic structures exist as methyl groups and possibly as side chains containing two or more carbon atoms, as connecting chains between aromatic nuclei, or as parts of hydroaromatic units.

This definition can be updated by substituting 'macromolecules' for 'substances' and by deleting 'possibly'. By 1961 the polymeric structure of coal was reasonably well established (van Krevelen, 1961). Nowadays this is expressed by regarding coal as having 'an extensive macromolecular structure' (Weller, Völkl and Wert, 1983) or as 'possessing macromolecular networks containing trapped molecules' (Given et al., 1983a).

The concept of a mobile or molecular phase within the macromolecular network of coals was debated by Given, Marzec, Barton, Lynch and Gerstein (Given et al., 1986), but the matter was not resolved. For example, treatment of coal with swelling solvents may affect parts of the macromolecular structure so that there is a contribution to the mobile hydrogen content of the coals. Evidence from experiments with some Australian and New Zealand brown coals (Lynch et al., 1988) indicates that 'the coals are composed of extract (guest) and residue (host) materials in differing proportions'. The formerly popular representation of coal by an average model structure (Given, 1960; Wiser, 1978; reviewed by Berkowitz, 1985) now 'becomes quite indefensible' and all such models should 'be pensioned off as being past their useful working life' (Given, 1984). Measurements of the internal friction spectra of several high-rank coals showed the polymer-like character of the coals, which is in keeping with 'the macromolecular character of the coal structure' (Wert and Weller, 1982). It is clear that modern coal scientists see 'coal as a macromolecular solid' (Larsen, 1985) or, expressed in another way, 'as a macromolecular organic rock' (Hatcher et al., 1983). The macromolecular nature of coal seems to mean that 'coals are macromolecular gels, i.e. have three-dimensionally cross-linked structures' (Green et al., 1982). Although Hayatsu et al. (1975) found that their experimental data supported the view 'that coal is predominantly an aromatic material', they suggested that there was 'a high degree of aliphatic or alicyclic linkages between aromatic units'. Nuclear magnetic resonance spectroscopy techniques have been used to detect aromatic and aliphatic groups in brown and in bituminous coals (Wilson and Pugmire, 1984; Wershaw and Mikita, 1987; Wilson, 1987). An investigation of the aromaticity of some high-sulfur US coals, using ^{13}C NMR, showed that the coals were more heterogeneous than expected (Neill et al., 1987), and it was concluded that 'there appears to be no really satisfactory means of determining forms of organic sulphur in a coal'. However, Hsieh and Wert (1985) have developed a method for determining organic sulfur directly in coal using a TEM technique on masses of coal of the order of 10^{-14} g. An X-ray absorption spectroscopic investigation of a bituminous coal from Kentucky, USA, led to the surmise that thiophene is the most likely main organic sulfur functional group (Spiro et al., 1984). Narayan, Kullerud and Wood (1988) suggested that a major part of 'the organic sulfur fraction in pristine coal contains sulfur in the form of a coal–polysulfide complex', which could produce some amorphous elemental sulfur during coalification. This elemental sulfur could react with alkyl aromatic compounds in coal to form thiophene compounds (White and Lee, 1980). Brooks (1936) had suggested that 'thiophenol and disulfide groups, together with heterocyclic sulfur in ring structures' could be present in coals. Measurements of the sulfur isotope composition of the Herrin (No. 6) coal from the Illinois Basin indicated that 'organic sulfur in the low-sulfur coals appears to reflect the original plant sulfur' (Westgate and Anderson, 1984). The statement made by Hauck (1975) that 'little is known with certainty about the N forms in coal' is probably still correct, although there is certainly 1–2% total nitrogen in most coals, the predominant forms being probably heterocyclic structures. Amino acids have been found in peats where they were probably adsorbed or linked to humic acid micelles (Swain, Blumentals and Millers, 1959), and these could have been the precursors of nitrogen forms in coal. Nitrogen in coal has probably been derived from plant and animal sources, the latter being regarded by some researchers as important (Stutzer, 1940).

In view of the relevance of humic acid to the early stages of coalification, it is surely worth while looking at advances in the chemistry and geochemistry of humic substances, for example there is a recent book which sets out to present an in-depth review of 'what we know, what we don't know, and what we think we know' about humic substances (Aiken et al., 1985). In the past, the main interest has been in humic acid (fraction insoluble in dilute acid, but soluble in dilute alkali) and fulvic acid (fraction soluble under all pH conditions), but more attention is now being given to humin, which is the fraction insoluble in water at any pH (Hatcher et al., 1985). It is well known that many metals have important roles in biological systems and hence are found in plants, animals and the like. It is not surprising that metals are found in sedimentary material, probably associated with various types of organic matter through oxygen, nitrogen and sulfur groups (Saxby, 1969). Studies of the interactions of metals with complex organic acids in soils have been reported by Kononova (1966). Hatcher (1985) has stated that 'humic acids can complex with inorganic cations and may be important in forming ore deposits'. It seems likely that several trace elements could have been removed from solution by humic substances during the early stages of coalification. Several trace elements were found in humic substances from marine-reducing environments (Nissenbaum and Swaine, 1976); and humic substances in municipal refuse material retained some Cu, Cd and Zn, probably because these elements were 'relatively unavailable to organisms' (Filip et al., 1985). Perhaps an analogous situation could have occurred in coal swamps. Differences in the nature and concentrations of trace elements associated with the organic matter, for example humic acid, could be caused by 'changes in the ability of the organic matter to fix metals', that is, by a preferential chelation effect, possibly related to 'the stage of diagenesis of both the organic and inorganic moieties', the pH and the oxygen concentration, as postulated for sediments (Nissenbaum and Swaine, 1976). At a later stage of diagenesis the metal–organic complexes could perhaps have broken down to set free trace elements which would then become available for incorporation in mineral phases (Hatcher, 1985). The insolubility of humin, and presumably of metal complexes formed with it, could have resulted in the longer persistence of such complexes than of humic acid–metal complexes. Results of a study of complexes formed between cupric ions and humic acids from four low-rank Spanish coals suggested 'that the geochemical concentration of heavy metals in coals by complexation reactions with humic acids is favoured in the early stages of coal formation' (Ibarra and Orduna, 1986). Despite the lack of detailed knowledge of coal structure, it is clear that certain specific organic groups are present and that some of these could bind some trace elements (Swaine, 1977a). The most important groups are carboxylic acid (−COOH) groups which are a feature of brown coals (Durie, 1961; Wolfrum, 1984) and lignites (Kube et al., 1984), possibly in some (Shinn, 1985) and perhaps in all subbituminous coals (Mraw et al., 1983), but certainly not in higher-rank coals (Shinn, 1985). Other groups that can bind some trace elements are phenolic (−OH), imino (=NH) and mercapto (−SH). Although a good deal is known about oxygen-containing groups in coal, 'evidence respecting the forms in which nitrogen and sulphur exist in coal is much more tenuous' (Berkowitz, 1985). Brooks and Sternhell (1957) estimated the −COOH and phenolic OH groups in four Victorian brown coals and an improved method for determining carboxylic acid groups in brown coal was developed by Schafer (1984). There is some analogy with the situation in water, where Malcolm (1985) believes that carboxylic acid and phenolic groups in humic substances are the main groups for associations with trace elements. A similar situation has been suggested for soils where humic and fulvic acids, mainly via carboxylic acid and hydroxyl groups, influence the properties of various trace elements (Stevenson, 1983; 1985). There is some enrichment of Ge in the surfaces of peat-rich soils in some parts of Scotland (Swaine, 1960).

It is clear that most of the trace-element associations with organic matter in coal occur during the early stages of coalification, say up to the subbituminous stage. Perhaps the same situation holds for mineral matter also, except that discrete minerals (major and minor amounts) would have been emplaced in cleats and cracks resulting from intrusions and tectonic effects. Although the question 'has coal a structure?' (Given, 1961) is not so applicable nowadays, there is still much to be done to elucidate further details of coal constitution (Green et al., 1982). It is salutary to recall the statement of Wender et al. (1981) that 'it is unfortunate that a great many of the publications on coal structure only serve to confuse, dealing all too often with studies on poorly defined coal specimens and leading to conclusions about coal constitution that, on inspection, are not warranted'. The same authors warn that 'perhaps the biggest block to progress in this field is that most scientists are accustomed to working with homogeneous specimens of good purity. There is no such thing as "pure" coal'. The difficulties of finding out the structure of coal are highlighted by Green et al. (1982) in their statement 'now let us take the tree, subject it to aerobic and anaerobic decomposition, mix its components thoroughly, bake at low heat for 200 million years, and determine the structure of the resulting mass. If one has mixed the debris from a variety of plants and added a variety of minerals, that is the problem facing the coal chemist.'

2.6 The history of trace elements in coal

It is clear that trace elements in coal have a diverse origin. Various influences, notably botanical, geological, biological, biochemical and chemical, interacted over long periods of time to increase or decrease the levels of different trace elements. Early suggestions by Goldschmidt (1935, 1937) involved three main stages for the enrichment of trace elements in coal, namely concentration during plant growth, during decay of vegetable matter and during mineralization. These suggestions were in keeping with a classification of mineral matter in coal by Lessing (1925). A coherent account of the origin of trace elements in coal was put forward by Swaine (1962a; 1975) who reviewed a wide range of publications relevant to the topic. More recent views are those of van Krevelen (1963), Nicholls (1968), Manskaya and Drozdova (1968), Saprykin, Kler and Kulachkova (1970), Finkelman (1981a), Bouska (1981), Stach et al. (1982), Finkelman (1982b), Hatch (1983), Rimmer and Davis (1984b), Given and Miller (1987) and Finkelman (1988).

On the basis of the above postulates, it seems reasonable to propose a sequence of events that could have influenced the trace-element contents of coal. The original vegetation of the peat swamp could have taken up trace elements by biological and physico-chemical processes, in some cases because the plants needed the elements for proper growth. Plants differ in their needs and tolerance for various trace elements (Underwood, 1977; Antonovics, Bradshaw and Turner, 1971) and an increased uptake of some elements is caused by conditions of impeded drainage (Swaine, 1951; Swaine and Mitchell, 1960). During the degradation of plant matter to form peat, it seems likely that many trace elements would have become associated with organic matter. At this stage there would have been intense bacterial activity and solutions entering the swamp would have contained ions in solution, together with fine mineral and rock particles. Some of these ions could have been removed from solution by interaction with organic peaty material and with mineral matter, for example clays, but there was also the possibility of loss of some ions from the peat and eventual removal from the swamp by outgoing solutions. This situation would depend on pH, the nature of the ion and the nature of the organic matter, the latter possibly depending on the stage of diagenesis (Nissenbaum and Swaine, 1976). Fine particles entering the swamp would be deposited in the peaty material and dispersed during the ensuing coalification (Alexseev, 1960). At various times, there could have been inputs of atmospheric material, for example volcanic dust, meteoritic dust, wind-borne dust and finely divided salts derived from lakes and seawater. During the early stages of coal formation there was surely intense biochemical activity, especially where the pH was above about 4, but such activity probably ceased at the end of the brown coal stage (Mackowsky, 1955), probably because of the subsequent increases in temperature brought about by metamorphic effects. At certain stages, compounds (minerals) could have been formed by precipitation from waters circulating in the swamp and by reactions between ions in solution and ions or minerals associated with peat. When pH conditions are around 6, not only are the conditions favourable for microbiological activity, but also there is less leaching of minerals and the precipitation of carbonate minerals can take place. Another source of trace elements could be the remains of microorganisms which took part in the early breakdown of plant matter.

At every stage of coalification there could have been geological disturbances, for example, arising from tectonic effects. In particular, subsidence would have allowed sediments to cover peat layers, thereby initiating burial processes leading to increases in rank. During compaction there would have been losses of water possibly containing trace elements. Once compaction has occurred, cleats and cracks would have formed, allowing the access of solutions, from which minerals could have precipitated. Such epigenetic minerals are generally larger than minerals formed earlier which are closely associated with macerals. After the brown coal–lignite stage there are probably some changes in the nature of the associations between trace elements and organic matter. Perhaps organically associated elements are at least partly 'flushed out of the lignite upon upranking' (Finkelman, 1982a), while elements associated with the mineral matter are not affected. It seems clear from studies by Finkelman (1981a,b) that detrital minerals can be a major source of mineral matter in many coals, depending on the nearness of a suitable source area of weathered rocks or the like. This view is in keeping with that of Ratynskii (1975) who stated that 'the influx of an adequate quantity of aqueous solutions rich in rare elements into regions of peat and coal formation is the main factor determining the concentration of several rare elements in fossil coals'. According to Kochenov and Kreshtapova (1967) trace elements in peat depend very little on the botanical composition of the plant community, but markedly on the geology of the nearby area which is the source of waters entering the peatland. However, Cecil et al. (1982) postulated that the inorganic matter from peat-forming plants was the main source of inorganic elements. Indeed, Cecil, Stanton and Dulong (1981) were of the opinion that the detrital origin of all mineral-rich bands 'does not have a sound physical or theoretical basis'. The relevance of the inorganic matter in plants cannot be completely discounted as a partial source of some major and trace elements for coal. For example, some epiphytic ferns had 3.9–11.6% Al in their ash (Dixon, 1881), biogenic silicon is well known and plants require several trace elements for their growth. However, most of the silicates (for example quartz and clay minerals) are expected to come from water-transported detritus, as postulated for organic-rich deposits in southern Louisiana (Bailey and Kosters, 1983). Perhaps plant inorganic matter could account for mineral matter in some coals, for example those distant from a detrital source, but this cannot be assumed overall. Studies by Rimmer and Davis (1984b, 1986) and Davis et al. (1984) showed the presence of minerals that were apparently derived from elements supplied from sources external to the swamp. In coals where there is no clear evidence for detrital minerals, perhaps there has been some indistinctness of mineral grains during diagenesis. It would seem reasonable to assume that minerals may be authigenic and detrital, depending on the various factors already discussed (Finkelman, 1984). The terms 'inherent ash' and 'inherent mineral matter', which were used to mean the inorganic material that had come from the original plant matter, are heartily discouraged because it is virtually impossible to isolate such material. It is a pity to see such terms still used, for example by Pike, Dewison and Spears (1989). It seems that plant-derived mineral matter is usually less than about 2% of ash yield for bituminous coals (Deul, 1955) and is only significant for coals with an ash yield of less than about 5% (Finkelman, 1982a).

There are some references that are pertinent to the above postulates. In connection with plants, Minguzzi and Naldoni (1950) concluded that As in coal was not predominantly plant-derived, whereas Janda and Schroll (1959) regarded plants as the source of B in some Austrian coals. Chlorine is taken up by plants and is therefore 'contained in coals', according to Correns (1956), while Bradburn (1958) supports the idea that Cl was present in coal measure plants, but adds that it could have entered coal during the early stages of its formation. Aubrey (1958) and Kizilshteyn and Syunyakova (1964) do not regard plants as the source of Ge, although Gordon and Motina (1959) believe that Ge was present in mosses at the peat stage. Germanium has been found in some Soviet peats (Ratynskii, 1946) and in some Scottish peats and organic-rich surface soils (Swaine, 1962a). Davidson and Ponsford (1954) do not support the idea that U was concentrated by pre-coal vegetation.

In discussing V, Buljan (1949) suggested that there may have been periods when organisms, including plants, utilized V more than they do nowadays, a postulate that could also apply to some other trace elements. Ourisson, Albrecht and Rohmer (1984) have suggested that various organic sediments, including coal, could have derived 'much of their organic matter from once unknown microbial lipids'. Taylor and Liu (1987) found the remains of microorganisms in several Australian coals and suggested that they could 'represent significant amounts of added organic matter'. It surely follows that these microorganisms, mainly fungi and bacteria, could have accumulated some trace elements.

It seems possible that plants have accumulated amounts of some trace elements of the same order as those found in some coals, as suggested by Zubovic, Stadnichenko and Sheffey (1961a), but there is no certainty that these plant trace elements have survived *in toto* into the coals. A full account of relevant aspects of plants and plant accumulation can be found in Bouska (1981).

Changes in the weathering of source rocks near coal swamps would have influenced changes in the amounts of some trace elements transported into the swamp (Zubovic, Stadnichenko and Sheffey, 1960a) as detrital minerals or in solution. The nature of the rocks being weathered would also influence the amounts of some elements, for example granites could have been a source of U for some Hungarian coals (Szalay, 1954) and of Ge for some Czechoslovakian coals (Zahradnik, Tyroler and Vondrakova, 1960). Stone (1912) suggested that the relatively high Au contents in some coals from the Black Hills, Wyoming, USA, were derived from Au-bearing sands overlying the coal swamp, the Au particles supposedly working down into the bog. Processes for the enrichment of trace elements in coal may well have taken place over long periods, for example Bouska and Stehlik (1980) estimated that the enrichment of Be, Ga, Ge and V in some Czechoslovakian coals by post-sedimentary processes took about 1.6 My. The role of adsorption in the concentration of some trace elements in peat and coal has been studied experimentally for Cu (Ong and Swanson, 1966), Ga (Eskenazi, 1967), Be (Eskenazi, 1970) and Ti (Eskenazi, 1972). Szalay and Szilagyi (1969) suggested, on the basis of laboratory experiments, that Mo and V migrated in solution as anions, which were reduced to Mo^{5+} and VO^{2+} when in contact with humic acids, and which were then fixed by cation exchange. The importance of reducing conditions for the accumulation of trace elements in coal has been stressed by Mason (1949) and Smirnov (1969b).

It can be concluded that in general 'the quantity of any minor element in a coal appears to have been controlled by the availability of that element to the swamp in which the coal was formed' (Zubovic, Stadnichenko and Sheffey, 1960b). Bouska (1981) put it more forcibly as 'it is unquestionable that the elements were supplied into the developing coal seam by solution percolating from the neighbouring rocks', and migrated partly in solution and partly in suspension as 'adsorbed cations on micelles' (Strakhov, Zalmanzon and Glagoleva, 1956), say, of clays. However, 'the question of how the elements originally become associated with the coal is a moot one' (Finkelman, 1981a). Given and Miller (1987) stated that 'the one generalization that we feel is fully warranted is that the organic matter in a peat or lignite, has in many ways, a profound influence in determining which inorganic matter, in what form, it will associate with'.

Chapter 3

Mode of occurrence of trace elements in coal

It is as important to know how each element occurs in coal as to know how much of the element there is.

(Finkelman, 1981a)

The history of trace elements in coals indicates that there are several possible ways in which an element may be present. In general, perhaps every element is present, partly associated with organic matter and partly associated with mineral matter, the relevant proportions of each form differing from element to element. It is reasonable to expect organic associations to predominate in low-ash coals, say those with less than about 5% ash yield (Finkelman, 1982a).

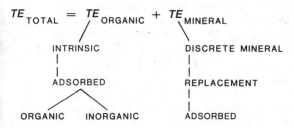

Figure 3.1 Modes of occurrence of trace elements in coal

The modes of occurrence are summarized in Figure 3.1. The term 'intrinsic' is used for chemical association with specific groups in the organic matter. 'Adsorbed' refers to the organic coaly matter and to organic matter on the surfaces of clay particles. The latter is in contrast to direct association with clays, which are regarded as part of the mineral matter. Mineral matter in bituminous coals is usually taken to mean minerals which mostly decompose on ashing to make up the major part of the ash yield, together with inorganic elements that may be associated with the coal organic matter. In other words, the mineral matter 'will embrace all components of a given coal mass which are not obviously organic in character' (Fowkes, 1978), an approach that is in keeping with the view of Given and Yarzab (1978) who stated that 'we treat mineral matter as consisting not only of minerals'. In terms of these views, all inorganic elements in coal are part of the mineral matter. In the following discussion, trace elements associated with organic matter will be dealt with initially, followed by those associated with minerals, and finally there are some comments on low-rank coals, especially, lignites and brown coals. The association of minerals with macerals has been investigated, for example by Smyth (1966), Finkelman (1981a, 1988), Cecil, Stanton and Dulong (1981) and Harvey and De Maris (1987).

3.1 Trace element–organic matter associations

There are certain specific groups that are part of the organic coaly matter, namely carboxylic acid (−COOH), phenolic hydroxyl (−OH), mercapto (−SH) and imino (=NH), and it has been suggested that trace elements may be associated with these groups (Swaine, 1977a). Evidence that some of the total content of certain trace elements in brown coals and lignites is ion exchangeable has been given by Morgan, Jenkins and Walker (1981) for Ba and Sr, by Swaine (1983) for Mn, Sr and Zn, and by Benson and Holm (1985) and by Miller and Given (1987b) for Ba, Mn and Sr. This indicates an association with carboxylic acid groups, as suggested for Mn and Sr by Brown and Swaine (1964), and possibly with phenolic hydroxyl groups. On the basis of fractionations of a Kosova Basin coal, Yugoslavia, using dimethyl sulfoxide (DMSO), Daci, Hoxha and Vujicic (1985) suggested that B, Ba, Cu, Mn, Sr, and Zn could possibly replace H atoms in carboxylic acid or phenolic hydroxyl groups.

The formation of trace-element complexes with organic matter in coal, probably by chelation, has been considered by Zubovic, Stadnichenko and Sheffey (1961b) who suggested that the stability of the complexes would depend on the charge and size of the ion, that is, the ionic potential, the bond configuration and coordination number, the tendency to form covalent rather than ionic bonds, and the tendency to combine with nitrogen rather than with oxygen or sulfur in the organic matter. According to Zubovic, Stadnichenko and Sheffey (1960c) 'it is significant that the elements whose ions are small and highly charged are generally associated with the organic fractions'.

Plots of the organic affinity (that is, the percentage associated with organic matter) and the ionic potential for 15 trace elements show similar configurations (Zubovic, 1976). The degree of organic complexing decreases with increase in rank, and carboxylic acid groups do not persist much beyond the brown coal/lignite stage. Metal–cystine complexes would not survive mild diagenesis (Saxby, 1973), but mercapto (−SH) groups may be expected to persist to the bituminous stage. Humic acids are probably 'the main complexing agents in most geological systems' (Stevenson, 1983) and together with fulvic acids they would have complexed several trace elements during the early stages of coalification. It has been proposed that Ge is retained by humic acids and coal by reaction with hydroxyl groups (Gordon, 1959a). Senesi, Sposito and Martin (1986) have studied humic acid complexes with divalent Cu in soils using infrared (IR) and electron spin resonance spectroscopy (ESR). Their results suggested that divalent Cu ions are bound in two ways, 'the stronger one involving functional groups that form complexes with a more covalent character, which are not easily dissociable and are stable against exchange by protons, the weaker one probably involving surface functional groups that adsorb highly hydrated metal ions through the formation of mainly electrostatic bonds, which are stable against exhaustive water washing but easily disrupted by protons'. The first of these two suggestions would seem applicable to coals during the early stages of coalification. Although Treibs (1935, 1936) reported metal porphyrin complexes in coal, it is only recently that trace amounts of Ga and Mn porphyrins have been satisfactorily proven to be present in several samples of lignite and bituminous coals (Bonnett and Czechowski, 1981), although 'nickel and vanadyl complexes have not as yet been identified'.

In view of the origins of trace elements in coals it is to be expected that most, indeed probably all, trace elements will occur partly organically associated and partly inorganically associated, as is inferred in Figure 3.1. The assessment of the proportions of each association is difficult, but several attempts have been made using differences in specific gravity to separate organic-rich and mineral-rich fractions. These so-called float–sink methods yielded a light (float) fraction which is organic-rich and a heavy (sink) fraction which is mineral-rich. Although the latter can be regarded as fairly representative of the mineral matter in coal, the float is not suitable for deciding the extent of organic association of trace elements in the parent coal, because it generally contains very fine particles of mineral matter embedded in an organic (coaly) matrix. This mineral matter is therefore in the float sample because its higher specific gravity is dominated by the lower specific gravity of the greater volume (and mass) of the coaly matter. Transmission electron microscopy shows the presence of fine submicron-sized particles in coal (Strehlow, Harris and Yust, 1978) and the use of scanning transmission electron microscopy (STEM) with an analysis facility allows the examination and identification of particles down to about 10 nm in diameter. Analyses of microareas of a sample of granular inertinite from a US coal showed the presence of Al, Si, Fe, S, Ca and K which indicated the presence of clays, quartz, pyrite, gypsum and calcite (Harris and Yust, 1981). The presence of Ti, Ni and S in other samples was indicative of rutile (TiO_2) and millerite (NiS). A study of some US coals by electron microprobe and STEM methods showed that most particles were 0.1–1 μm minerals (Minkin et al., 1983). Recent work on three US coal samples, using STEM, has shown the presence of several particles less than 200 nm in diameter. The predominant particles were Fe-rich and Ba-rich in a lignite, Ti-rich in a bituminous coal and Ca-rich and Ti-rich in a semi-anthracite (Allen and van der Sande, 1984). Finkelman et al. (1979) recognized that 'substantial amounts' of several trace elements may be retained in the light fractions (s.g.<1.50) because of the close association of fine-grained minerals with the organic coaly matter. Despite the comment by Given and Miller (1987) that 'there is no reason to suppose that this is generally a major phenomenon', I believe that it is a definite limitation on the use of the float fraction to estimate the association of trace elements with the organic coaly matter. However, the sink fraction is generally useful for estimating the association of trace elements with mineral matter. Another aspect of the organic association of trace elements in coal is the association of trace elements with different macerals. Again there is the problem of minute particles of mineral matter in the macerals. Inagaki (1951a) did not find any special differences in trace elements in different petrographical components and Hashimoto (1953) found no correlation between the distribution of several elements and petrographical components. However, Eskenazi, Mincheva and Ruseva (1986) found that lignites from the Elhovo Basin showed three distributions of trace elements, namely those strongly affected by lithotypes (Co, Ge, Mo, V, Y), those depending on lithotype composition (As, Cr, Cu, Mn, Ni, P, Pb, Sc, Ta, Th, Ti, U, W) and those fairly equally distributed among the lithotypes (Ba, Ga, Hf, Sb, Sr, Zn, Zr). Some modern analytical methods permit the detection and identification of several trace elements in very small areas (volumes) of sample, for example Chen et al. (1981) and Minkin et al. (1982) used the proton (PIXE) microprobe to obtain information on the distribution of some elements in specific macerals in some US coals, the target area being a few μm^2. Their results suggested the highest concentrations of various elements occurred 'most frequently in vitrinites'. Samples of vitrinite, fusinite and exinite from US bituminous coals were examined by McIntyre et al. (1985) using secondary ion mass spectrometry (SIMS) and X-ray photoelectron spectroscopy (XPS). They found some regions in vitrinite 'where inorganic ions dominate the ion images' and these were separate from 'regions rich in organic materials'. Perhaps the most convincing illustration of elemental variation was shown by the sharp change in the content of organic sulfur from vitrinite to sporinite in a specimen of sporinite embedded in vitrinite which was measured by Wert et al. (1987) using TEM. Cecil et al. (1982) grouped several trace elements statistically into those associated with vitrinite, fusinite, semifusinite and macrinite (34 elements) and those associated with sporinite and micrinite (As, Ge, Hg). In view of the known occurrence of ultrafine mineral particles in coal (in the nm range) caution should be exercised in assigning specific trace elements to macerals. Even with the most modern techniques there is the possibility of missing an element present in an ultrafine mineral embedded in the maceral. Caswell, Holmes and Spears (1984a) found a correlation between coal type and total Cl, bright coals having higher contents than dull coals, probably due to the vitrinite, with its greater porosity, being the major source of the Cl. Perhaps part of the total Cl is combined with organic matter (Caswell, Holmes and Spears, 1984b). On the basis of changes of lustre, Kitaev (1970) has suggested that Ba, Be, Cu, Mn, Mo, Pb and W are closely related to organic substances in coals of the Bureya and Tyrminsk synclines.

Another approach to assessing the organic or inorganic association of trace elements in coal is to plot the content in coal or in ash against the ash yield. For example, values of B in ash plotted against ash yield for coals from the Sydney Basin, Australia (Clark and Swaine, 1962a), showed an inverse relationship, especially for coals from the northern areas, and hence it was inferred that 'it seems most likely that boron is mainly organically bound in the coal substance'. In the absence of direct evidence, it seems reasonable to infer that B is at least partly organically bound in

Australian bituminous coals (Brown and Swaine, 1964; Swaine, 1967). Nicholls (1968) used these plots for published data for coals from the Sydney Basin, NSW, from the Sydney Coalfield, Nova Scotia, and from the Svea seam, Spitsbergen. Nicholls concluded that B 'is largely, almost entirely, associated with the organic fraction in coals', while Ba, Co, Cr, Pb, Sr and V are usually associated with the inorganic fraction. Some elements (Cu, Ga, Ge, Mo and Ni) may be associated with either or both fractions. Results for the contents of several trace elements in clean coal composites from the Sydney Basin, NSW (that is, composite samples, prepared from subsection samples with ash yields < 35%), given in Clark and Swaine (1962a), were plotted against ash yields. The correlation coefficients (r) varied greatly and were often low (<0.40). However, some r values were useful as indicators of the nature of the association of the elements. Organic association was indicated for B from the northern area ($r = 0.60$), Co from the southern area ($r = 0.45$) and Ni from both areas ($r = 0.42, 0.65$). Inorganic association was indicated for B from the southern area ($r = 0.65$), Be from both areas ($r = 0.56$), Co from the northern area ($r = 0.49$), Cu from both areas ($r = 0.43$), Ga from both areas ($r = 0.68$), Mn from both areas ($r = 0.57$), Mo from both areas ($r = 0.57$), Sc from both areas ($r = 0.48, 0.65$) and Zr from both areas ($r = 0.53$).

Despite the limitations on the use of float–sink fractions for assessing organic and inorganic associations of trace elements in coal, they may give useful general information. An idea of the variability of such assessments can be gleaned from Table 3.1, which summarizes information on the percentage of organic association (Horton and Aubrey, 1950), on the percentage of organic affinity (Zubovic, 1966b; based on Zubovic, Stadnichenko and Sheffey, 1960c and Zubovic, Stadnichenko and Sheffey, 1961b), and on the order of organic affinity (Ruch, Gluskoter and Shimp, 1974; Gluskoter, 1975). There is broad agreement on the predominantly organic association for Ge, Be, Ga, Ti, B and V as assessed by Horton and Aubrey (1950) and Zubovic (1966b). Kirkby (1950) reported that Horton and Aubrey also found Mn, P and Sn to be 'associated entirely with the adventitious mineral matter'. The assessments for the four coals performed by Ruch, Gluskoter and Shimp (1974) show general similarities with the other two assessments, for example B, Ge, Ga, Ti, Be. The predominantly inorganic association for Zr, Mo, Mn, Pb, As, Hg, Cd and Zn is in keeping with the known presence of these elements in mineral matter. On the basis of contents of trace elements in density fractions from six NSW and Queensland coals, Ward (1980) grouped B, Ni, V, Zr, Co and Be as having organic affinity, Cu, Pb, Zn, Mo, Sr and Cr as having inorganic affinity and Ge as having variable affinity. Float–sink data for some German bituminous coals led Otte (1953) to suggest that Ge, Ga, Be, Ti, V, Ni, Cu, Mo, Zr and Cr were organically bound, that Mn was inorganically bound and that Co, Pb, Zn, Sb and Sn were partly organically and partly inorganically bound. Warbrooke and Doolan (1986) examined several density fractions from seven NSW bituminous coals and were able to establish interelement correlations, but 'organic bonding of elements could not be discriminated, but is probably of minor significance'. However, they found some degree of organic 'affiliation', albeit different

Table 3.1 Organic and mineral matter associations of trace elements in coal, as indicated by float sink methods

% organic association (Horton and Aubrey, 1950)		% organic affinity (Zubovic, 1966b)	Order of organic affinity (Ruch et al., 1974)			
			Davis	De Koven	Colchester	Herren
Ge	100	87	B	Ge	Ge	Ge
Be	75–100	82	Ge	Ga	B	B
Ga	75–100	79	Be	Be	P	Be
Ti	75–100	78	Ti	Ti	Be	Sb
B	75–100	77	Ga	Sb	Sb	V
V	100	76	P	Co	Ti	Mo
Ni	0–75	59	V	P	Co	Ga
Cr	0–100	55	Cr	Ni	Se	P
Co	25–50	53	Sb	Cu	Ga	Se
Y	–	53	Se	Se	V	Ni
Mo	50–75	40	Co	Cr	Ni	Cr
Cu	25–50	34	Cu	Mn	Pb	Co
Sn	0	27	Ni	Zn	Cu	Cu
La	–	3	Mn	Zr	Hg	Ti
Zn	50	0	Zr	V	Zr	Zr
			Mo	Mo	Cr	Pb
			Cd	Pb	Mn	Mn
			Hg	Hg	As	As
			Pb	As	Mo	Cd
			Zn		Cd	Zn
			As		Zn	Hg

for different coals, for Ge, Mo, Ni, Be, Br and some other elements. Kojima and Furusawa (1986) used float–sink fractionations from some Japanese coals to group elements into degrees of organic affinity from strong (Ti) to quite strong (V, Sr, B) to moderate-to-strong (Ba, Sn, Cr, Ni, Sc, Y, Be, Co, Cu, Zn, Zr, As) to weak (Mn). Studies of Illinois, US, coals showed that B, Be, Br, Ge, Ni, Sb, U and V had predominantly high organic associations (Harvey et al., 1983), while As, Ba, Cd, Mn, Mo, Pb, Tl and Zn had predominantly high inorganic associations; other elements had mixed associations. Using float–sink fractions, acid treatments and ion-exchange treatments, Kuhn et al. (1980) studied coals from eight states in the United States, and assigned B, Be, Br, Ge and Sb as organically associated, Zn, As and Cd as inorganically associated, and several others (for example Ga, Ni, P and Ti) as intermediate. In general, they found that most of the trace elements were in mineral form, a conclusion also reached by Finkleman and Gluskoter (1983). These deductions were general and variations must be expected for different coals. Density fractionation of lignite samples from the Began deposit in the Soviet Union indicated that Be, Mo, B, U, As and Ag were accumulated in the organic matter, Ti, Ga, Ba, Sr and Mn were in the inorganic matter and other elements, namely Cr, V, Zr, Co, Ni, Cu, REE, Zn, Pb and Sn, occurred in both forms (Smirnov, 1969b). Ratynskii and Glushnev (1967a,

1967b) have given a metal–organic affinity series, namely Be > W > Ga > Be > Nb > Mo > Sc > Y > La > Zn > Pb where an affinity factor varied from 2.6 to 0.02, values >1 indicating that the element is 'combined principally with coal organic matter'. On the basis of an enrichment coefficient, that is, the ratio of the highest concentration of an element in coal ash to its concentration in high ash rock interlayers, the following series was proposed by Minchev and Eskenazi (1966) to give the degree of binding with organic matter in Bulgarian coals: As, Ge, U, Sb, Ga, Mo, Co, Ni, Cr, Sc, V, Be, Cu, Yb, Zn, Y, Mn, Pb. On the basis of trace-element contents in ashes from some Bulgarian lignites, Minchev and Eskenazi (1972) suggested that Ge, As and Ag would be found only in organic matter, Y, Mo, Yb, Sr, Ba, V, Mn, Cu, Ni, Sn, Zn and Co would be mainly associated with organic matter, while Be, Sc, Zr, Ti and Cr would be associated mainly with mineral matter. Eskenazi and Chubriev (1984) found that Ge, W, Mo and Be were related to 'humic substances and fulvic acids' in brown coals from the Pirin deposit, Bulgaria. For coals from Belogradcik, Bulgaria, Minchev and Eskenazi (1963) suggested that Ge, Ga and Sb are bound to organic matter, that Cu, Zn, Ni and As are bound mainly to organic matter, and that Mn, Ba and Ti are bound mainly to inorganic matter. Kortenski (1986) suggested that the trace elements in coals from the Sofia Pliocene Basin, Bulgaria, could be grouped into those associated totally with organic matter (Ag, As, Ge, Mo, Ni, W), those strongly associated with organic matter (Ba, Co, Cr, Cu, Pb, Ti, V. Zr), those strongly associated with inorganic matter (Mn, Zn) and those associated totally with inorganic matter (Bi, Sn, Sr, Tl). However, for some Hungarian coals Odor (1967) suggested that Mn, Mo and Sr are predominantly organically bound. Valeska, Malan and Kessler (1967) found that Ge, Be, P, B and V contents in some Czechoslovakian coals increased with decreasing ash yield and hence these were regarded as organically associated. Correa da Silva Filho (1982) and Correa da Silva et al. (1984) suggested that B, V and Cr were associated with the organic matter in a Brazilian coal, whereas Ga was inorganically associated. In another Brazilian coal, Azambuja, Formoso and Bristoti (1981) classified Cu, Co, Ni, Cr, V, Zn and Pb as mainly associated with the organic fraction, but Mn, together with Pb and Zn in high-ash coals, were inorganically associated. Kitaev (1971) regarded Cu, Pb, Mo, W, Be, Ba and Mn as preferentially organically bound, while Zn, Cr, V, Ga, Zr, Ti and Sn were predominantly bound to mineral matter. Goodarzi (1987b, 1988) assigned As, B, Br and Cl to the organic fraction of lignite–subbituminous coals from Hat Creek, Canada, and of bituminous coals from British Columbia, while Goodarzi and Cameron (1987) assigned Br, Cl and Mn to the organic fraction and Ti, Cr, Hf, Ta, Th and V to the inorganic fraction of the bituminous to semi-anthracite coals from the Kootenay Group, Mount Allan, Alberta, Canada. Scholz (1987) has developed a graphical–statistical method for dealing with results from float–sink separations, thereby obtaining estimates of the proportions of an element that are organically and inorganically associated. Studies of North Dakota lignites showed organic affinity for Br, Mn, Sr, Y and Ba (Karner et al., 1984), mainly organic affinity for Rb, Sr, Mn, Ba, V and P (Benson, Falcone and Karner, 1984) and organic affinity for Sr and Ba

(Karner et al., 1986). Ion-exchangeable Ba, Sr and Mn were regarded as associated with carboxylate groups in samples of lignite from North Dakota, lignite from Texas and subbituminous coal from Montana (Benson and Holm, 1983, 1985). Studies of lignite from the Moose River Basin, northeast British Columbia, led to the conclusion that 'association with the organic matter is the most common mode of occurrence of trace elements' in this lignite (van der Flier-Keller and Fyfe, 1987a).

The above discussion has been based on studies of several elements for particular coals, but there are also references to work on, and conclusions for, one or a few elements. The following comments will be grouped for individual elements.

There are several references to the organic association of Ga, Ge, U and V, so these elements will be discussed first. A small part of the total Ga in some coals has been isolated as a Ga porphyrin (Bonnett and Czechowski, 1980, 1984; Bonnett, Burke and Czechowski, 1987). The geochemistry of metal complexes, mainly metalloporphyrins, in fossil fuels has been reviewed by Filby and van Berkel (1987). Some Ga in Russian coals is organically associated (Volodarskii, Zharov and Ratynskii, 1976; Alekseev, 1960; Ratynskii and Zharov, 1979) and Ratynskii and Zharov (1977) suggested that Ga reacted with carboxylic acid and phenolic groups in brown coals. Much attention has been given to Ge in Soviet coals. On the basis of the inverse relationship between Ge content and ash yield, Vistelius (1947) and Volkov (1958) assigned Ge to the organic coaly matter, and other workers regarded this as the prime association for Ge in other Soviet coals (Gordon, Volkov and Mendovskii, 1958; Menkovskii and Aleksandrova, 1963; Shpirt and Sendulskaya, 1969; Syabryai, Kornienko and Kuzminskaya, 1971; Sharova, Gertman and Semerneva, 1973). Studies of Ge in British coals (Aubrey, 1952) and in Czechoslovakian coals (Hak and Babcan, 1967) favoured an organic association. On the basis of the enhancement of Ge by vitrains, several workers regarded Ge as mainly organically bound (Inagaki, 1951b, 1952a; Minchev and Eskenazi, 1963–1964; Khanamirova, 1963; Ryabchenko and Bochkareva, 1968; Minchev and Eskenazi, 1969). Hallam and Payne (1958) regarded Ge as associated with organic matter in lignite. Zahradnik, Tyroler and Vondrakova (1960) postulated that Ge is bound to coal mainly by chemical adsorption, while Sofiyev, Gorlenko and Semasheva (1964) favoured adsorption rather than chemical bonding. In low-ash coals from Assam, India, Ge was reported as being 'in the form of organo metallic complexes' (Banerjee, Rao and Lahiri, 1974). Although it is difficult to establish the nature of the association between Ge and organic coaly matter, some Soviet workers have made suggestions. For example, Ge could be bonded to coal through non-reactive oxygen (Ryabchenko, Grishaeva and Lefeld, 1968), possibly by ether-type bridges O–Ge–O (Ryabchenko, 1968) and O–Ge–C (Ryabchenko, Alekhina and Lisin, 1968), the latter being as a connecting link between structural components in coal (Ryabchenko and Lisin, 1965). Saprykin (1965) proposed that Ge is bound to coal as organic Ge compounds, as Ge salts (e.g. phenates) of humic acids and by adsorption by organic matter. Bekyarova and Rouschev (1971) agree generally with the latter suggestions, but add an inorganic form, namely silicogermanates, i.e. solid solution of GeO_2 in SiO_2. According to Shpirt et al. (1984) 'it has been established that germanium does not form individual compounds in coals, but is attached to various fractions of the macromolecule of the coal substance by Ge–C bonds or with the formation of closed rings of the type of complex humates'. Whatever the nature of the association, it seems likely that Ge is generally associated with the organic coaly matter in most coals.

Since the early studies by Szalay (1964) and others which established the close association between U and humic acids, it has been recognized that this is a likely association for U in coals. According to Morales and Gasos (1985) 'U occurs trapped in the lignite network, not as discrete minerals' in Spanish lignites. In some West German organic-rich materials, it has been suggested that U bonding is based on ion exchange reactions with acid groups (von Borstel and Halbach, 1982), while U in a Texas lignite is 'most probably coordinated to the oxygen-containing functional groups' (Mohan et al., 1982). Studies of West German brown coals indicate that there is 'complexation of UO_2^{2+} by carboxylate groups which act as bidentate ligands' (Koglin, Schenk and Schwochau, 1978). In addition to organically bound U in a south Texas lignite, Ilger et al. (1987) suggest that there could be some inorganic U, possibly as the mineral coffinite. However, Mohan et al. (1982) pointed out that very small inclusions of minerals in organic material are difficult to distinguish from organically bound U, as was found by Rhett (1979) who detected 'submicron-sized crystals of coffinite encapsulated in an inert organic matrix' present in a carbonaceous U ore from New Mexico.

Although Treibs (1935) found metal porphyrins in coals, there is no firm evidence for vanadyl porphyrins in any coals. The careful work of Bonnett and coworkers (Bonnett, Burke and Czechowski, 1987) on metalloporphyrins has not led to the detection of vanadyl porphyrins in coals. The examination of a sample of V-rich material (about 2% V) with about 25% organic matter, from Mina Ragra, Peru, did not show V porphyrin, the limit of detection being about 0.2 ppm V as porphyrin (C.J.R. Fookes, 1989, pers. commun.). However, there is the possibility that some V porphyrin, albeit a very small amount, may be resistant to the extraction techniques used, perhaps because it is bound in a complex framework. Wong et al. (1982) did not find any evidence for porphyrins in their tests on coal using high-resolution X-ray spectroscopy. Marczak and Lewinska-Ochwat (1987) found about 2% of the total V in some Polish coals in the organic matter. There are reports of V associated with organic coaly matter in coals from Hungary (Almassy and Szalay, 1956), from Bulgaria (Uzunov and Karadzhova, 1968; Uzunov, 1976, 1980) and from Brazil (Correa da Silva et al., 1984).

The association of Sb with organic matter in coals from Witbank, South Africa, was proposed by Hart, Leahy and Falcon (1982), while As in Texas lignite is probably in organic combination (Clark et al., 1980). There are suggestions that Be is organically associated in coals from the Urgal deposit, USSR (Alexseev, 1960), and from the Sokolov Basin, Czechoslovakia (Hak and Babcan, 1967). Although there is no direct evidence, it has been suggested that B is 'strongly related to the organic matter' in coals from the Santa Rita Basin, Brazil (Correa da Silva et al., 1984), that it is 'mostly associated with the organic matter' in coals from the Angara–

Chulym depression, USSR (Timofeev *et al.*, 1967), that it 'may be organically bound, at least in part' in NSW bituminous coals (Brown and Swaine, 1964), and partly bound to organic matter in coals from the Gissar Range, USSR (Valiev, Pachadzhanov and Adamchuk, 1977), where the amount of silicate-bound B increased with increase in rank. Bromine is regarded as organically bound in coals from the Witbank Basin, South Africa (Hart, Leahy and Falcon, 1982), and from the Kootenay Group, British Columbia (Goodarzi, 1987a). It has been stated that Cl is organically combined in some South African coals (Technologist, 1959) and that it may be held on the coal substance as Cl^- ions (Edgcombe, 1956), possibly through ion exchange (Daybell and Pringle, 1958). There is possibly some organic Cl in Victorian brown coals from Yallourn but not in those from Morwell (Durie, 1961). In some coals from the United Kingdom most of the Cl is in vitrinite (Saunders, 1980), and it seems that the Cl occurs 'most probably as ions associated with the coal matter' (Gibb, 1983). A study of bituminous coals from the Asturian central coalfield, Spain, showed that 'Cl is mainly associated with the organic matter' (Martinez-Tarazona, Palacios and Cardin, 1988). Ladner (1984) has doubted the general applicability of the method of determining organically bound Cl, which is based on heating a sample of coal in a stream of hydrogen and ammonia (Chakrabarti, 1978).

Some Co has been reported as organically bound in coals from the Witbank coalfield, South Africa (Hart, Leahy and Falcon, 1982), and from Ghugus coals, India (Singh, Singh and Chandra, 1983). Some peats in the Tantramar swamp, New Brunswick, Canada, contain 3–10% Cu, which is organically bound, possibly as a Cu chelate (Fraser, 1961). Functional groups ($-OH, -COOH$) can chelate Cu in peats and low-rank coals (NAS, 1979). Using a SIMS technique, McIntyre *et al.* (1985) found evidence that 'indicated the F was sited organically' and further work with XPS showed that the F was 'largely directly bonded to carbon'. In most Bulgarian coals, Hf is regarded as inorganically bound, the exceptions being some low-ash coals where part of the Hf seems to be organically bound (Eskenazi, 1987a). Indium is bound to the inorganic and organic parts of Bulgarian lignites (Eskenazi, 1982a). In brown coals from the Burgas Basin, Bulgaria, Li is partly associated with the organic matter. Manganese is predominantly inorganically associated in bituminous coals, but Bonnett and Czechowski (1981) isolated a small amount of Mn porphyrin from a Polish coal. Manganese in brown coals is regarded as associated with the organic matter via carboxylic acid ($-COOH$) groups (Brown and Swaine, 1964; Swaine, 1967). According to Idzikowski (1960) Mn is not limited to the inorganic matter in coal. Organically associated Hg has been reported by Dvornikov (1981a, 1981b). In Hungarian coals Mo has been reported as bound to organic matter (Almassy and Szalay, 1956; Szilagyi, 1971). It has been suggested that Ni is at least partly associated with the organic coaly matter (NAS, 1979; Singh, Singh and Chandra, 1983). Smith (1941a) has stated that organic P can only be present in South African coals 'in negligibly small amounts'. There is indirect evidence that Australian bituminous coals may contain 40–95% of total Sc organically bound (Swaine, 1964), the corresponding figure for Soviet coals being up to 87% (Borisova *et al.*, 1976). Clark *et al.* (1980) postulated that Se is probably organically combined in Texas lignite, while Uman et al. (1988) suggested that organically associated Se 'may be the most common form' in subbituminous coals from the Powder River Basin, USA. A high-Se coal from the Soviet Union has approximately 50% of total Se 'adsorbed on coal' according to Savelev and Timofeev (1977). There are references to organic Ti in coals from Bulgaria (Eskenazi, 1972–1973), USA (Wong et al., 1983; McIntyre et al., 1985) and Japan (Kojima and Furusawa, 1986). Ershov (1961) suggested that Y and REE of the Y group occur in coals as organometallic compounds. On the basis of adsorption and desorption experiments with peat humic acids and cation exchange resins, Eskenazi (1977) suggested that W can form organometallic complexes, which may explain enhanced concentrations of W in some coals. In Bulgarian coals, some of the total Zr is organically bound, especially in low-rank coals (Eskenazi, 1987a).

3.2 Methods for determining the total amount of mineral matter

Minerals that occur in coals as discrete particles may be disseminated, in layers, as nodules, in fissures and as rock fragments (Harvey and Ruch, 1984).

It is appropriate to refer to methods of determining the amount of mineral matter in coal, as distinct from the ash yield, because ash is not a property of coal *per se*. Brief reference will also be made to methods of ascertaining the nature of the constituent minerals in coal. It is important to stress that the term 'ash' should not be used for mineral matter. Of course, there is a relationship between mineral matter and ash yield, the latter being usually slightly lower than the former, mainly because of losses of water and gaseous products of decomposition of some minerals, depending on the temperature of ashing. Minerals are found in coals in a range of sizes from submicrometre to millimetre or more. Mineral impurities can often be seen directly, depending on their size, for example by X-radiography of pieces of coal (Briggs, 1929; CSIRO, 1957), petrological microscopy and TEM, the limits of detection for width or diameter being 1–2 mm, about 5 μm and about 20 nm respectively.

The two main approaches that have been used for the determination of the total amount of mineral matter are low-temperature oxidation and acid digestion. Based on published information about the effects of heat on minerals commonly found in coal, Brown, Durie and Schafer (1959) carried out experiments on Australian coals, thereafter proposing a method which found the total mineral matter content. In order to avoid local hot spots during the oxidation of coaly matter, the coal sample was heated at about 200 °C for 2–3 h prior to the final heating at 370 °C for about 100 h. This method was also used by Brown, Belcher and Callcott (1965). An improvement was introduced by Gluskoter (1965) who applied the technique of plasma ashing which uses oxygen in a radiofrequency field to form an activated gas plasma to oxidize the coaly matter at around 150 °C. This method was found to be suitable for Australian coals (Frazer and Belcher, 1973) and is recommended in an Australian Standard (AS, 1983a), together with the 370 °C air oxidation method. The oxygen plasma method, now usually termed LTA, was investigated by Miller, Yarzab and Given (1979) who confirmed Frazer and Belcher's observations that some pyrite may be oxidized and that some organic sulfur may be fixed as sulfate. Miller, Yarzab and Given (1979) found that the oxygen plasma method is not applicable to lignites and subbituminous coals, although it can be modified for use with brown coals from Victoria, Australia (AS, 1983a), and Felgueroso *et al.* (1988) recommended extraction with NH_4Ac, followed by LTA, for determining the 'true' mineral matter in Spanish brown coals. Miller, Yarzab and Given (1979) 'urge users of LTA for any purpose to control carefully the conditions of ashing'. The methodology of LTA has been discussed in detail by Miller (1984). A new method, termed 'glow discharge-excited LTA', has been used by Adolphi and Störr (1985) who reduced the ashing time to $4 h g^{-1}$ for lignites and to $8 h g^{-1}$ for bituminous coals, with no 'distinct influence on preexisting minerals'. The acid digestion approach, using HCl and HF, proposed by Radmacher and Mohrhauer (1955), aimed at removing the mineral matter without affecting the coaly matter. This approach with some modifications was applied to UK coals (Bishop and Ward, 1958), US coals (Tarpley and Ode, 1959) and Australian coals (Brown, Durie and Schafer, 1959, 1960). The preferred approach would seem to be an LTA method, which has the added advantage of providing a relatively unchanged sample for the identification of its constituent minerals.

It is of historical interest to recall that the total mineral matter content used to be calculated from the empirical KMC formula, namely mineral matter = 1.09 ash + 0.5 (pyrite) + 0.8 CO_2 + 2.5 S (sulfate) + 0.5 Cl − 2.75 S (ash). This formula can give good values, for example the ranges of total mineral matter contents for 18 Queensland coals were 15.7–33.8% by direct determination and 16.1–34.5% by the KMC formula, the largest difference being 1.1% (CSIRO, 1962).

3.3 The identification of minerals in coal

It is very difficult to examine fine-grained minerals that are intimately mixed with the coaly matter, although cleat minerals and the like may be sufficiently large to be easily removed for identification. Hence, the use of the LTA method to prepare samples that are substantially free from organic matter. There are reviews of the various methods required to identify minerals in coal, for example Gluskoter, Shimp and Ruch (1981), Jenkins and Walker (1978) and Finkelman and Gluskoter (1983), the latter being concise and helpful. For details of methods, including background theory, special texts should be consulted, but there are some references especially relevant to coal and these will be considered.

An important method for identifying minerals is X-ray diffraction (XRD), which has been widely used for coal mineral matter, some relevant references being Rekus and Haberkorn (1966), O'Gorman and Walker (1971), Rao and Gluskoter (1973) and Renton (1984). Although claims have been made about quantitative XRD, it is generally regarded as semiquantitative at best.

The scanning electron microscope (SEM), plus an energy dispersive X-ray detector (EDX), is a most useful and versatile means of studying mineral matter, as has been shown by Russell (1977), Finkelman and Stanton (1978) and Russell and Rimmer (1979). Indeed, the thorough study of inorganic matter in coal (Finkelman, 1981a) was greatly aided by SEM–EDX, especially for examining fine particles of the order of a micrometre or so and by electron microprobe analysis (EMPA: Raymond and Gooley, 1979). The results of Finkelman's work can be found in Finkelman (1978, 1979, 1982b, 1986, 1988). The association of inorganic elements with particles of mineral matter can also be ascertained by SIMS. Scanning transmission electron microscopy is being used to identify minerals smaller than about 100 nm (Allen and van der Sande, 1984; Palmer and Wandless, 1985). Other useful methods for identifying some minerals in coal are differential thermal analysis (DTA: Warne, 1979), infrared (IR: Estep, Kovach and Karr, 1968; O'Gorman and Walker, 1971), Fourier transform infrared (FTIR: Painter *et al.*, 1978, 1981) ion microprobe mass analysis (Finkelman *et al.*, 1984a), optical microscopy (Stach *et al.*, 1982), Mössbauer spectroscopy (Montano, 1979; Huggins and Huffman, 1979; van der Kraan, Gerkema and van Loef, 1983; Pankhurst, McCann and Newman, 1986) and solid state NMR (Barnes *et al.*, 1986). Several of the above methods were used in an interlaboratory comparison of the mineral constituents in a sample of an Illinois, USA, coal (Finkelman *et al.*, 1984b). Procedures for the megascopic–microscopic examination of bituminous coals are given by Chao, Minkin and Thompson (1982).

3.4 Trace element–mineral matter associations

The occurrence of trace elements associated with mineral matter may be as discrete minerals, as replacement ions in minerals, and adsorbed on minerals, as shown in Figure 3.1 (Swaine, 1967, 1977a). Elemental grains may occur in some coals, for example Au particles have been found in some US coals (Finkelman, 1981a). Examples of these occurrences will be given below. Although the comment that 'perhaps every mineral known to mankind could be found in coal, if one looked hard enough' (Wert *et al.*, 1987) is an overstatement, there are certainly more than a hundred minerals found in coals and careful investigation, especially of micron and smaller sized particles, should lead to the identification of many more minerals. Any trace element may occur in a coal in more than one form and, in general, it seems that most elements are partly organically associated and partly mineral associated. Trace elements in coals with ash yields of more than a few percent are likely to be predominantly mineral associated.

Some mineral associations have necessarily been mentioned in Chapter 3.1. There are several collected papers and reviews which are relevant to mineral matter in coal. The occurrence, form and distribution of mineral matter in peat are discussed in detail in 19 papers collated by Raymond and Andrejko (1983). Early studies on mineral matter in US coals were carried out by Gauger, Barrett and Williams (1934), Ball (1935), Sprunk and O'Donnell (1942) and Thiessen (1945). Recent useful publications include Finkelman (1981a), Renton (1982), Mraw et al. (1983) and the book edited by Vorres (1986) which covers 38 papers presented at a Conference (ACS, 1984). Mineral matter in coal has been reviewed by Nelson (1953), Ode (1963), Watt (1968), Mackowsky (1968), Gluskoter (1975), Gluskoter, Shimp and Ruch (1981), Mackowsky (1982) and Raask (1985b). Information on mineral matter in Australian coals is presented by Kemezys and Taylor (1964), CSIRO (1965a), Brown, Belcher and Callcott (1965), Ward (1978), Doolan, Mills and Belcher (1979) and Corcoran (1979), in New Zealand coals by Soong and Gluskoter (1977) and in South African coals by Falcon (1978). Geological factors that influence mineral matter in coal are discussed by Cecil et al. (1979, 1982) and the distribution of mineral matter in lithotypes from some Indian coals has been studied by Singh (1987).

Statistical approaches have been used to indicate minerals in coal and trace-element associations. Roscoe, Chen and Hopke (1984) used target transformation factor analysis for some US coals and confirmed their results by XRD. Relationships between elements may be suggested by the dendograph, which is a method of general cluster analysis (Labonte and Goodarzi, 1985). Glick and Davis (1987) applied principal component analysis and cluster analysis to data for 335 US coals, with some interesting results. After commenting that these techniques are useful for 'suggesting the geologic sources of variability in the inorganic content of coals' they rightly cautioned that 'the approach should be followed by careful investigations of individual samples to confirm the validity of the statistically observed relationships'. Mraw et al. (1983) examined the use of correlation coefficients and concluded that 'it is important to remember that there could be some instances where significant statistical correlations may mean nothing geochemically and, conversely, that a significant geochemical mechanism may be operating which does not show the statistics'. The above comments by Glick and Davis (1987) and Mraw et al. (1983) endorse the view that the identification of mineral matter and related trace-element associations is the province of direct scientific measurements which cannot be replaced by statistical techniques.

In most coals, the dominant mineral groups are silicon-rich, especially clays, carbonates and sulfides. Trace elements are associated with these major minerals, as will be discussed below, followed by sections on other minerals containing trace elements.

3.5 Silicon-rich minerals

Among the commonest silicon-rich minerals found in coals are the clays, kaolinite, illite, montmorillonite, mixed-layer (for example illite-montmorillonite) and chlorite. Other silicon-rich minerals that have been reported include halloysite, muscovite, biotite, sericite, feldspars, topaz, zircon, tourmaline and garnet. Several of these minerals that have been found in the Okefenokee Swamp and Everglades Basin can be regarded as detrital, authigenic or diagenetic (Andrejko, Cohen and Raymond, 1983). It has been suggested by Davis et al. (1984) that clay minerals in coals could be derived from detrital input, transformation and neoformation (authigenesis). Transformation involves the alteration of detrital minerals, for example Staub and Cohen (1978) showed that smectites can change to kaolinite under conditions of low pH and the loss of cations in the presence of organic matter. Other changes that could occur after the peat stage include the alteration of smectites to illite, an increase in crystallinity of kaolinite and the loss of biogenic silica (Bailey and Kosters, 1983). Neoformation involves the precipitation of clay minerals from solutions or gels. Under conditions of impeded drainage in coal swamps, surface properties of clays may change to give increased reactivity as reported for gleyed soils (Mitchell *et al.*, 1968). Clays may adsorb ions from aqueous solutions, the capacity varying from about 100 meq kg^{-1} for kaolinite to about 1000 meq kg^{-1} for montmorillonite (Pickering, 1986). During the early stages of coalification there may be clay–organic matter associations, including organic coatings on clay particles. Thus adsorption of ions may be either by clay particles *per se* or by the organic surface layers on clays. Some trace elements may have migrated as cations adsorbed on clays, as postulated for sediments by Strakhov, Zalmanzon and Glagoleva (1956). Finkelman (1981a) suggested that Be, Cr, Cs, F, Ga, Li, Rb, Ti, V and possibly Ni and Sc may be associated with clays in bituminous coals. A study of illites, kaolinites and mixed-layer clays from the Upper Freeport coal, USA, showed that Ti was strongly associated with the illites (Minkin, Chao and Thompson, 1979). Leutwein and Rösler (1956) found 80 ppm Ga and 40 ppm V in two kaolinites from German coals, while data are given by Kronberg *et al.* (1987) for 48 trace elements in clays associated with a Canadian lignite. Various authors have added other elements to this list, namely B, Ba, Co, Mn, Mo, Sb, Sr and W, while Ba, Cs, Pb, Rb and Sr would seem likely to be in orthoclase feldspars found in some coals. A study of 27 coals from eight areas in the United States led to the suggestion that at least 20 trace elements are commonly associated with clays (Kuhn *et al.*, 1980). Quartz, chalcedony and cristobalite are found in coals, but are unlikely sources of trace elements. Zircon, topaz, tourmaline and garnet are only found in traces in some coals, perhaps because their detection is limited by difficulties at trace levels. Several trace elements may be associated with these minerals namely F (tourmaline, topaz), B and Li (tourmaline), Cr (garnet), Hf (zircon) and, of course, Zr (zircon). Several other elements may be associated with clays in sediments, namely, Ag, Cu and Th. It has been suggested that allanite (a rare earth aluminium iron silicate) may be present in Waynesburg, USA, coals (Finkelman, 1978).

3.6 Carbonate minerals

Carbonate minerals are commonly found in bituminous coals, the main ones being calcite, siderite, dolomite and ankerite. Aragonite is rarely found, probably because its formation needs a high salinity and a ratio of Mg to Ca of greater than 12. Although magnesite has been reported occasionally, it is likely to be present in coal only rarely. It would seem that the available Mg was used to form dolomite or ankerite. Calcite, siderite and ankerite were found in some Queensland Triassic coals (Cook and Taylor, 1963) and in some NSW Permian coals (Smyth, 1968), while one or more of these carbonates was also found together with dolomite in some other NSW Permian coals (Taylor, 1968).

In some coals barium carbonate (witherite) and strontium carbonate (strontianite) have been identified. For example, witherite has been found in some Australian coals (Day, Jones and Belcher, 1979).

It has been known for some time that Mn is often relatively high in carbonate minerals, including those found in coal. For example, 78 samples of ankerite with varying amounts of Ca, Fe and Mg contained <50–15 800 (with a mean of 6200) ppm Mn (Pringle and Bradburn, 1958), while 17 samples of ankerite from NSW Permian coals contained 500–17 700 (with a mean of 7500) ppm Mn (Brown, Durie and Schafer, 1960). Sinnatt, Grounds and Bayley (1921) had found 2400–8600 ppm Mn in some ankerites in coals from Lancashire, UK. The presence of ankerite, which is a ferroan variety of dolomite, poses the question 'is Mn replacing some Fe or Ca or both?'. Goldschmidt (1937) proposed that ions could substitute for other ions in crystals, depending mainly on the size and charge of the ions. Although there are exceptions to these rules (Burns and Fyfe, 1967), it was considered that Mn could substitute for some Fe in siderite and for some Ca in calcite. Samples of hand-picked siderite and calcite from Australian bituminous coals were found to contain 3000 and 10 000 ppm Mn respectively (Brown and Swaine, 1964). Samples of calcite from other Australian seams and from a Japanese coal contained 400–19 200 ppm Mn. The sample of calcite from a fissure in a Queensland coal (Figure 3.2)

Figure 3.2 A sample of calcite from a fissure in a Queensland coal seam (actual size)

contained 2240 ppm Mn and 1620 ppm Sr. 'Indirect evidence for the association of Mn with siderite comes from the good linear correlation ($r = 0.89$) between carbonate iron (that is, total iron less pyritic iron) and Mn for 56 samples of coal from 7 locations in the Lithgow seam, New South Wales.' (Swaine, 1986b.) A positive correlation between Mn and Ca was found by Stricker (1974) who analysed 82 coal samples from the Illinois Basin, USA, and by Hatch, Avcin and van Dorpe (1984) for bituminous coals from Iowa.

The replacement of Ca^{2+} ions by Mn^{2+} ions in calcite in a coal from Pittsburgh was shown by an electron paramagnetic resonance (EPR) study (Malhotra and Graham, 1985). However, for some Canadian coals Goodarzi and Cameron (1987) found that Mn should be one of several elements 'with the greatest affinity for the organic fraction'. Some samples of dolomite from Upper Carboniferous coals from the Ruhr region, West Germany 'contain high amounts of manganese' (Littke, 1987). However, 12 samples of dolomite-rich material from NSW coal seams were relatively low in Mn, namely 10–150 ppm Mn (D.J. Swaine, unpublished). Strontium is also associated with calcite, for example calcites from several NSW seams contained 350–15000 ppm Sr, with aragonite present in some Bulgarian coals (Eskenazi and Minceva, 1989). In a study of some coals from eastern United States, Palmer and Wandless (1985) found that Zn was associated with carbonate minerals, as well as with pyrite.

3.7 Sulfide minerals

Trace-element aspects of the abundance and origin of sulfide minerals in coal have been reviewed by Swaine (1984b), and there is relevant information on sulfide minerals in Czechoslovakian coals (Bouska, 1981). 'Reduced sulfur compounds are ubiquitous components of anaerobic sediments.' (Morse et al., 1987.) Hence it is expected that sulfide minerals would occur in coals. Pyrite is found in most coals, while several other sulfides have been reported, albeit at trace levels. Although optical microscopy is often sufficient to identify small particles, say down to 5–10 μm diameter, success with smaller particles requires SEM plus EDX (Finkelman, 1981a), EMPA (Raymond and Gooley, 1979) and STEM (Hsieh and Wert, 1981). Pyrite and marcasite (FeS_2) are found in various coals in a range of forms and sizes (Mackowsky, 1982) which may be related to different stages of syngenesis and epigenesis (Querol, Chinchon and Lopez-Soler, 1989). Framboidal pyrite, which was named by Rust (1931) because of its raspberry-like appearance under the microscope, is made up of a mass of tiny micelles (around 1 μm diameter) forming a cluster with a diameter of, say, 30–50 μm. The resulting large surface area would seem to favour the adsorption of ions. The rare iron sulfides, pyrrhotite ($Fe_{1-x}S$: Hsieh and Wert, 1981) and greigite (Fe_3S_4: Harvey and Ruch, 1984), have been reported. Greigite is usually referred to as melnikovite in Soviet publications. Some trace elements may substitute isomorphously in iron sulfides and may occur also as mineral sulfides in the main sulfide mineral, say pyrite. Sedimentary sulfides, and this is probably the case with sulfides in coal, do not permit much lattice substitution, simply because lattice expansion is not favoured at low temperatures (Degens, 1965). 'It is often difficult to distinguish between matter in true solid solution in the pyrite structure and impurities contained in discrete minerals.' (Morse et al., 1987.) Among trace elements that have been reported as associated with pyrite are As, Hg and Se (Palmer and Filby, 1984). Minkin et al. (1984) found some As in pyrite which seemed to be in solid solution, and some As and Se which could have been mineral inclusions in pyrite. Arsenic has been reported as arsenopyrite associated with pyrite (Crossley, 1946). Finkelman (1981a) suggested that Ag, Bi, Ga, Ge, In, Mn, Sb, Sn and Tl may be present as sulfides or associated with sulfides in some coals. Arseniferous pyrite has been referred to by Bayet and Slosse (1919), Crossley (1946) and Hokr (1977) and a pyrite crystal from a North Dakota lignite had 37 ppm Mn (Miller and Given, 1986). Goldschmidt (1954) reported pyrite with 3000 ppm As. A sample of pyrite from the Lithgow seam, NSW, had 70 ppm Tl and 1000 ppm As (D.J. Swaine, unpublished). In a study of coals from Iowa, USA, Hatch, Avcin and van Dorpe (1984) found correlations that strongly suggested the association of As, Hg and Sb, and to a lesser extent Co, Cr, Cu, Mo, Ni, Pb, Se, U and V with pyrite. In some West German brown coals, Brinkmann (1977) found Co and Ni usually bound to marcasite. An investigation of coals, mostly from the Illinois Basin, USA, indicated that 'some pyrites contain As and perhaps Sb' (Gluskoter et al., 1977). Samples of pyrite and marcasite from coals from several areas in the Soviet Union contained up to 200 ppm Tl and As, Cu, Mo, Pb and Zn were also present (Voskresenskaya, 1968). Some pyrites from Donbass coals, USSR, contained Ag, As, Co, Cu, Hg, Ni, Pb, Sb and Zn (Dvornikov and Tikhonenkova, 1968). Lauf (1981) found that Cd, Co, Cr, Cu, Ni and V appeared 'to be associated with microscopic framboidal pyrite' in some US coals. Wandless (1958) found 70 ppm Se in a pyrite from a UK coal. In the Au-bearing coal from Wyoming, South Dakota, USA, the Au may be associated with pyrite (Stone, 1912). Pyrites from coals in the Donets Basin, USSR, contained Ag, As, Co, Cu, Hg, Ni, Pb and Zn (Dvornikov, 1967a, 1967b). Elemental sulfur has been reported in coal (Yurovskii, 1960), but in some cases this may have been formed by oxidation on exposure to air (Duran, Mahasay and Stock, 1986). Trace elements are not likely to be associated with elemental sulfur. Goldschmidt and Peters (1931) did not detect Ga (<4 ppm) in pyrite from bituminous coals or in marcasite.

Sphalerite has been reported in many coals, early references being to a sample from the Netherseal colliery, Leicestershire, UK (Binns and Harrow, 1897), and to a sample from Bicknell, Indiana, USA (Dove, 1921), where the sphalerite was in nodules of pyrite–marcasite in shale adjoining the coal beds. Perhaps the most widely studied occurrences of sphalerite are those in the coals of the Illinois Basin, USA (Hatch, Gluskoter and Lindahl, 1976; Cobb et al., 1979; Cobb et al., 1980). Sphalerite occurs in NSW coals (Taylor and Warne, 1960; Smyth, 1968), in Queensland coals (Kemezys and Taylor, 1964), in German coals (Mackowsky, 1982), in Czechoslovakian coals (Bouska, 1981), in coals from the West Midlands, UK (Spears and Caswell, 1986), in South African coals (Falcon and Snyman, 1986), in various US coals (Parratt and Kullerud, 1979; Finkelman, 1981a; Whelan, Cobb and Rye, 1988), and elsewhere. As well as being the prime site of Zn in most bituminous coals, sphalerite is also an important site of Cd. Binns and Harrow (1897) suggested that 'a minute yellow precipitate, too small for identification', which was found during the analysis of a coal-derived sample of sphalerite, was probably Cd. Gluskoter and Lindahl (1973) found Cd in sphalerites from Illinois coals, where up to about 1% Cd has been reported (Hatch, Gluskoter and Lindahl, 1976).

Binns and Harrow (1897) found galena (PbS) in the roof rock of a coal seam in Leicestershire, UK. This galena contained some Ag, Cu and Zn. Galena has also been found in coals from Central Missouri, USA (Hinds, 1912), New South Wales (Kemezys and Taylor, 1964), Germany (Mackowsky, 1982), the Illinois Basin (Chou and Harvey, 1983), various parts of the United States and several other countries (Finkelman, 1981a), West Midlands, UK (Spears and Caswell, 1986), Czechoslovakia (Bouska, 1981) and Leicestershire, UK (Cressey and Cressey, 1988). Although galena has not been observed frequently, it is clearly a source of Pb in some coals. A rare source of Pb, mainly in coals from the Appalachian region, USA, is the lead selenide, clausthalite (Finkelman, 1981a), which was also found in a fracture in a seam from Leicestershire, UK (Cressey and Cressey, 1988).

Chalcopyrite ($CuFeS_2$) was found in the Netherseal colliery, Leicestershire, UK (Binns and Harrow, 1897), and, together with bornite (Cu_5FeS_4), in various other coals (Stutzer, 1940). Other findings were in an Australian coal (Kemezys and Taylor, 1964), in various US and other coals (Finkelman, 1981a), in German coals (Mackowsky, 1982) and in Leicestershire, UK, coals (Cressey and Cressey, 1988). Bouska (1981) records several copper sulfide minerals found in Czechoslovakian coals, namely, chalcopyrite, bornite, azurite ($Cu_3(CO_3)_2(OH)_2$), chalcocite (Cu_2S), malachite ($Cu_2CO_3OH_2$) and tetrahedrite (($Cu,Fe_{12})Sb_4S_{13}$). In some coals from western Canada, Goodarzi, Foscolos and Cameron (1985) 'identified CuS with some isomorphic substitution by Zn and As for Cu'. Coals from Tirana, Albania, contained some chalcopyrite, bornite, chalcocite and covellite (CuS) which were irregularly distributed (Mullai, 1984). Malachite occurs in coals from Zwickau, East Germany (Stutzer, 1940). According to Moore (1947), stibnite (Sb_2S_3) may occur in coal and it was indeed found in a coal from Czechoslovakia (Bouska, 1981).

Arsenopyrite (FeAsS) has been suggested as the form of As in UK coals (Wandless, 1959) and realgar (As_2S_2) and opiment (As_2S_3) were found in some Czechoslovakian coals (Cech and Petrik, 1972), while others contained arsenopyrite (Bouska, 1981). The statement by Chapman (1901) that there is 'little doubt that arsenic in coal is present as arsenical pyrites and that little exists in any other form' may well be correct, but the nature of the association between As and pyrite is probably not easily ascertained. Careful work by Minkin et al. (1984) on a Pennsylvanian coal indicated that most of the As was probably in solid solution in pyrite, but some As-bearing mineral inclusions may also be incorporated in pyrite. Linnaeite (($Co,Ni)_3S_4$) has been reported in a coal from Wales (Des Cloizeaux, 1880; North and Howarth, 1928) and from Czechoslovakia (Bouska, 1981). It also occurs in concretions in the South Wales coalfield (Firth, 1973). Correlations between Ge and S from 400 samples of coal from the Soviet Union indicated some association between Ge and S, but the form of S was not given (Lazebnik, Grinvald and Dolgopolov, 1967). In view of the general opinion that Ge is organically associated, perhaps the above correlation is really with organic S. The work of Stadnichenko et al. (1953), together with their analyses of 12 pyrites from US coals showing <20 ppm Ge, makes it unlikely that Ge is in pyrite or other sulfide minerals, with the possible exception of sphalerite. Cinnabar (HgS) has been found in several coals from the Soviet Union (Vasilevska and Shcherbakov, 1963; Dvornikov, 1967c; Dvornikov, 1971; Dvornikov and Tikhonenkova, 1974; Dvornikov, 1981a,b; 1985) and in coals from Bavaria (Stutzer, 1940). There is some evidence for Mn in pyrite (Cambel and Jarkovsky, 1967; Finkelman, 1981a), but this is not considered to be the usual situation because manganese sulfides are much more soluble than iron sulfides and they have a very restricted stability field (Logvinenko, 1972).

The occurrences of Ni minerals in coal have been summarized by Swaine (1980b). Millerite (NiS) has been found in concretions in the South Wales coalfield (North and Howarth, 1928; Firth, 1973), following early discoveries there by Miller (1842), and by Des Cloizeaux (1880) and Grosjean (1943) in Belgian coal areas. There are also reports of millerite in NSW coals (Lawrence, Warne and Booker, 1960) and in Czechoslovakian coals (Bouska, 1981). There are reports of ullmanite (NiSbS) from the Durham area, UK (Spencer, 1910), and of bravoite ((Ni, Fe)S_2) from the Kladno–Rakonitz area, Czechoslovakia (Bernard and Padera, 1954). For the Upper Freeport coal, USA, it has been suggested that a minor proportion of the Se may be as tiny mineral inclusions 'incorporated into the pyrite at the time of crystallization' (Minkin et al., 1984).

3.8 Oxides

Limonite (FeOOH.nH_2O) and hematite (Fe_2O_3) were found in some US coals (Gauger, Barrett and Williams, 1934) and limonite and goethite (FeOOH) in some NSW coals (Kemezys and Taylor, 1964). Goethite has also been found in some coals from New Zealand (Soong and Gluskoter, 1977) and in some Albanian coals (Mullai, 1984), and hematite in two US coals (O'Gorman and Walker, 1971). Magnetite (Fe_3O_4) occurs in some US coals (Gauger, Barrett and Williams, 1934; Finkelman, 1978; Alexander, Thorpe and Senftle, 1979), in some NZ coals (Soong and Gluskoter, 1977), in some Indian coals (Mehdi and Datar, 1960) and in some Albanian coals (Mullai, 1984). Rekus and Haberkorn (1966) observed goethite, hematite and magnetite in various coals.

Anatase (TiO_2) was found in some NSW coals (Ward, 1978; Day, Jones and Belcher, 1979), in an Indian coal (Mehdi and Datar, 1960) and, together with rutile (TiO_2), in US coals (Palmer and Wandless, 1985; Steinmetz, Mohan and Zingaro, 1988). Rutile has been found in various US coals (Gauger, Barrett and Williams, 1934; O'Gorman and Walker, 1971; Finkelman, 1978; Palmer and Filby, 1984), in NZ coals (Soong and Gluskoter, 1977), in South African coals (Falcon and Snyman, 1986) and in German coals (Mackowsky, 1982). Anatase and brookite (TiO_2) were found together in Leicestershire coals (Cressey and Cressey, 1988). All three forms of TiO_2 have been seen in coal (Hsieh and Wert, 1983). An Fe–Cr oxide was found in a US coal (Finkelman, 1978). There is a report of titanomagnetite and chromite (FeCr$_2$O$_4$) in an Albanian coal (Mullai, 1984). It has been suggested that Nb and Sn may be present in coal as columbite ((Fe,Mn)Nb$_2$O$_6$) and cassiterite (SnO$_2$) respectively (Merritt, 1988).

3.9 Sulfates

Barite (BaSO$_4$) was detected in a Leicestershire colliery (Binns and Harrow, 1897) and later reported by Stutzer (1940). It has been found in NSW coals (Kemezys and Taylor, 1964; Doolan, Mills and Belcher, 1979), in Queensland coals (Shibaoka, 1971), in Illinois coals (Chou and Harvey, 1983), in Upper Freeport coals (Palmer and Wandless, 1985), in Czechoslovakian coals (Bouska, 1981) and in South African coals (Falcon and Snyman, 1986). Barytocelestine ((Ba,Sr)(SO$_4$)$_2$) was found in a Leicestershire coal (Cressey and Cressey, 1988) and celestine (SrSO$_4$) was detected in some Soviet coals (Sokolov and Vidishev, 1958) and East German coals (Adolphi and Störr, 1985). The presence of iron sulfates is usually the result of oxidation of pyrite, but there is an unusual occurrence of szomolnokite (FeSO$_4$.H$_2$O) in a Kentucky, USA, bituminous coal, where some pyrite has been oxidized in the unexposed, unweathered coal (Raymond, Bish and Gooley, 1983). Gypsum (CaSO$_4$.H$_2$O) is found in some coals, and may contain some Sr, for example a sample from Leigh Creek, South Australia, had 400 ppm Sr (D.J. Swaine, unpublished).

3.10 Phosphates

Apatite is found in various coals from the United States (Ball, 1935; Finkelman and Stanton, 1978; Chou and Harvey, 1983; Straszheim et al., 1988), from South Africa (Smith, 1941a; Falcon and Snyman, 1986), from Australia (Cook and Taylor, 1963; Day, Jones and Belcher, 1979), from New Zealand (Soong and Gluskoter, 1977), from Germany (Mackowsky, 1968) and from the United Kingdom (Spears and Caswell, 1986). Apatite in coals is usually fluorapatite, that is, some OH has been replaced by F, and there are references to fluorapatite in UK coals (Crossley, 1944a; Cressey and Cressey, 1988) and in NSW and Queensland coals (Cook, 1962; Durie and Schafer, 1964). The mention of phosphorite in coals (Stutzer, 1940) probably means fluorapatite. Fluorite (CaF$_2$), was found in some Czechoslovakian and Canadian coals (Bouska, 1981; Fyfe, Kronberg and Brown, 1982). Gorceixite (Ba Al$_3$(PO$_4$)$_2$(OH)$_5$H$_2$O) occurs in some Australian coals (Doolan, Mills and Belcher, 1979), in an Illinois coal (Finkelman and Simon, 1984), in a South African coal (Roberts and van Rensburg, 1985) and in some Alaskan coals (Brownfield, Affolter and Stricker, 1986). Goyazite (SrAl$_3$(PO$_4$)$_2$(OH)$_5$H$_2$O) is found in some Australian coals (Ward, 1974, 1978; Doolan, Mills and Belcher, 1979) and in some Alaskan coals (Brownfield, Affolter and Stricker, 1986). Monazite (CePO$_4$) and xenotime (YPO$_4$) have been reported in some US coals (Finkelman, 1978; Finkelman and Stanton, 1978; Hsieh and Wert, 1983; Palmer and Wandless, 1985).

3.11 Lignites and brown coals

Although lignites and some brown coals may contain much the same major-element minerals – for example quartz, clays, sulfides – and some trace-element containing minerals – for example barite in North Dakota lignites (Paulson, Beckering and Fowkes, 1972; Fowkes, 1978; Rindt *et al.*, 1980; Huffman and Huggins, 1984; Yeakel and Finkelman, 1984; Schobert *et al.*, 1987), rutile (Finkelman and Bhuyan, 1986; Schobert *et al.*, 1987; Störr, Adolphi and Kasbaum, 1987), galena (Schobert *et al.*, 1981), clausthalite (Gueven and Lee, 1983), As strongly associated with pyritic iron (Yeakel and Finkelman, 1984) and an occurrence of the sulfides of Co, Cu and Ni (Ratynskii *et al.*, 1982), there is a fundamental difference from bituminous coals, namely the occurrence of several trace elements associated with carboxylic acid and phenolic hydroxyl groups. Among elements associated with low-rank coals via the above groups are Ba, Mn, Sr, Zn (Brown and Swaine, 1964; Morgan, Jenkins and Walker, 1981; Miller and Given, 1987b) and perhaps Rb (Benson, Falcone and Karner, 1984). As well as chelation with divalent cations, ion pairing may occur with univalent cations. Although it had been stated that the usual LTA technique (Gluskoter, 1965) was not applicable to low-rank coals (Miller, Yarzab and Given, 1979), it can be used after removing ion-exchangeable cations with 1N NH_4Ac at pH 7 (Given and Yarzab, 1978). There is a new glow discharge-excited LTA technique which 'makes it possible' to ash low-rank coals in about 4 h (Adolphi and Störr, 1985). As with other LTA techniques, some artifacts are formed, for example bassanite ($CaSO_4.0.5H_2O$). Although carbonate minerals have been reported in several lignites (for example Fowkes, 1978; van der Flier-Keller and Fyfe, 1988), in German brown coals (Gumz, Kirsch and Mackowsky, 1958) and in some peats (Kivinen, 1936) Burns, Durie and Swaine (1962) presented evidence that carbonate minerals are not present in Victorian brown coals and probably some other low-rank coals. This view is supported by an investigation of a US lignite by extended X-ray absorption fine structure and X-ray absorption near-edge spectroscopy which showed that the Ca 'is present as salts of carboxylic acids' (Huggins and Huffman, 1983). Further work confirmed this finding and indicated that calcite could not be present at more than 2% (Huffman and Huggins, 1984). As with bituminous coals, minerals occur in low-rank coals authigenically, epigenetically and as detritals. However, there is some difference of opinion about the relative importance of the association of cations with organic matter and mineral matter in low-rank coals. Van der Flier-Keller and Fyfe (1988) state that 'association with the organic matter appears to be most important', at least for the Canadian lignite studied by them. It would seem reasonable for organic association to predominate for low-ash coals with less than, say, about 5% ash yield, which is in keeping with qualifications mentioned by Finkelman (1982a). Victorian brown coals merit special mention as they differ from most lignites. They have high bed moisture (49–~65%), low ash (0.5–~4% ash yield) with corresponding low mineral-matter content (Kiss, 1982). The main minerals are quartz, kaolinite and an aluminium hydroxide (AlOOH), with minor amounts of pyrite/marcasite and minerals containing K and Ti (Kiss and King, 1977). For brown coals from the Latrobe Valley, Victoria, Kiss and King (1977, 1979) suggested that the mineral matter should be divided into minerals and inorganics, the former being discrete mineral particles and the latter being exchangeable and acid extractable cations. Because some subbituminous coals may contain carboxylic acid groups, the above comments will also apply to them.

It is probable that 'mineral matter in peat may be detrital (transported or reworked older sediments), authigenic (produced *in situ*) or diagenetic (altered) in origin' (Andrejko, Cohen and Raymond, 1983), providing sources of trace elements for later incorporation in coal, organically and inorganically. Hatch (1983) has summarized the factors that control the distribution of elements in coal, namely, 'geochemical conditions (primarily pH) in the peat swamp, composition and depositional environment of coal roof rocks, thermal maturity (rank), nature and intensity of any epigenetic mineralization, composition of ground waters that come in contact with coal and degree of weathering'. Hence, it is to be expected that each trace element has been subjected to several changes that would affect how it is ultimately present in coals. The main deduction from the review of published information on trace element–organic matter associations (Section 3.1) is that there is no concensus about the predominance of organic matter associations for any trace element in bituminous coals. Perhaps some of the differences are related to differences in coal properties, but many are the products of deductions based on density separations and on statistical operations. Although direct evidence is lacking, it is reasonable to assume that several trace elements are organically combined, probably as chelation complexes, in low-rank coals. As coals mature and increase in rank some of these complexes will change, for example complexes of alkali and alkaline earth elements are not found in bituminous coals, where there is an increase in the number and diversity of minerals. Indeed, trace elements in coals with more than about 5% ash yield are predominantly associated with minerals (Finkelman, 1981a), and as a result of a detailed study of US and other coals, Finkelman has stated that 'the bulk of most trace elements in bituminous coals are associated with fine-grained accessory minerals' (Finkelman, 1988). The extent of trace-element–organic matter associations remains elusive, but many early suggestions, based on float–sink separations, are suspect, because trace elements in fine micrometre-sized mineral particles, closely associated with macerals, are thought to be directly associated with the organic coaly matter, i.e. the float fraction. It seems reasonable to postulate that each trace element in coal occurs in more than one form and that the mineral matter is a major site. It is not possible to predict how trace elements *will* occur in a particular coal, but published work does permit suggestions of how they *may* occur. Local conditions may change the habitat of an element from that commonly found, for example when pyrite is very low, As, Hg and Se may be associated with clays and carbonates (Palmer and Wandless, 1985). There is no substitute for carrying out direct examinations of coals, in particular the use of SEM–EDX as carried out by Finkelman (1981a), together with STEM, XRD and EMPA. As Lindahl and Finkelman (1984) stated 'we encourage analytical chemists and geologists to interact closely' thereby leading to improvements in the quality of data and its interpretation.

Chapter 4
Methods of analysis

> The determination of trace inorganic elements in coal is one of the most difficult problems facing the analytical chemist.
> Guidoboni (1973)

The determination of trace amounts of any element in any material is a demanding task, but good results can be obtained by attention to details, proper sampling, the awareness of possible errors from contamination, the choice of relevant reference standards and, of course, the use of the proper analytical method. It is salutary to note that 'any suggestion that more sophisticated instruments remove the need for analytical chemists is a myth. Indeed, there is a danger that the figures produced by such instruments may be taken unquestionably for results!' (Swaine, 1985a.)

There are two books dealing with coal analysis, including the determination of trace elements: Babu (1975) and Karr (1978a, 1978b, 1979) and two general reviews by Gluskoter, Shimp and Ruch (1981) and BCRA (1988, 1989). The specialized review of the determination of trace elements in bituminous coals (Swaine, 1985a) gives an approach to the total problem, as well as an appraisal of methods. Factual reviews of methods for determining trace elements in coals are published biennially in *Analytical Chemistry* in the section entitled 'Solid and gaseous fuels', up to 1985 in the April issue, thereafter in June. Methods used by the US Geological Survey for the sampling and inorganic analysis of coal are given by Golightly and Simon (1989).

Although the method of analysis may well depend on the availability of particular equipment and experience in its use, there is a place for standard methods and some are available for several elements (ASTM, 1983a; AS, 1986, 1988). However, such methods may require modifications for some purposes (Lindahl, 1985c). The total variance (V_T) of an analytical procedure is the sum of the sampling variance (V_S), the preparation of the analytical sample variance (V_P) and the variance of the analytical determination (V_A), that is, $V_T = V_S + V_P + V_A$. At each stage, contamination is possible and must be minimized to a tolerable level, as shown by blanks. There are various interpretations of the term 'detection limit', which have been discussed prior to suggesting a better phrase, namely the 'detection limit of an analytical system' (AMC, 1987). A useful definition of 'detection limit' is the minimum concentration that can be measured with a stated degree of confidence. In practice, the limit of detection, x ppm, can be used to give some meaning to terms, such as 'present', 'found', 'trace', 'not detected' or 'absent', by stating these as <x ppm (Swaine, 1955). The mineral matter dictates the major element composition of most coals and hence trace elements are determined in a matrix of several per cent of Si, Al and Fe plus a few per cent (or even more) of Ca, Mg, Na and K, the relative amounts depending on the type of coal.

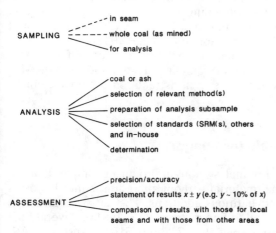

Figure 4.1 The role of the analytical chemist in obtaining meaningful data for trace elements in coal (reprinted with permission from *CRC Crit. Rev. Anal. Chem.*, **15**(4), 315–346, 1985)

The overall approach to the attainment of meaningful results for trace elements in coals is shown in Figure 4.1 (Swaine, 1985a) and it is suggested that the analytical chemist should be involved, at least indirectly, at all stages, including the initial planning, as suggested by Iyengar (1982) for biological systems.

4.1 Bulk sampling

It has been said that 'sampling is frequently a weak link in the chain of analytical operations. Unglamorous and often inconvenient, although essential to the result, the sampling preparation may be given far less thought and care than it deserves.' (Hume, 1973.) The successful attainment of good results depends on proper sampling and this can be achieved, as shown by the results of replicate analyses.

The initial sampling in a mine may be carried out by pillar (or block) sampling, by channel sampling or by strip sampling, preferably by standard procedures (AS, 1983b). Coal samples may be taken from conveyor belts bringing coal from underground mines, from trucks removing coal from surface mines and from drill cores. There are standard methods for sampling coal from stationary situations (AS, 1984a) and from moving streams (AS, 1984b). Various aspects of sampling coal are covered in the procedures used by the US Geological Survey (Swanson and Huffman, 1976), and in standard procedures (ISO, 1975; ASTM, 1982; SABS, 1977a). Relevant information about samples should be given, for example the mine, seam, location in seam, the method of sampling, the state of the coal (fresh or weathered) and the presence of dirt-bands or inclusions. Coal is frequently sampled away from the field situation, for example from stockpiles, railway wagons, trucks, barges and ships, where continuous mechanical samplers may be used (Berchtold, 1983). In every case, samples should be kept away from contamination and oxidation should be minimized, especially in coals containing pyrite, where rapid oxidation could form sulfuric acid which could dissolve some trace elements and cause their translocation within the sample (Swaine, 1980a).

Because of its heterogeneous nature coal 'presents one of the most difficult challenges in the world of sampling' (Berchtold, 1983). This challenge is even greater when the sample is required for trace-element analysis.

4.2 Preparation of sample for analysis

The preparation of samples for analysis aims at producing samples that are representative of the original material without alteration, either by losses of volatile elements or by adventitious additions. The initial requirement is the reduction of lumps of coal to fine material (AS, 1984c). Several final sieve sizes have been suggested, namely $-75\,\mu$m (CSIRO, Australia; AS, 1986), $-150\,\mu$m or finer (Illinois State Geological Survey, Gluskoter et al., 1977), $-180\,\mu$m (US Geological Survey, Swanson and Huffman, 1976), $-200\,\mu$m (South African Bureau of Standards, SABS, 1977b) and $-250\,\mu$m (American Society for Testing and Materials, ASTM, 1978a). Although most coals can tolerate a few ppm Mn introduced from crushers or presses, the hardened manganese steel face plates should be very low in Cr and other alloying elements. At CSIRO, 'lumps of coal are reduced to a few millimeters in size in a tungsten carbide jaw crusher or press, then ground in a tungsten carbide (or agate ring) grinder to $-75\,\mu$m particle size' (Swaine, 1985a). The general use of nylon sieves may be avoided by carrying out some experiments to ascertain the time required to grind a particular weight of sample to the necessary mesh size.

There are several ways of preparing a coal sample for analysis, namely ashing, acid dissolution, bomb decomposition, fusion or sintering and slurry formation. Some methods use finely ground whole coal, for example neutron activation analysis (NAA), X-ray fluorescence spectrometry (XRFS) and methods for the determination of volatile elements.

Ashing removes carbon, thereby achieving a concentration of trace elements, and has been widely used, perhaps because of the dominance of emission spectroscopic methods for determining trace elements. There are several reported ashing temperatures, the highest being 1050°C (Clark and Swaine, 1962a). Commonly reported temperatures are 850°C (Pollock, 1975), approximately 740°C (Headlee and Hunter, 1955a), 600°C (O'Gorman and Walker, 1971), 500°C (Watling, Watling and Wardale, 1976; ASTM, 1983a), 450°C (Zubovic, Stadnichenko and Sheffey, 1961a) and

400 °C (Mukherjee and Dutta, 1950). At CSIRO, samples are ashed at 450 °C in aluminium dishes. Cole et al. (1987) found that the oxidation of coal organic matter and the oxidation of iron minerals were rapid at 250–450 °C which is 'also the temperature range where the concentration of organic free radicals reached a maximum' (Cole et al., 1985). An LTA technique used for biological samples (Gleit and Holland, 1962) was successfully applied to coal samples by Gluskoter (1965) and is suitable for retaining trace elements in the ash (Ruch, Gluskoter and Shimp, 1974; Purchase, 1987), although it is slow (up to about 70 h: Nadkarni, 1984). The new technique of glow discharge-excited LTA (Adolphi and Störr, 1985), which produces ash in 4–8 h, may be useful for preparing ash for trace-element determinations. Much has been written about losses of some trace elements during ashing, for example Watling and Watling (1976), Egorov, Laktionova and Popinako (1979), Doolan et al. (1984a; 1985b). References to studies of volatilization losses are given by Gluskoter, Shimp and Ruch (1981) for As, Be, Ge, Hg, P, Ti, U and Zn. Although some authors dislike ashing prior to the determination of Ge, no detectable losses were found by Waring and Tucker (1954), Aubrey and Payne (1954), Campbell, Carl and White (1957) and Corey et al. (1959), while Menkovskii and Alexsandrova (1963) found that losses were insignificant if the heating rate was $3.5\,°C\,min^{-1}$. Indeed, Medvedev and Akimova (1964), after recommending 700 °C, stated that 'it has been repeatedly shown that no germanium is lost when coals are ignited . . . at either 500–600 °C or even 700–800 °C'. However, Nurminskii (1971) carried out a Kjeldahl digestion (1–1.5 g coal plus mixture of H_2SO_4, K_2SO_4 and $CuSO_4$) to avoid losses of Ge, said to be inevitable during ashing. Egorov, Laktionova and Popinako (1979) recommend 500 °C, but state that B, Mn, Pb and V 'must be determined directly in the coals without ashing', although there does not seem to be sufficient experimental evidence for this conclusion. Mikhailova, Larina and Vlasov (1978) found losses of 40–47% of Ni and 8–12% of Ti during 800 °C ashing of some Soviet coals. Ashing at up to 800 °C is regarded as suitable for Zn by Thürauf and Assenmacher (1963) and Weaver (1967). Guidoboni (1973) provides evidence about ashing at 850 °C for the determination of several elements, including Cd, Cu, Mn, Ni, Pb, V and Zn, and concludes that 'this would seem to indicate that none of these particular elements is lost during the dry ashing of coal'. In general, ashing at up to about 500 °C is suitable for most trace elements in most coals, provided that the carbon is oxidized slowly and that the coal is in relatively thin layers. It seems that the amounts of alkali metals (Ca, Mg) may influence the retention of some trace elements, and that large amounts of halogens, notably chloride, may cause volatilization of some elements (AMC, 1960; Gorsuch, 1962). It is important to maintain oxidizing conditions at all stages of ashing, since reducing conditions may lead to losses of some elements, for example Ge may form volatile carbonyl compounds (Gordon, 1959b). There is no doubt that losses of halogens and Hg occur, and that some Se may be lost during ashing. However, Se is retained by LTA (say, about 150 °C), although there are losses of halogens (Ruch, Gluskoter and Shimp, 1973). There are differences of opinion about As and Sb, which may well require Ca and Mg, often present as carbonates in coals, for retention in ash.

Wet-ashing techniques are often used for destroying organic matter. These utilized mixtures of H_2SO_4, HNO_3, H_2O_2 or $HClO_4$ (Tölg, 1972), but some elements may be lost and the procedure may take up to 48 h according to Watling, Watling and Wardale (1976). Sim and Lewin (1975) used a mixture of HNO_3-H_2SO_4 at lower than 350 °C to wet-ash some NZ coals. Advantages and disadvantages of wet decompositions in closed systems are discussed by Knapp (1984). A modern approach to wet ashing is to carry out acid digestion in a microwave oven (Nadkarni, 1984; Papp and Fischer, 1987; Alvarado et al., 1988). Full information on the theory and practice of microwave dissolution, together with safety guidelines, is available (Kingston and Jassie, 1988).

Dissolution of coal or ash in an oxygen bomb has been used (Nadkarni, 1981), but care must be taken to avoid losses of volatile elements in the vapour that may escape when the pressure is released, and the possible loss of some elements by adsorption on the walls of the bomb. Hartstein, Freedman and Platter (1973) treated coal with fuming HNO_3 in a Teflon-lined bomb for 2.5 h at 150 °C, then, after cooling, added HF for a further 15 min at 150 °C. Sample weights of 1–50 mg were satisfactory for several elements. Decomposition in open and closed systems by fusion and by sintering is dealt with by Dolezal, Povondra and Sulcek (1966), by Sulcek, Povondra and Dolezal (1977) and by Sulcek and Povondra (1989). Lithium metaborate ($LiBO_2$) fusion (Suhr and Ingamells, 1966) was used by Karacki and Corcoran (1973) for the dissolution of coal ash. Information on a wide range of decomposition methods and container materials is given by Bock (1979). There is a review of methods of sample preparation, including borate and other fusion techniques (Hickman, Rooke and Thompson, 1987), and detailed information on the preparation of coal for analysis has been given by Mills and Belcher (1981).

4.3 Contamination

'Contamination is one of the major problems hampering analyses at the trace (1–100 µg/g) and ultratrace (<1 µg/g) level' (Zief and Mitchell, 1976), and must always be considered at each stage of a total analysis. Due care is a *sine qua non* and 'this means the proper choice of implements, storage in suitable containers, and the avoidance of contamination during handling (ashing, chemical treatments, preparation for instrumental analysis). Checks must be made on materials used for dishes and on chemical reagents.' (Swaine, 1985a.)

Most of the problems of contamination incurred during the analysis of coals for trace elements are common to trace analysis of other materials and hence extensive studies carried out in connection with rocks, soils and biological materials are relevant to coals. Among such useful works are those by Thiers (1957), Pinta (1970), Scott and Ure (1972) and Zief and Mitchell (1976), which give a thorough coverage of most aspects. The initial breaking down of lumps of coal usually involves hardened steel with the consequent addition of several ppm Fe and probably a few ppm Mn (Senftle et al., 1982). Final reduction to -75 µm is achieved satisfactorily in a tungsten carbide ring mill, but contamination with W and with Co, used as a binder, must be expected, albeit in very small amounts. According to Lynskey, Gainsford and Hunt (1984) Ni is sometimes used as a binder instead of Co. Agate mortars and pestles are probably the ultimate clean material for sample preparation. Contamination from grinding and sieving, using a wide range of materials, showed differences as expected (Thompson and Bankston, 1970). There are many references to specific sources of contamination, for example B from paper containers (Huffman, 1960), and paint, tobacco ash, cosmetics and cellulose tape (Hamilton, Minski and Cleary, 1972) are other sources of some trace elements. An unexpected source of Zn could be human skin, which is high in Zn (Davies, 1977).

For many elements there will be some contamination, albeit hopefully at a very low level. Meaningful trace-element determinations depend on making proper corrections for contamination. This may be done by carrying out the analysis, either without the coal sample or using a sample virtually free of the elements in question, thereby obtaining a 'blank' value which is equivalent to the contamination. The blank should be precise, that is, its value should be constant within very narrow known limits, something only attainable by reducing the sources of contamination and depending 'on the competence of the analytical chemist in using the methods' (Swaine, 1985a).

4.4 Reference materials

Reliable reference materials are essential for checking new methods and for producing credible data which can be compared with other data. The role of reference materials in chemical analysis, various aspects of the production of standard reference materials (SRM) and how the values are obtained are outlined by Uriano and Gravatt (1977) of the National Bureau of Standards (NBS), who state that 'NBS SRMs carry the full weight and authority of NBS and the U.S. Department of Commerce as they are specifically authorized by federal legislation.' Great care is taken to state the uncertainty of a certified value for any SRM. The real need for reference standards became painfully apparent in 1951, when results of trace-element contents in two rocks showed great discrepancies, although they had been analysed by about 30 well-known analytical chemists. This early, perhaps the first round-robin, produced what Abbey (1981), describes as 'heterogeneous data' and eventually led to the NBS program of reference standards. In 1974, several laboratories reported poor results for trace elements in samples of coal, flyash, fuel oil and gasoline, which stressed the need for more attention to analytical methods and for reference standards (von Lehmden, Jungers and Lee, 1974).

The high reputation of the NBS is reflected in the quality of their SRMs. The first one for coal was SRM 1630, Hg in coal, issued in 1972. Others soon followed, namely 1632, 1632a, b and c (bituminous coal), 1635 (subbituminous coal), 1633 and 1633a (coal flyash). Results for a wide range of trace elements are given as certified or for information only, usually for about 20 elements. SRM 1632 and 1633 were analysed by four laboratories using INAA, their results being in satisfactory agreement with those of the NBS (Ondov et al., 1976). Data for a few or many trace elements are to be found in a myriad of publications, for example Germani et al. (1980) determined 51 elements in SRM 1632a and 43 elements in SRM 1635; Chu et al. (1981) determined 30 elements in SRM 1632a; and Scholz et al. (1985) reported the results of a round-robin on SRM 1632a and 1633a. Special INAA methods are still being used to analyse coal and flyash SRMs (Obrusnik and Posta, 1983; Gladney, Jurney and Curtis, 1976; Kendrick, Kyle and Kuellmer, 1988). The homogeneity of SRM 1633a has been examined by Korotev and Lindstrom (1982) and Korotev (1987), who found that 10 mg samples were satisfactory as standards for most elements, but smaller samples (0.8–5.5 mg) indicated some lack of homogeneity for several elements (Filby et al., 1985). However, Gladney (1972) found a precision of better than ±10% of the amount of Be (1.52 ppm) in SRM 1632, using only 0.3–2.0 mg subsamples.

There are three South African reference standard coals, namely SARM 18 (bituminous), SARM 19 and SARM 20 (each subbituminous to bituminous). It is commendable that full details of the preparation of the samples, brief mention of methods of analysis, all results from 28 participating laboratories in ten countries, certified values and uncertified values are given by Ring and Hansen (1984). There are three European coal reference standards with certified values for six to eight elements (Griepink, Scholz and Wilkinson, 1988) and six new US coals, including five from the Argonne National Laboratories (Vorres, 1987), which have been analysed for 70 elements (Palmer et al., 1987). Data are also available for six Illinois coals (Kruse, Harvey and Rapp, 1987; Harvey and Kruse, 1988) and another Illinois coal (Finkelman, 1984) which could be useful as standards. There are sharp differences of opinion on the treatment of outlying results and on how to arrive at usable values. There is a frank discussion in Abbey and Rousseau (1985) of the 'select laboratories' approach advocated by Abbey, and of the 'Dybczynski' method (Dybczynski, 1980) which is based on statistical treatment, and the result is that the matter remains unresolved. However, a useful practical approach is advocated by Gladney and Burns (1983) who deduce a 'consensus value' from published data for ranges and means of many standards. They are not dogmatic about this approach and suggest that 'the responsibility for the end use of these data lies with the individual investigator'. There are several useful compilations that should be consulted for trace-element data on coal, flyash and other relevant materials, namely Gladney (1980a,b) and Gladney et al. (1981, 1982). Constant updating of values in reference standards can be found in *Geostandards Newsletter*. Some trace elements have lacked suitable values in reference standards until recently, notably F (Godbeer, 1987). The need for more good coal reference standards is ongoing, so that more accurate results can be obtained, thereby improving the value of comparative data for different coals.

4.5 Chemical methods

Most trace elements in coal are determined by instrumental methods, although Cl, F and P are still usually determined by chemical methods.

Colorimetric or spectrophotometric methods are available for most elements and may be used where sundry analyses are required or where the relevant instruments are unavailable. It must not be assumed that the replacement of spectrophotometric methods by instrumental methods confers more precision or accuracy. The success of chemical methods depends very much on the skill of the analytical chemist, and scientists trained in other disciplines should be wary of assuming that good results can be obtained by following directions given for a particular method, especially if the method was developed for material other than coal. The many special books on chemical analysis give much relevant information, and there are useful sections on quantitative ultratrace analysis, laboratory conditions, materials for containers and the purification of reagents in Zief and Mitchell (1976). From the vast literature of the analytical chemistry of trace elements, a selection of relevant references, especially those for coal, will be dealt with.

A widely used spectrophotometric method of determining As in coal is based on the blue colour developed with ammonium molybdate (Edgcombe and Gold, 1955; Abernethy and Gibson, 1968; AS, 1980a) but this is not satisfactory at the levels of 1 ppm or less which are found in many Australian coals. The Australian standard method for B in coal (AS, 1988) uses curcumin, and Pollock (1975) uses carminic acid for B, the colours being measured spectrophotometrically.

There are standard methods for determining Cl in coal. One approach is to mix the coal sample with Eschka mixture (2 parts MgO plus 1 part Na_2CO_3) and heat in an oxidizing atmosphere at about 675 °C (BS, 1977; ASTM, 1978b; AS, 1980b); another is to burn a mixture of coal plus Eschka mixture in a bomb with oxygen under pressure (ASTM, 1978b); and there is also a high-temperature combustion method in which the coal is burnt in a stream of oxygen in a tube furnace at 1350 °C (AS, 1980b). The liberated chloride is finally determined titrimetrically by the classical Volhard method or potentiometrically, as suggested by ASTM (1978b) and by Gibb and Mayne (1980). Another method is to burn a coal sample plus Na_2CO_3 solution in an oxygen bomb and measure the liberated chloride by ion-selective electrode (ASTM, 1983b).

Early determinations of F in coal were usually based on a bleaching effect on a zirconium–alizarin indicator (Crossley, 1944b) or a thorium–alizarin reagent (McGowan, 1960). The ASTM standard method (ASTM, 1979) is based on the method of Thomas and Gluskoter (1974) in which the coal is combusted in an oxygen bomb with a dilute base. However, another approach is now advocated, namely pyrohydrolysis, which has been used for several years for rocks and is based on studies by Warf, Cline and Tevebaugh (1954). The pyrohydrolysis method involves heating a coal sample in a stream of moist oxygen in either a tube furnace (Godbeer and Swaine, 1987; Doolan, 1987) or an induction furnace (Bettinelli, 1983a; Rice, 1988). Godbeer and Swaine (1987) mix the coal sample with finely ground silica and heat at 1200 °C, while Doolan (1987) uses a catalyst of SiO_2 plus WO_3 plus V_2O_5 at 1100 °C, and Rice (1988) uses a catalyst of silica gel plus WO_3 plus KH_2PO_4. In each case, the liberated fluoride is measured by an ion-selective electrode. Other catalysts have been used, for example U_3O_8 (Riepe, 1986) or MoO_3

(Conrad and Brownlee, 1988). For rocks, $Na_2W_2O_7$ or $Li_2W_2O_7$ were chosen by Berns and van der Zwaan (1972). Godbeer and Swaine (1987) found that lower results for F in Australian coals were obtained by the ASTM (1979) method than by their pyrohydrolysis method. The work of the three Australian groups (Doolan, 1987; Godbeer and Swaine, 1987; Rice, 1988) is incorporated in a new Australian standard method for F in coal (AS, 1989).

There are many references to the colorimetric determination of Ge in coal, mostly based on the complex formed between Ge and phenylfluorone, which was suggested as a specific reagent for Ge by Gillis, Hoste and Claeys (1947). Cluley (1951) adapted this reagent for use with coal, and some later modifications refined the method (Kunstmann and Müller, 1959). The phenylfluorone method is still used and has been suggested recently as a 'rapid and simple method' (Sager, 1984). Early work on Ge in Australian coals was carried out by Pilkington (1957) and by Durie and Schafer (1958) using phenylfluorone methods. Among other spectrophotometric methods are the use of Beryllon P for Be (Vasileva and Vekhov, 1972), the use of malachite green (Urbanek, 1961; Nadkarni, 1982a) or rhodamine B for Ga (Xu et al., 1980), the oxidation to permanganate for Mn (Lustigova, 1976), the use of dithizone for Hg (Vasilevskaya, Shcherbakov and Klimenchuk, 1962; Vasilevskaya, Shcherbakov and Karakozova, 1964), the use of Arsenazo for Th (Pakalns, 1972), the use of ferrocyanide (Szonntagh, Faraday and Janosi, 1955; Ujhelyi, 1955) and bromo-PADAP for U (Fukushima and Nakaoka, 1984), the use of diphenylcarbazide and cetyltrimethylammonium bromide (Li, Guo and Hu, 1986) and N-m-tolyl-N-phenylhydroxylamine for V (Inoue, Hoshi and Matsubara, 1986), and the use of dithizone for Zn (Thürauf and Assenmacher, 1963).

The determinations of P and Ti are often carried out as part of the major-element analysis of coal ash, the measurements being done spectrophotometrically (ASTM, 1984). For P the colour may be developed with ammonium molybdate (Cosstick and Schafer, 1959) or better with molybdivanadate solution (ASTM, 1984). For Ti the colour may be developed with H_2O_2 (ASTM, 1984) or with Tiron containing sodium dithionite to remove any colour from ferric iron (Durie, Schafer and Swaine, 1965).

4.6 Atomic emission spectroscopy

The various techniques for determining trace elements by atomic emission spectroscopy (AES) depend on the measurement of emission by atoms which have been energized to boost their valence electrons above their normal stable levels. Good background information on emission spectrochemical analysis is given by Strock (1957) and by Ahrens and Taylor (1961), while special methods for coal are outlined by Dreher and Schleicher (1975). Most of the early data for trace elements in coal were obtained by methods usually termed 'optical emission spectrography', for example Clark and Swaine (1962a), Abernethy and Gibson (1963), Ruch, Gluskoter and Shimp (1974). There are several methods of excitation, including open direct current (DC) arc, controlled atmosphere DC arc, alternating current (AC) and other forms of spark, and inductively coupled plasma. The open DC arc has been used widely by Zubovic, Stadnichenko and Sheffey (1961a), Clark and Swaine (1962a, 1963a), Peterson and Zink (1964), Abernethy, Peterson and Gibson (1969) and Laktionova, Egorov and Popinako (1978). The early work by Goldschmidt and his coworkers at Göttingen used DC arc techniques with the sample in a carbon cathode, under conditions which utilized the cathode layer effect (Strock, 1936) to increase sensitivity. However, subsequent work on rocks, coal ash and the like has used the sample as anode, thereby shortening the time of excitation and giving smoother burning 'probably due to the high temperature of the anode, which results in the uniform melting of the substance and a rapid and very steady burning of the sample' (Kvalheim, 1947). There has been a change from the use of carbon electrodes to graphite electrodes which are easier to produce. However, at CSIRO the sample is used in a graphite anode which is arced against a pure carbon cathode, which seems to lessen wandering of the arc (Ahrens and Taylor, 1961). Controlling the atmosphere in the arc is another approach to stabilizing arcing conditions, a mixture of 80% Ar plus 20% O_2 being suitable (Dreher and Schleicher, 1975; Harvey *et al.* 1983; Fletcher and Golightly, 1985) for depressing CN band formation. The use of buffers to reduce differences between samples and standards has been in vogue for a long time and many compounds have been suggested, for example Li_2CO_3 (Skalska and Held, 1956), $Na_2B_4O_7$ (Radmacher and Hessling, 1959), SnO_2 (Bronshtein, Sendulskaya and Shpirt, 1960) and Sb_2O_5 (Kekin and Marincheva, 1970). Most buffers are compounds of elements that one wants to determine and hence they can be a limitation and can contaminate the laboratory and equipment (e.g., agate mortars). The use of such buffers ceased at CSIRO about 20 years ago. Although the presence of CN band spectra may limit the use of some lines, especially in the visible region around 358–425 nm, this is not a real problem. However, to avoid CN bands, the determination of some elements has been carried out in Cu electrodes (Konieczynski, 1960) or in Al electrodes (Zimmer and Haypal-Galocsy, 1970). An integral part of the success of AES methods is the choice of spectral lines. Every element gives a spectrum with many lines and a choice must be made of the most suitable line or lines for each element. This choice is based on the sensitivity of the lines and on freedom from interference by lines of other elements, especially elements present in concentrations greater than about 1%. In most coal ashes, the line-rich elements, Fe and Ti, and sometimes Zr, must be considered for possible interferences, and reference should be made to tables of spectral lines, for example Harrison (1939), Zaidel *et al.* (1970). The competence of an emission spectrographer depends largely on knowledge of the suitability of spectral lines for particular types of samples. Spectra are taken on a photographic plate which may be read visually using a comparator to compare lines against those from standard spectra (semiquantitative), or the density (blackness) of lines may be measured in a microdensitometer (quantitative). Some instruments use direct reading, where sensitive detectors receive radiation at preset wavelengths for elements required, for example Gluskoter *et al.* (1977). Another approach is to evaluate the spectral lines on a photographic plate by computerized methods, such as the automated plate reader (Zubovic et al., 1980).

For the two techniques of semiquantitative AES used at CSIRO since 1959 (Clark and Swaine, 1962a; Brown and Swaine, 1964), samples of coal are ashed at 450 °C in Al dishes for 16 h. This means that there will be losses of volatile elements, notably halogens, Hg and Se, but this is not important because these elements are not sensitive by DC arc methods. The two techniques are (a) total burn in which the coal ash sample is mixed with graphite powder and (b) selective volatilization in which the coal ash is mixed with Al_2O_3 plus $CaCO_3$ plus K_2CO_3 (Tennant, 1967). In method (a) during the main phase of arcing, the field strength is 4.5 to about 7 v per mm increasing to 11–12 v per mm towards the end of the burn, i.e. after the volatilization of the alkalis and similar elements. These values are of the same order as those for a source approximately in thermal equilibrium, i.e. electron excitation is not excessive.

The total burn method (a) is used to determine involatile elements including B, Be, Co, Cr, Cu, Ge, Mn, Ni, Sc, Ti, V, Y and Zr. The selective volatilization method (b) is used to determine volatile elements, namely Ag, B, Bi, Cu, Ga, Ge, In, Mo, P, Pb, Sn, Tl, W and Zn. Some elements are determined by each method which means that there is a check on sampling and arcing. The limits of detection vary from about 0.1 ppm for Ga by method (b) and about 1 ppm for Mo by method (a) to several hundred ppm for some insensitive elements. The wavelengths of the spectral lines which were chosen for coal ash are given by Clark and Swaine (1962a) and by Swaine, Godbeer and Morgan (1984b). Wavelengths for several other elements are given by Peterson and Zink (1964). Although it has been stated that the V line at 318.5 nm cannot be used in samples high in Ca (Shaw, 1958) because of a coincidence of V 318.5 with a weak Ca line, which was reported by several workers before Shaw's statement (Scott and Swaine, 1959), V 318.5 can be used for bituminous coal ash samples. Interference by Fe on the Mo line at 317 nm was found by McKenzie (1962), but 'it can be used, without correction, in the presence of relatively high concentrations of Fe under certain conditions of excitation in a DC arc' (Swaine, 1963a). The usefulness of a particular line depends on the dispersion of the spectrograph, the conditions of excitation and the concentration of the element, usually a major constituent, that is suspected of causing interference. Hence, categorical statements about line interferences should not be made, but should only be based on experimental evidence. Constant checks should be made against samples, preferably SRMs, analysed by other methods.

The above CSIRO methods have provided acceptable results when checked against standards (Clark and Swaine, 1963b) and when checked against other methods, as shown in Tables 4.1, 4.2. In Table 4.1 the AES results are from Clark and Swaine (1962a) and the AAS results are by W.C. Godbeer (1984, pers. commun.), while in Table 4.2 the AES and AAS results are by N.C. Morgan (1985, pers. commun.) and the INAA results are by J.J. Fardy (1985, pers. commun.). These comparative results attest to the validity of the CSIRO results obtained by AES. Even in the present age of insistence on quantitative results, there is still a place for semiquantitative AES methods to provide useful information prior to

Table 4.1 Comparisons of results for NSW bituminous coals by semiquantitative (AES) and quantitative (AAS) methods (ranges and mean values, as ppm in dry coal)

	Be	Cr	Cu	Mn	V
AES	<0.4–7(1.6)	2–20(5)	6–30(13)	2.5–500(114)	5–50(19)
AAS	0.4–9.6(2.0)	2.0–19(5.6)	8–26(15)	2–637(102)	5–47(20)

Table 4.2 Comparisons of results for Swiss coals by semiquantitative (AES) and quantitative (INAA, AAS) methods (ranges and mean values, as ppm in dry coal)

	Co	Ga	Mn
AES	<1–60(9)	0.9–30(10)	5–300(84)
INAA	0.3–53(9.5)	1.5–45(10.3)	3.2–291(84)
AAS	–	–	5–322(85)

carrying out quantitative determinations, perhaps even to justify the latter. It could be argued that there may be some advantages in having a large number of semiquantitative results, rather than a few quantitative results. In special cases, where there is a shortage of sample, there are methods using only a 1 mg (Rait, 1981) or a 2 mg sample (Swaine, 1963b). The precision for semiquantitative AES is often given as ±30%, but for most results the precision is better than this (Zubovic, Stadnichenko and Sheffey, 1961a; Clark and Swaine, 1962b).

Among other methods of excitation that have been used for coal ash are AC spark (Monnot, 1953), high-voltage spark (Hawley and Rimsaite, 1954), point-to-plane high-voltage spark (Mills, 1983), shock tube (Kosasa and Nakajima, 1987) and a ruby laser (Swaine, 1968). Unusual methods of introducing samples into an arc were reported by Rusanov and Bodunkov (1940) who used powdered coal ash on paper strips and by Safonov (1940) who fed powdered coal through an axial canal in an electrode. There are reports of using coal, without ashing, in a DC arc, in a condensed spark (Hegemann, Giesen and Kostyra, 1959) and in 'a modified emission spectrographic technique' (Watling, Watling and Wardale, 1976). Fletcher and Golightly (1985) have developed a method for determining 28 elements in whole coal using DC excitation in an argon or argon–oxygen atmosphere (80% Ar, 20% O_2), the gases being introduced through an alumina nozzle arrangement developed by Helz (1964). Problems caused by rapid gas evolution, ejection of some sample and erratic flaming of the organic phase were overcome by mixing the coal with Li_2CO_3 buffer.

4.7 Inductively coupled plasma atomic emission spectrometry

An improved excitation source, namely the inductively coupled plasma, is being widely used in an emission spectrometric method (ICPAES), which is applicable to trace elements in coal. Following studies on plasmas for AES by Greenfield, Jones and Berry (1964) and Wendt and Fassel (1965), ICPAES soon became established for rocks and the like and was eventually used for coals. Excitation is achieved by a high-temperature plasma formed from an electrodeless discharge in argon, which is maintained by inductive coupling to a radiofrequency electromagnetic field. Solutions for analysis are injected as aerosols at the bottom of the plasma. ICPAES has a high multielement capability, low chemical interferences, moderate spectral interferences and automatic data handling (Ihnat, 1984). The detection limits for most elements are better than for flame atomic absorption spectrometry (FAAS), but not as good as for electrothermal atomic absorption spectrometry (EAAS), except perhaps for B and Sr. It is also possible to operate ICPAES sequentially (Rice and Bragg, 1983), rather than in the more common simultaneous mode. Among useful reference books are Thompson and Walsh (1983) and Boumans (1987a,b) which cover all aspects, the latter being offered 'to act as a handbook and textbook for the novice and the expert'. The proper operation of an ICPAES instrument requires that the plasma should be run long enough for stabilization (say, about 1.5 h), together with strict control of the argon flowrate (sample and coolant) and of the power to the plasma. It is advantageous to keep the content of total solids to about 1% in solution. Multielement analysis of coal by ICPAES has been carried out by Nadkarni (1980), Heinrichs (1982), Que Hee et al. (1982), Satoh (1984), Pearce, Thornewill and Marston (1985), Sato and Sakata (1985) and Haraguchi, Kurosawa and Iwata (1985). Marquardt, Luederitz and Grosser (1983) developed a method for brown coal. Solutions for ICPAES are usually prepared from coal ash, several methods being reported. Nadkarni (1980) treated coal ash with a mixture of aqua regia plus HF in a Parr acid-digestion bomb, similar to that designed by Bernas (1968). Boric acid was then added to mask excess fluoride and to complex insoluble fluorides, so that a glass nebuliser system could be used for solution injection. Satoh (1984) and Sato and Sakata (1985) used $HF+HNO_3$, while Pearce, Thornewill and Marston (1985) used HF alone and Que Hee et al. (1982) used HNO_3-HClO_4 followed by HF. Botto (1980) found that complete dissolution by LiB_4O_7 fusion could only be achieved by using an automatic fusion device (Claisse Fluxer). According to Meyberg, Berger and Dannecker (1986), conditions of measurement must be right for each type of sample and this requires the relevant matrix or correction. Mahanti and Barnes (1983) used a poly-dithiocarbamate chelating resin to concentrate 14 trace elements from coal prior to digestion, to give a solution for ICPAES. Another approach is to determine trace elements in coal using a DC argon plasma spectrometer (DCP) with an echelle grating (Suhr and Gong, 1977). This excitation source is based on early work by Margoshes and Scribner (1959). Samples of coal have been introduced into a DCP for the determination of several trace elements, an important factor being fine grinding, at least to 1–23 μm (McCurdy, Wichman and Fry, 1985). Slurry atomization, in which an aqueous suspension of powdered coal with a dispersing agent is sprayed into an ICPAES (Wilkinson, Ebdon and Jackson, 1982; Parry and Ebdon, 1988), 'seems to show promise' (Ebdon and Wilkinson, 1987). Several methods of introducing powdered solids, including coal, directly into the plasma have been tried, including electrothermal vaporization (Hull and Horlick, 1984). Thompson (1985) discussed various needs for improvements to ICPAES, and made suggestions which could be very helpful to skilled users; ICPAES is a good method for determining B in various geological materials, including coal, so it is not surprising to find several relevant methods, namely, Owens, Gladney and Knab (1982), Mills (1986), Pougnet and Orren (1986) and Zarcinas and Cartwright (1987). Special techniques have been reported for Ge in coal ash (Nadkarni and Botto, 1984) and for Sc in coal ash (Bettinelli, Baroni and Pastorelli, 1987).

Currently the fastest growing new technique in trace element analysis is the ICP–mass spectrometer combination (ICPMS) in which the ICP is really the source of ions for measurement by a mass spectrometer. Advantages of this method are the simplicity of the spectra and low limits of detection for a wide range of elements. Several applications have been discussed by Date and Gray (1989) and there is a useful review by Riddle, van der Voet and Doherty (1988), which deals with advantages and disadvantages and makes a pertinent comment, namely 'as with any new technique, the initial optimism generated by the advantages apparent in the early development of ICPMS became tempered as problems were encountered in its application to "real" samples'. It seems clear that ICPMS may well become the favoured method for determining rare earth elements (REE). Hickman, Rooke and Thompson (1987) regard this as 'an area of considerable interest' and Jarvis (1988) has analysed some geological materials for REE with promising results. 'The field of plasma source/mass spectrometry is in a state of substantial flux . . . the technique will certainly grow' and it may become by the end of this century 'a dominant method for elemental analysis.' (Hieftje and Vickers, 1989.)

4.8 Atomic absorption spectrometry

Following independent research by Walsh (1955) and Alkemade and Milatz (1955), which showed that most elements that give free atoms in flames can be detected by atomic absorption, the new, versatile, sensitive technique of atomic absorption spectrometry (AAS) soon became established. Since most of the pioneering work on AAS was done in Australia, it is not surprising that the first publications on the use of AAS for elements in coal came from Australia (Durie, Schafer and Swaine, 1963; Belcher and Brooks, 1963). There are many books and reviews on AAS (Ihnat, 1984) and there are relevant sections on the determination of trace elements in coal in the review by Mills and Belcher (1981). An Australian standard (AS, 1970) gives practical details of FAAS. Differences in the type of excitation delineate four methods of AAS commonly used, namely flame AAS (FAAS), electrothermal AAS (EAAS), hydride AAS and cold vapour AAS. In FAAS the sample solution is passed through an atomizer into a flame, commonly air–acetylene, sometimes nitrous oxide–acetylene, which is in the path of radiation from a light source (hollow cathode lamp or discharge-type spectral lamp). The resonance line of a particular element is isolated by a monochromator and detected by a photomultiplier–amplifier system. Flame atomic absorption spectrometry is sensitive for many elements and is the simplest of the four AAS methods. Following early studies by L'vov (1961) on electrothermal heating as a source of excitation, Massman (1968) and West and Williams (1969) developed graphite furnaces which were adapted for use in AAS. The lower detection limits for many trace elements make EAAS a very attractive method; indeed it is probably the most sensitive method for many trace elements. There is a good review of EAAS, often termed graphite furnace AAS, by Noller, Bloom and Arnold (1981). Early uses of EAAS for coal were the determinations of As, Sb and Se (Aruscavage, 1977) and of Cd (Godbeer and Swaine, 1979), the coal samples being decomposed by heating with HNO_3–H_2SO_4–$HCLO_4$ and the 450 °C coal ash being treated with aqua regia ($1HNO_3$–$3HCl$, + HF) respectively. Interferences in the determination of As, Cd, Co, Cr, Ni and Pb in various environmental samples, including coal, were overcome by using suitable matrix modifiers and conditions of atomization (Berger, Meyberg and Dannecker, 1986). Hydride methods depend on the reduction of dissolved compounds by $NaBH_4$ to form gaseous hydrides that are fed into a flame for atomization. This method has been used for several elements, for example As, Bi, Pb, Sb, Se, Sn and Te in coal (Nadkarni, 1982b). Lower detection limits with hydride methods can be gained by using an electrically heated quartz cell instead of a flame (Lindahl, 1985a).

There is a good review of the generation of covalent hydrides in AAS (Godden and Thomerson, 1980) which discusses interferences (background absorption, interelement effects and others) and suggests that 'the reduction or elimination of interference effects' would improve hydride generation. Van der Sloot *et al*. (1982a) comment on complete volatilization of hydrides and state that 'in many cases analysts are satisfied with "constant" yields without knowing the causes of incomplete volatilization. The highest response is supposed to represent complete volatilization. Relying on a constant yield without thorough previous investigation may imply systematic errors.' One would not expect competent analytical chemists to make such mistakes.

Cold vapour AAS involves reduction to atomic vapour and transfer by inert carrier gas to an absorption cell; this is especially useful for Hg, where an amalgam collector is desirable.

It is usually possible to choose suitable resonance lines to avoid spectral interference, but Riley (1982) found that Al, which is a major constituent of most coals, interferes with the As line at 193.7 nm by producing 'significant nonspecific absorption which is not eliminated by using continuum source background correction'. However no spectral interference occurs with the As line at 197.2 nm. The determination of As in coal using Haynes' (1978) method, which involves heating the coal sample at 450–500 °C with nickel and magnesium nitrates, then digesting the ash with HNO_3–HF, is free from Al interference on the As line at 193.7 nm. This is probably because most of the Al precipitates as $MgAlF_5.xH_2O$, thereby reducing Al in the graphite furnace to an insignificant level (Riley, 1984).

Atomic absorption spectrometry methods mostly require that the sample is in solution and several dissolution techniques have been used. At CSIRO, samples are ashed at 500 °C and the ash is treated sequentially with HF, HNO$_3$ and HClO$_4$ in a PTFE (polytetrafluoroethylene) crucible or with aqua regia and HF in a closed polypropylene bottle, which is heated on a boiling water bath for 2 h. Similar techniques were used by others, for example Nakashima et al. (1983), Nakashima, Kamata and Shibata (1984), Yamashige et al. (1984) and Papp and Harms (1985). Kamada et al. (1982) used HClO$_4$–HIO$_4$ and Rigin (1985) used autoclave fluorination with XeF$_4$, but these two techniques are not in general use. Lindahl and Bishop (1982) developed a fairly rapid method involving combustion of a coal sample in a standard Parr oxygen bomb, with or without a quartz liner, depending on the need for adding dilute HNO$_3$ to the sample. Nadkarni (1981) used an oxygen bomb technique for determining volatile trace elements. Fusion techniques have been used in which coal ash is treated with Li$_2$B$_4$O$_7$ (Bettinelli, 1983b) or LiBO$_2$ (Boar and Ingram, 1970). In some cases, trace elements have been extracted prior to the AAS determination, with, say, ammonium pyrrolidine dithiocarbamate (Bailey, 1975; Tamura, Inue and Fudagawa, 1986), but such extra treatment is not necessary with most coals. The two standard methods for up to 11 trace elements (ASTM, 1983a; AS, 1986) use 500 °C ash treated with aqua regia–HF in a sealed plastic bottle heated on a steam bath, the solutions being analysed by FAAS with background correction. In a study of Victorian brown coal, Bone and Schaap (1980) determined a wide range of trace elements by several AAS methods, namely FAAS, FAAS after chelation with sodium diethyldithiocarbamate, hydride generation and Hg by cold vapour AAS. The initial digestion of the coal was by HNO$_3$ followed by H$_2$O$_2$, except for Hg.

In order to by-pass the dissolution of coal by acids or by fusions, efforts have been made to use solid coal directly in a graphite furnace (EAAS), for example for Be (Gladney, 1977), for Cu, Ni and V (Langmyhr and Aadalen, 1980) and for several trace elements (Doolan, Mills and Turner, 1980). An early approach was to spray aqueous suspensions of finely ground (at least $-44\,\mu$m) rocks or soils directly into the flame of a conventional FAAS (Willis, 1975). Aqueous slurries of finely ground whole coal have been used by O'Reilly and Hale (1977), O'Reilly and Hicks (1979) and Ebdon and Pearce (1982), and results which are sufficient for some purposes have been obtained for several elements. Another aspect of AAS has been considered in detail by Harnly, Miller-Ihli and O'Haver (1982), namely the simultaneous determination of up to about 30 elements by FAAS or EAAS, using a continuum source and a dedicated microcomputer. Atomic fluorescence spectrometry (AFS), has been used for the determination of Hg (Rigin, 1981; Ebdon, Wilkinson and Jackson, 1982) and for several trace elements (Rigin, 1984, 1986). Although AFS gives lower detection limits for some elements than FAAS, it does not generally match EAAS in this respect, and hence 'AFS procedures are unlikely to be widely adopted' (Mills and Belcher, 1981).

Special techniques have been developed for some trace elements, especially volatile ones. The determination of Sb can be done satisfactorily by the method of Haynes (1978) in which the coal sample is mixed with Mg(NO$_3$)$_2$ and Ni(NO$_3$)$_2$ prior to ashing at 450–500 °C. Arsenic has been determined after decomposition of the coal sample with HClO$_4$–HIO$_4$ (Spielholtz, Toralballa and Steinberg, 1971), by hydride generation

(Lindahl, 1985b) and by slurry injection of whole coal (Ebdon and Parry, 1987). Gladney (1972) determined Be by EAAS after treating the coal with HNO_3–HF–H_2SO_4, then $HClO_4$. For Cd, Fudagawa and Kawase (1985) treated the low-temperature ash with m-xylene or chloroform solution of dithizone, prior to EAAS determination. Instead of measuring the excess Ag with KCNS, as in the classical Volhard technique, Martinez-Tarazona and Cardin (1986) determined the excess Ag by FAAS for doing Cl in some Spanish coals. Shan, Yuan and Ni (1985) developed a method of determining Ga in coal using EAAS (graphite furnace) after the addition of $Ni(NO_3)_2$ plus $(NH_4)_2SO_4$ in solution to an acid-treated coal, thereby improving sensitivity. Germanium has been determined in some Spanish lignites by a hydride method (Castillo, Lanaja and Aznarez, 1982). There are several methods for determining Hg in coal (Huffman et al., 1972; Heinrichs, 1975a; Wimberley, 1975; Gardner, 1977; Uchikawa, Furuta and Mihara, 1982); in some cases, the Hg is collected on Ag or Au prior to determination by flameless AAS (either EAAS or cold vapour AAS). The standard method (ASTM, 1983c) involves combustion of the coal sample in an oxygen bomb with dilute HNO_3 to absorb Hg which is measured by the cold vapour method. A method for determining La and Sc in coal by FAAS or EAAS depends on coprecipitation with calcium oxalate, then hydrated ferric oxide, after decomposition with HNO_3 and with HF (Sen Gupta, 1982). There are two general methods for Se, perhaps the common one being a hydride generation technique (Imai, Terashima and Ando, 1984; Agterdenbos et al., 1986), but the other one, using separation by distillation with HCl–HBr followed by EAAS, is also satisfactory (Woo et al., 1985). Kellerman, Haines and Robert (1983) ashed coal at 400 °C, fused the ash with Na_2O_2 and determined As, Se and Te by hydride generation. Another method for determining Se and Te, based on coprecipitation with a trivalent As carrier and EAAS, was developed by Woo et al. (1987). The generally low concentrations of Tl in coal require EAAS (Shan, Yuan and Ni, 1986), sometimes after preconcentration (Berndt et al., 1981). Among the methods for background correction, the technique of on-line correction, based on the Zeeman effect, is commended because nonatomic background absorption is corrected at the same wavelength as the line being measured.

The determination of trace elements in coal will continue to rely on AAS methods in many laboratories. Koirtyohann (1980) has cautioned that modern instrumentation using microprocessors should not be accepted uncritically, in particular, single-point calibration is not to be recommended.

4.9 Neutron activation analysis

Neutron activation analysis is a well-tried method which is suitable for whole coal and related materials. The most commonly used technique involves thermal neutron irradiation of small encapsulated samples, followed by instrumental measurements of the emissions from the radioactive isotopes (radionuclides) formed by the various elements. This direct approach (INAA) is not suitable for some elements in some materials, but such elements may be determined by radiochemical separation techniques (RNAA). Among the advantages of INAA are its multielement capability by which 30–40 elements can be determined simultaneously with low limits of detection for many elements. Minimal sample preparation reduces the risk of contamination and interferences can usually be overcome by improvements in gamma-ray detectors and sophisticated computer programs. Some measure of the advances in the past decade or so is the reduction in the number of systematic errors and disadvantages listed by Zief and Mitchell (1976). The attainment of high sensitivity requires a nuclear reactor with a high neutron flux. Although pure metals or chemical compounds may be used as standards, it would seem desirable nowadays to use SRMs (Nadkarni and Morrison, 1978), which would improve accuracy. Becker (1977) has endorsed the use of reference standards in his comments on achieving accuracy with INAA. Among the many references to the determination of trace elements in whole coal or coal ash by INAA are Block and Dams (1973), Nadkarni (1975), Abel and Rancitelli (1975), Sheibley (1975), Lyon (1977), Millard (1977), Ruch et al. (1977), Filby, Shah and Sautter (1977), Weaver (1978), Steinnes (1979), Harvey et al. (1983), Ward, Kerr and Otsuka (1986), Gladney, Garcia and Newlin (1986) and Bellido and Arezzo (1986). There are two good reviews, one on INAA in coal research (Tripathi, 1979) and the other on the techniques of NAA (Faanhof, 1986). As with all analytical methods, some trace elements in coal are not readily amenable to thermal INAA, notably B, Be, Cd, Cu, F, Hg, Mo, Ni, Pb and Tl. For some elements, including Mo and Ni (Rowe and Steinnes, 1977a), irradiation with epithermal neutrons improves the sensitivity of INAA by enhancing 'the formation of a radionuclide relative to interfering ones' (Steinnes, 1979). For epithermal irradiation, samples are wrapped in Al foil and placed in a cadmium box ready for irradiation. Details of the method and comparisons of results from this method with those from normal INAA have been given by Rowe and Steinnes (1976, 1977a,b).

In some cases, for example where there is interference from other radionuclides or where there is a very low concentration of an element, separations may be used after irradiation, the so-called RNAA method. Several examples are given in detail by Frost et al. (1975) for 11 trace elements in coal, namely As, Br, Cd, Cs, Ga, Hg, Rb, Sb, Se, U and Zn. Among the separation techniques that have been used are distillation, precipitation, adsorption, solvent extraction, ion exchange and chromatography. Probably the first use of RNAA for coal was the determination of Cs and Rb, ferric hydroxide being the initial separation medium followed by chemical precipitations (Smales and Salmon, 1955). Orvini, Gills and LaFleur (1974) developed a method for As, Cd, Hg, Se and Zn, the elements being precipitated as sulfides. Precipitation of HgS was used by Kostadinov and Djingova (1979) to determine Hg in Bulgarian coals. Two methods for Se were based on separations using a Hg carrier and the formation of HgSe (Rook, 1972) and reduction to elemental Se (Pillay, Thomas and Kaminski, 1969). Column chromatography enabled Perricos and Belkas (1969) to separate U, while anion exchange chromatography was the choice of Casella et al. (1981) for separating Pb, Th and U.

Several special methods have been reported, for example prompt gamma-ray spectrometry for B and Cd (Gladney, Jurney and Curtis, 1976), neutron-capture prompt gamma-ray activation analysis for up to 17 elements (Failey et al., 1979) and photo activation analysis for up to 28 elements in coal (Galatanu and Engelman, 1981; Robertson et al., 1983). Another technique, proton-induced gamma-ray emission (PIGE), has special application for determining F (Roelandts et al., 1986) and has been applied to Australian coals (Clayton and Dale, 1985).

An early use of RNAA was the determination of Hg and Se in Australian coals, separations being effected by distillation followed by precipitation of HgS and of elemental Se (Porritt and Swaine, 1976). Up to 32 trace elements were determined in a range of Australian coals (Fardy, McOrist and Farrar, 1984), using irradiation facilities at the Lucas Heights Laboratories and the methods set out by Carr and Fardy (1983). The same methods were used to study variations in trace-element concentrations in a NSW coal seam (Fardy and Swaine, 1985). Some Victorian brown coals have been analysed by INAA (Bone, Schaap and Hughes, 1981).

4.10 X-ray fluorescence spectrometry

There are two main approaches to the analytical use of X-ray emission, usually termed X-ray fluorescence spectrometry (XRFS), namely wavelength dispersive and energy dispersive, the former being the more sensitive and the more expensive. The wavelength dispersive instrument uses diffraction by crystals to separate the excited secondary X-rays according to their wavelengths, whereas the energy dispersive instrument uses a solid state detector to separate the X-rays according to their photon energies. Relevant general information on XRFS can be found in Dzubay (1977). Early applications of XRFS to coal analysis were by Sweatman, Norrish and Durie (1963) at CSIRO and by Kiss (1966) for Victorian brown coals, and by Kuhn (1973), Kuhn, Harfst and Shimp (1973, 1975), and Kuhn and Henderson (1977) for Illinois bituminous coals. Although XRFS is a good method for determining major and trace elements in coal ash the main attraction is whole coal, which usually requires grinding to $-75 \mu m$ (1–2 g sample), although grinding to $-45 \mu m$ can improve precision (Ruch, Gluskoter and Shimp, 1974). The procedure used by Ruch, Gluskoter and Shimp (1974) starts with grinding the coal sample mixed with 10% of a binding material, then preparing a disk at 275 MPa (40 000 psi) which is dried in a vacuum oven before the disk is submitted for analysis. Following a study of three modes of excitation for XRFS, namely proton beam, Mo X-ray tube and radioactive sources (^{57}Co and ^{109}Cd), Valkovic et al. (1984) suggested the most appropriate technique for a particular purpose. For elements heavier than Fe the best sensitivity was given by X-ray tube excitation.

Matrix effects are important and may be dealt with by having a range of suitable standards, by modifying samples to minimize the effects, or by applying mathematical correction procedures (Kuhn and Henderson, 1977). In some cases SRMs may be available, but often synthetic standards must be made from spectrographically pure compounds and briquetting graphite. The preparation of such standards requires much care (Knott, Mills and Belcher, 1978; Wheeler, 1983; Dewison and Kanaris-Sotiriou, 1986). Ways of modifying samples to minimize matrix effects include dilution with an inert substance, addition of a heavy absorber or preparation of a very thin sample, thereby allowing all the generated X-rays to escape. The disadvantage of dilution and addition of a heavy absorber is that the limit of detection is increased. The thin sample technique is used, although great care must be taken to produce a uniform representative sample. Mathematical procedures for correcting matrix effects depend on empirical measurements of mass absorption coefficients or the use of Compton scatter to estimate the mass absorption coefficients. Garbauskas and Wong (1983) applied fundamental mathematical corrections using doped graphite standards to overcome matrix effects on the determination of Ti in coal. Willis (1983) has summed up the situation succinctly: 'the quality of the data depends critically on the accuracy of the mass-absorption coefficients and the accuracy with which the background and line-overlap corrections can be calculated'. For light element matrices, Giauque, Garrett and Goda (1979) found that incoherent scattered radiation, corrected for matrix absorption, could serve as an internal standard. Heinrich and Foscolos (1984) devised a method of correcting for interference from spectral lines of elements which occur in targets; their method does not require prior knowledge of the major element composition of samples being analysed. Among the many studies of trace elements in coal or coal ash by wavelength dispersive XRFS are those on Illinois coals (Kuhn, 1973), on Victorian brown coals (Dimitrakakis, Rankin and Schaap, 1976), on US coals (Prather, Guin and Tarrer, 1979), on Australian coals (Mills and Turner, 1980; Turner, 1981; Mills et al., 1981), on South African coals (Willis, 1983, 1986; Willis and Hart, 1985), on Italian coals (Farne et al., 1983), on UK coals (Thorne et al., 1983), and on Soviet coals (Vyalov and Stepanosov, 1986). The methods worked out by Mills and Turner (1980) and Willis (1983) are the results of thorough investigations of the applicability of XRFS to Australian and South African coals, respectively, and show the usefulness of XRFS and its limitations. Despite some limitations of sensitivity, energy-dispersive XRFS has been the method of choice for investigations of some US coals (Giauque, Garrett and Goda, 1977; Cooper et al., 1977; Prather, Tarrer and Guin, 1977; Wheeler and Jacobus, 1980; Harris, Barrett and Kopp, 1981). X-ray fluorescence spectrometry has been used for specific trace elements, for example As in Polish coals (Rozkowska, Orlowska and Zyczkowska, 1982), As in Czechoslovakian brown coals (Simon and Hally, 1984), Cl in Polish coals (Kusmierska and Badura, 1983), Ge in US coals (Campbell, Carl and White, 1957), Ge in lignites from Umbria, Italy (Abbolito, 1960) and Se in Soviet coals (Belopolskaya and Serikov, 1969). Methods of preconcentration have been devised to cope with low concentrations of some trace elements. For example, the Coprex method of Luke (1968), in which about 60 elements are concentrated by specific reagents and coprecipitated with ferric iron, could be applicable to coal. Another approach is preconcentration by cation exchange resin filters which was applied to coal prior to energy-dispersive XRFS (Kingston and Pella, 1981). The proper operation of instruments for XRFS depends on keeping the instrument in optimum condition and there is a standard for ascertaining the precision of wavelength dispersive machines (AS, 1982).

A new rapid, simultaneous method for determining up to 75 elements, utilizing X-ray emission from excitation by protons, is proton induced X-ray emission (PIXE). There is a good review of the general principles of the analytical use of ion-induced X-rays (Folkmann, 1975). Simms, Rickey and Mueller (1977) and Kullerud et al. (1979) give details of a PIXE method which is applicable to coals. Small samples (0.2–2 mg) of finely powdered coal (<20 μm) are deposited on a thin plastic backing, which is the target for a proton beam. About 75 elements have been determined in Indiana coals, and Simms, Rickey and Mueller (1977) determined 29 trace elements in the NBS SRM 1632. Raj (1986) carried out an investigation of some US coals using a PIXE technique, as did Bujok et al. (1980) for Polish coals. Detection limits depend on the time of exposure to the proton beam, background radiation, presence of high concentrations of some elements and energy interference (Simms, Rickey and Mueller, 1977). For many elements, 1 to a few ppm can be detected, for REE about 10 ppm and for P about 250 ppm. The concentrations of several elements associated with vitrinite, exinite and inertinite in six samples of coal from the United States and one from China were determined by a modified PIXE technique (PIXE microprobe) which uses a fine beam (Chen et al., 1981). Synchrotron radiation was used by Chen et al. (1984) to determine some trace elements in vitrinites. The latter two methods are not practicable for bulk analysis, but are suitable for examining minerals in coal or microsections of coal. Proton induced X-ray emission is useful for thin samples, when the amount of sample is limited and for the analysis of surface layers, whereas XRFS is suitable for bulk samples, including coal. Willis (1988) gives a good appraisal of XRFS and PIXE.

4.11 Spark source mass spectrometry

In spark source mass spectrometry (SSMS) the sample is ionized in a vacuum by igniting a spark between two electrodes made from the sample. The ions are accelerated into a magnetic field, where splitting of the ion beam takes place and ions are separated according to their mass-to-charge ratios. The mass spectrum may be recorded on a photographic plate or electronically. Basic information especially relevant to geological samples is given by Taylor (1965, 1971) and Taylor and Gorton (1977), while recent advances in analytical SSMS are dealt with by Adams (1982). Beske et al. (1981) have reviewed and evaluated SSMS as an analytical method, special emphasis being given to the main applications, detection limits, precision, accuracy and to comparisons with INAA and AES. Details of the determination of trace elements in coal ash are given by Guidoboni (1973, 1978) and Carter et al. (1978). In an early use of SSMS for coal ash, 36 trace elements were detected including several REE (Sharkey, Schultz and Friedel, 1963), while Kessler, Sharkey and Friedel (1973) and Sharkey, Kessler and Friedel (1975) determined 56 trace elements in raw coal samples from ten seams from five US areas. This use of raw coal is unusual, as coal ash is normally used, being mixed with pure graphite to form an electrode. Canadian coal ashes have been examined by SSMS, 52 trace elements being determined (Kronberg et al., 1981; Fyfe, Kronberg and Brown, 1982) and several trace elements were determined in some Japanese coals (Koseki, Ogawa and Hasuda, 1986). Although the photographic plate is often used as the ion detector, electronic detection is available as an alternative which, together with computerized data handling, saves time. SSMS is very sensitive for a wide range of elements, detection limits being about 0.02 ppm or better. Under suitable conditions with highly skilled scientists, SSMS can produce good results, sometimes quantitatively, but the lack of suitable reference standards is a limitation. Comparisons between SSMS (two laboratories), INAA, AAS and XRFS (three laboratories) for results for a sample of volcanic ash (Kronberg et al., 1988) showed large variations for several trace elements, in some cases even for the two results by SSMS. Better results are obtainable by semiquantitative AES, although not as many trace elements would be detected. However, a variant of SSMS, known as isotope dilution spark source mass spectrometry (IDSSMS) enables better precision and accuracy to be achieved. In IDSSMS, an enriched isotope (spike) in pure graphite or silver powder is mixed with coal ash, and the new isotopic ratios are measured after establishing that there is complete isotopic equilibrium between the coal ash sample and the spike. Either SSMS or thermal emission mass spectrometry (TEMS) can be used with isotope dilution, as has been carried out by Carter et al. (1978), the latter method having high precision but being limited to elements that ionize easily from a heated filament. Details of IDSSMS and TEMS applied to coals are given by Carter, Walker and Sites (1975) and Carter et al. (1978). In order to avoid dissolving samples, Carter, Donohue and Franklin (1977) used a dry matrix comprised of graphite or high purity Ag containing the necessary isotopes, which had been applied in solution and dried. Koppenaal et al. (1980) treated LTA of coal with mixed acids in a Teflon-lined Parr bomb, the isotope spikes being added before the decomposition procedure. Another application of IDMS is for the accurate determination of Pb in coal, biological and environmental materials (Machlan et al., 1976), the Pb being separated by ion exchange, purified by electrodeposition and then measured by TEMS.

The association of elements with mineral matter and macerals in coals can be ascertained by SIMS, which ionizes the sample by bombardment with a primary ion beam (Martin, McIntyre and MacPhee, 1985). However, SIMS is intended primarily for surface analysis and is not suitable for bulk analysis.

The multielement capacity of SSMS surpasses that of other methods, but quantitative analysis is difficult, perhaps unattainable, for several trace elements, and Beske et al. (1981) have warned that 'high accuracy and low detection limits are in every case only achievable by painstaking measurements'.

4.12 Other methods

In addition to the more commonly used methods already discussed, trace elements in coals have been determined by other methods, for example the electroanalytical techniques of polarography and anodic stripping voltammetry, fluorimetry and chromatography. The electroanalytical techniques are very sensitive. Weclewska (1960) determined Cu, Ge, Pb and Zn in Polish coal ash by a polarographic method, while Somer, Cakir and Solak (1984) developed a method for determining As, Cd, Cu, Mo, Pb, Sb, Ti and Zn in Turkish coals after digestion of the coal samples with mixed acids in a Parr bomb. Polarographic methods have also been used for Ge (Weclewska and Popanda, 1958; Khizhnyak, Tsebrii and Lenkevich, 1971), Mn (Kessler and Dockalova, 1955), Ni (Wang, 1978) and U (Korkisch, Farag and Hecht, 1958). Another technique for determining U in Greek lignite utilized an ion-exchange and extraction procedure prior to polarography (Ochsenkuhn-Petropulu and Parissakis, 1985). Kaiser and Tölg (1986) determined Bi, Cd, Cu, Pb, Se, Te, Tl and Zn, with detection limits of a few ppb, by anodic stripping voltammetry. Fluorimetry has been used for Se, utilizing the complex with 2,3-diaminonaphthalene (Guo, Cao and Zhuo, 1986) and for U (Korkisch, Farag and Hecht, 1958). Talmi and Andren (1974) developed a method for determining Se in coal, based on gas chromatography with a microwave emission spectrometric detection system, which gives high sensitivity (to about 15 ppb). Another chromatographic technique was used by Szonntagh, Farady and Janosi (1955) for determining U in some Hungarian coals.

4.13 Consideration of results

The end product of sampling and analysis is the analytical result, which is dependent on careful attention to detail at every stage of the process. It is difficult to assign the relative magnitude of errors, but van Loon (1986) regards the errors as greatest for sampling and least for the determination, with sample preparation errors being intermediate. This could be so for coal, which is a heterogeneous solid. Trace-element results should always be stated as the element, not as an oxide or other compound, and usually referred to air-dried or to dry coal (105–110 °C) or sometimes to as-received coal. Concentration or content is commonly expressed as parts per million (ppm), which is often given as $\mu g\,g^{-1}$ by analytical chemists. Other equivalent forms are mg per kg and g per tonne. The assignment of a value of zero when an element is not detected is meaningless and should be given as a value of less than the limit of detection. Values should be given to the number of figures in keeping with the method of analysis. With semiquantitative results there should be a clear statement, so that the user, albeit innocently, cannot achieve a seeming increase in accuracy, when the results are used for another purpose or transcribed. The user of results should not round off values without consulting the analytical chemist.

The analytical chemist is concerned with the precision and accuracy of results. The precision is the measure of the reproducibility of a method, while the accuracy is the closeness to the true value. The total error (E_T) is the sum of systematic (E_S) and random (E_R) errors, that is, $E_T = E_S + E_R$. Systematic errors 'are those which tend to give results that are always higher or lower than the true figure', while random errors 'lead to results that are sometimes higher and sometimes lower than the true result' (Pickering, 1977). Precision is a measure of the purely random errors, whereas accuracy measures $E_S + E_R$, and hence a decrease in E_T, i.e. $E_S + E_R$, means an improvement in accuracy. A decrease in the scatter of repeated results leads to an increase in precision. Ultimately the attainment of accuracy depends on the use of good standard reference materials, which are discussed in Section 4.4. In stating the accuracy of a result a realistic approach should be used and the temptation to quote too many figures should be resisted! For trace elements in coal or other heterogeneous materials, an accuracy of ±10% of the content should be acceptable, particularly at ppm and sub-ppm concentrations. Perhaps good precision is difficult to achieve at low concentrations (Taylor, 1983), but the situation is usually better than that shown by the 'Horwitz trumpet', where the coefficient of variation is plotted against concentration for the range 0.001 ppb to 10% (Horwitz, Kamps and Boyer, 1980); at the ppm level the coefficient of variation is about ±15%, increasing to about ±45% at the ppb level. There is useful information on precision, accuracy and bias in ASTM (1986). In order to maintain satisfactory results in a particular laboratory or group of laboratories, programs of analytical quality assurance may be helpful. There are two aspects of such programs, namely internal checks that are carried out by any reputable analytical chemist, and external checks that involve the round-robin approach, that is, the analysis of unknown samples prepared by an independent laboratory. Such programs help to overcome the satisfaction to be gained from obtaining precise results that could be inaccurate; in other words the analytical chemists may be alerted to systematic errors, which may be different in each laboratory.

> It is important to state results in a way that is meaningful for the user, rather than just for other chemists. Various statistical operations may be carried out on the data in order to obtain the standard deviation, the relative standard deviation, the coefficient of variation, or other measure of precision or accuracy. However, trace-element contents in coal may be stated simply as a range of values, for particular seams, locations, geological areas (e.g., a coal basin) or geographical areas. This enables the user to ascertain the variability of results for various trace elements and the presence of abnormal, especially high, results.
>
> (Swaine, 1985a)

Mean values may be useful for comparing results and the mean for 90% of values avoids the bias from the too low and too high values. An estimate of the most probable concentration (mode) can be calculated and is designated as the geometric mean (GM). Variations from GM can be calculated to give the geometric deviation (GD = antilog of standard deviation of logarithm of concentration). Since the data for trace elements in coals often show positively skewed frequency distributions, the use of the GM and GD is a valid means of normalizing the data, as is done by Swanson et al. (1976) and Zubovic et al. (1980).

Results that are less than the limit of detection, or unfortunately given as zero, must be considered in the calculation of mean values. Such values may be treated as 0.5 or 0.7 times the limit of detection, but there are proper procedures for limiting bias in estimating mean values (Cohen, 1959). Sometimes trace-element contents can be compared usefully by comparing GM values or arithmetic means for 90% of values.

For some years, the trend has been towards multielement methods which may determine more than about 40 elements simultaneously. In practice, one method cannot be expected to determine all elements with the required sensitivity and freedom from interference, although SSMS probably comes closest to this ideal. Zief and Mitchell (1976) have listed their criteria for an ideal analytical method, namely extreme sensitivity for many elements, high specificity yet capable of simultaneous multielement operation, nondestructive of samples, independent of matrix effects and interferences, free from contamination problems, inexpensive, simplicity of operation, capable of automatic operation, high precision and accuracy, reasonably independent of operator error and capable of giving absolute values independent of standards. The last requirement is no longer applicable because of the current availability of standard reference samples. All these criteria are not met and cannot be expected to be met by any one method, but they are useful as guidelines. Magyar (1982) commented that 'such an ideal method does not exist, but AAS has characteristics pointing in this direction'. When large numbers of trace elements need to be determined it is necessary to have several methods available. For example, the US Geological Survey used AES, AAS, INAA, XRFS and wet chemical methods (Zubovic et al., 1980), as did the Illinois Geological Survey (Cahill et al., 1976). Compromises can be made according to circumstances, for example it is scarcely worth while making special efforts to determine some trace elements, usually present at sub-ppm levels, notably Au, Bi, In, Re and Te, although they may be detected during an analysis by SSMS for other elements. The choice of methods is often governed by the availability of particular equipment, so that if INAA was available, there would hardly be a need for, say, XRFS or SSMS. Sometimes semiquantitative results are sufficient and the choice should then be AES or SSMS. When there is a need for high precision, special methods may be required, for example EAAS, RNAA, IDSSMS, or a particular spectrophotometric or other chemical technique.

There is still a place for semiquantitative AES methods which give ballpark or better values for a wide range of elements. In many cases there is merit in having numerous semiquantitative results rather than a few quantitative results, and such information is useful for planning later quantitative analysis. It is salutary to recall the recent statement by West (1986) that 'in capable hands in specialist laboratories, emission spectrography is, and always has been, a most powerful and flexible technique of multichannel elemental analysis'. The current status of ICPAES establishes it as a multielement quantitative method, with the minor disadvantage of requiring solutions. The determination of trace elements in coal is still the domain of AAS, which has high sensitivities for many elements, but with the disadvantages of sequential operation and the need for solutions. Perhaps the most useful multielement method for many trace elements is INAA, which uses solid samples and determines 30–40 elements simultaneously. A disadvantage is the lack of results for B, Be, Cd, Cu, F, Hg and Pb, which are of environmental interest in coal. Despite limitations of sensitivity for some elements, notably B, Be, Cd, F, Hg and Sb, XRFS is a good multielement method with the advantage of using solid samples. The wide coverage of elements plus high sensitivity are special attributes of SSMS, but quantitative analysis is not easily attained. Very small solid samples are used, but this is not necessarily an advantage for analysing bulk coal samples. There is a good evaluation of SSMS (Beske et al., 1981) which indicates research that could lead to improvements. There is still a place for chemical methods in some cases. Without wishing to be dogmatic, it is suggested that AES, AAS and INAA are perhaps the most suitable methods, an opinion supported by the fact that most published results have been obtained by these methods. Chemical methods are suggested for Cl, F and P. Suitable methods for determining trace elements in coals are shown in Table 4.3 for those elements of environmental interest. The pros and cons of the various methods of removing organic matter and the dissolution of coal samples have been discussed already, but although dry ashing seems to be viewed unfavourably in some quarters, this reputation 'appears to be largely undeserved, at least for biological materials' (Thiers, 1957).

It is imperative that trace-element determinations are carried out by competent and experienced analytical chemists *au fait* with the nuances of the various methods, chemical or instrumental. As Abbey (1981) has stated 'the reliability of a result depends more on <u>who</u> produced it than on <u>how</u>, it was done', a sentiment also echoed by Willis (1988). The

Table 4.3 **Suitable methods for determining trace elements in coals** (×, suitable for most coals; O, limited by sensitivity or interference; −, usually unsuitable)

Trace element	AES	AAS	INAA	XRFS	SSMS	CHEM	ICPAES
As	O	×	×	×	×	×	O
B	×	−	−	−	×	×	×
Be	×	×	−	−	×	×	×
Cd	O	×	−	O	×		O
Cl	−	−	×	×	−	×	−
Co	×	×	×	×	×		×
Cr	×	×	×	×	×		×
Cu	×	×	O	×	×		×
F	−	−	−	−	×	×	−
Hg	−	×	O	−	×		−
Mn	×	×	×	×	×		×
Mo	×	×	O	×	×		×
Ni	×	×	O	×	×		×
P	−	−	−	×	×	×	×
Pb	×	×	−	×	×		×
Sb	O	×	×	O	×		O
Se	−	×	×	O	×		O
Sn	×	×	×	×	×		O
Th	−	−	×	×	×		O
Tl	×	×	−	−	×		O
U	−	−	×	×	×		O
V	×	×	×	×	×		×
Zn	×	×	×	×	×		×

importance of the researcher is stressed by Ihnat (1977) who stated that 'constant vigilance and efforts by analytical chemists are essential to ensure that analytical data generated are sufficiently reliable to form a solid base for sound decisions'.

Results for the contents of trace elements in coals involve much work in sampling, preparation of samples for analysis and the final determination. Hence, the statement of results should give relevant information, for example on precision and accuracy, to enable the user to judge the value of the results for his purposes.

Chapter 5

Contents of trace elements in coals

No matter how perfect a bird's wing may be, it could never lift the bird to any height without the support of air. Facts are the air of science.

I.P. Pavlov

5.1 General comments

The data for the contents of trace elements in coals are found in a wide range of journals and reports. Where multielement methods of analysis have been used, there may be results for 30–40 elements per sample. In other cases, perhaps only one element was determined. It is not considered worth while to include every published result, especially if only one sample was analysed. A bibliography of literature on trace elements in coal up to 1980 has been prepared by Valkovic (1981) and there are two earlier bibliographies confined almost entirely to coals from the United States (Averitt et al., 1972, 1976). There are many references to trace-element contents in coal in Clark and Swaine (1962a), Bethell (1963), Abernethy and Gibson (1963), Swaine (1975), Finkelman (1981a) and Valkovic (1983). Data for trace elements in coal ash will not be given, although in some cases where ash yields were stated values were recalculated to a coal basis.

The following publications give multielement data for a relatively large number of coals, together with relevant information about the coals. Data for Australian coals are in Clark and Swaine (1962a), Brown and Swaine (1964), Swaine (1967, 1975, 1977a, 1979), and more recently in Cahill and Mills (1983), Knott and Warbrooke (1983), Knott et al. (1984), Doolan et al. (1984b) and Fardy, McOrist and Farrar (1984). There is specific information on Victorian brown coals (Bone and Schaap, 1980, 1981a, 1981b), Queensland coals (CSIRO, 1960, 1962), on Leigh Creek (South Australia) hard brown coals (CSIRO, 1963, 1964, 1965b, 1966) and on Collie (Western Australia) subbituminous coals (Davy and Wilson, 1984). Relevant information on Canadian coals is given by Landheer, Dibbs and Labuda (1982), Evans et al. (1985) and van Voris et al. (1985), while results for NZ coals are summarized by Purchase (1985). South African coals are featured by Willis (1981, 1983), Hart, Leahy and Falcon (1982) and Hart and Leahy (1983). Taylor (1973) and Hislop *et al.* (1978) cover UK coals, and US coals are dealt with by Swanson *et al.* (1976), Gluskoter *et al.* (1977), Zubovic, Hatch and Medlin (1979) and Zubovic *et al.* (1979, 1980). There are thousands of results in CSIRO Investigation and Location reports which have been incorporated in the ranges of values given in the tables. A similar mass of data resides in the US Geological Survey's data bank and in Open-File reports, and there may well be such sources in other countries.

For some countries there are several publications covering data for many samples and often earlier data are updated. In these cases in the tables, only one reference will be given, but it will cover data from several publications. For example, for Australia, Clark and Swaine (1962a) covers Brown and Swaine (1964), Swaine (1967, 1975, 1977a, 1979), and CSIRO (unpublished). For Queensland coals, Brown and Swaine (1964) includes Swaine (1977a), Swaine (1979) and CSIRO (unpublished). For Leigh Creek coals, CSIRO (1966) covers CSIRO (1963, 1964, 1965b). Knott and Warbrooke (1983) includes Cahill and Mills (1983), Knott *et al.* (1984) and Doolan *et al.* (1984b). Many thousands of results for coals from the United States are given in Swanson *et al.* (1976), Zubovic, Hatch and Medlin (1979) and Zubovic *et al.* (1979, 1980). Most of these results had been gleaned from Open-File and similar reports, too many to list here.

5.2 Detailed information for individual trace elements

Information on 47 trace elements, the platinum group elements (PGE) and the rare earth elements (REE) will be given for each trace element, arranged alphabetically. The information for each element will include mode of occurrence, methods of analysis, environmental and health aspects and data. Comments on environmental and health aspects are restricted to effects from low concentrations that are generally associated with coal mining and usage. It is unwise to categorize elements as toxic *per se* because toxicity depends on several factors, including concentration, chemical structure and availability. Several trace elements, for example Cu and Zn, that are essential for good health, are harmful in excess. In discussing analytical methods, SSMS will not be mentioned frequently, as most elements can be determined by SSMS. Other relevant matters are discussed in Chapters 3 and 4. Data will be tabulated under five headings, namely sample details (country, location of seam and other relevant information), rank of coal, number of samples, content (range of values or mean value in parentheses) and reference. The following abbreviations will be used for rank: BR (brown), L (lignite), H-B (hard brown), S (subbituminous), BI (bituminous), S-A (semi-anthracite), A (anthracite). All results will be stated as ppm of the element in air-dried or oven-dried coal, unless stated otherwise. Usually the lower end of the range of values will be given as $<x$, where x is the limit of detection, but it cannot be inferred that this lower value is necessarily near x ppm.

5.2.1 Antimony

In about the mid-nineteenth century Daubrée (Briggs, 1934) reported the presence of Sb in some coals and it is often determined nowadays, although the low concentrations generally found in coals are not considered to be environmentally hazardous. The main sources of Sb in the atmosphere are metal smelting, waste incineration and coal combustion (Austin and Millward, 1988), while Sb has also been detected in emissions from Hawaiian volcanoes (Cadle *et al.*, 1973). It is not clear how Sb occurs in coals, but it is likely that an organic association prevails in many coals, together with a sulfide association, which may predominate in others. Despite the report of ullmanite in a Durham, UK, coal (Spencer, 1910), this cannot be regarded as a common mineral in coal. In the high-S coals from Greta, New South Wales, Sb is <0.1 ppm. In some coals, Sb may be present as tiny grains of a sulfide, possibly stibnite, closely associated with the organic coaly matter (Finkelman, 1981a) and hence seemingly having an organic association.

The determination of Sb may be carried out by AAS and INAA, the high sensitivity of INAA being noteworthy. However, a method based on EAAS (Haynes, 1978) with a limit of detection of 0.1 ppm, is sufficient for most Australian coals. A similar method was used by Aruscavage (1977), who extracted Sb from the coal with mixed acids, then removed the Sb from a H_2SO_4-iodide solution into toluene, the final determination being carried out by EAAS. A method based on hydride generation–AAS was used by Nadkarni (1982b).

Results for the contents of Sb in coal are given in Table 5.1. Apart from a few high values, that are local and atypical, the range for most coals is about 0.05–10 ppm Sb and the corresponding range for Australian coals is <0.01–1.2 ppm Sb. Approximate mean values in ppm Sb are 0.5 (Australia), 1 (West Germany), 2.5 (United Kingdom), 0.7 (United States), and 2.3 for coals burnt in nine EC countries (calculated by Sabbioni *et al.*, 1983).

It would seem that there is no reason to regard Sb as in any way deleterious during coal winning and utilization.

5.2.2 Arsenic

Arsenic was found in some European coals before there was much interest in trace elements (Daubrée, 1858; Percy, 1875), but the first real investigation was carried out by Goldschmidt and Peters (1934). It is not surprising to find that there has been increased interest in As in coals, together with work on rocks, soils, plants and waste materials, probably because of possible adverse health effects of high concentrations. There is a wide range of values from less than 1 ppm to several hundred ppm, sometimes enrichment being related to nearby As-rich ores, as is probably the case with some peats from Finland with up to 340 ppm As (Minkkinen and Yliruokanen, 1978). As with some other environmentally sensitive elements, coal tends to be seen as a major source of As, whereas it only contributes 1.8% of the total emissions to the atmosphere (Walsh, Duce and Fasching, 1979), which is about the same as wood fuel. Volcanic

Table 5.1 Results for antimony (as ppm Sb)

Sample details	Rank	Samples	Content	Reference
AUSTRALIA:				
Latrobe Valley, Vic.	BR	20	<0.02–0.03	Bone and Schaap (1981a)
Latrobe Valley, Vic.	BR	28	<0.01–0.5	Fardy, McOrist and Farrar (1984)
St Vincent Basin, SA	L	5	0.08–0.87	Fardy, McOrist and Farrar (1984)
Leigh Creek, SA	H-B	3	0.3–2.6	K.W. Riley (1983, pers. commun.)
Leigh Creek, SA	H-B	2	<1	Knott and Warbrooke (1983)
Collie, WA	S	22	<0.4–2	Davy and Wilson (1984)
Collie, WA	S	5	<0.1	Brown and Swaine (1964)
Northern area, NSW	BI	20	<0.1–0.8	K.W. Riley (1983, pers. commun.)
Southern area, NSW	BI	10	0.4–1.3	K.W. Riley (1983, pers. commun.)
Western area, NSW	BI	60	<0.1–1.4	Swaine, Godbeer and Morgan (1984b)
NSW and Qnsld	BI	65	<0.1–9	Knott and Warbrooke (1983)
NSW and Qnsld	BI	45	0.1–4.7	Fardy, McOrist and Farrar (1984)
Bowen Basin, Qnsld	BI	10	<0.1–0.8	K.W. Riley (1983, pers. commun.)
Fingal, Tasmania	BI	2	<1	Knott and Warbrooke (1983)
BELGIUM:				
From 12 mines	BI, A	48	0.3–4.0	Block and Dams (1975)
BRAZIL:				
Rio Grande do Sul	BI	13	1.6–13	Bellido and Arezzo (1987)
BULGARIA:				
'Some coals'	–	–	(1.3)	Kostadinov and Djingova (1980)
CANADA:				
Saskatchewan	L	8	0.5–3.3	Landheer, Dibbs and Labuda (1982)
Hat Creek No.1, BC	L, S	14	0.08–2.4	Goodarzi (1987b)
Hat Creek No.2, BC	L, S	32	0.2–1.5	Goodarzi and van der Flier-Keller (1988)
Alberta and BC	S	23	<0.3–1.5	Landheer, Dibbs and Labuda (1982)
'Typical'	S, BI	59	0.08–0.23	Jervis, Ho and Tiefenbach (1982)
Fording mine, BC	BI	22	0.2–3.6	Goodarzi (1988)
Nova Scotia and New Brunswick	BI	7	<0.3–5.1	Landheer, Dibbs and Labuda (1982)
CHINA:				
'Selection'	–	15	<0.06–124	Sun and Jervis (1986)
from 110 mines	–	–	0.047–29	Chen et al. (1986)
FRANCE:				
2 areas	–	2	(5.2)	Sabbioni et al. (1983)
GERMANY WEST:				
Mostly Ruhr	BI	27	0.4–2.0	Kautz, Kirsch and Laufhutte (1975)

Table 5.1 (contd)

Sample details	Rank	Samples	Content	Reference
Various	BI	10	0.14–2.0	Heinrichs (1975b)
For power production	BI	6	1.8–5.0	Sabbioni et al. (1983)
NEW ZEALAND:				
Westland	S-BI	–	0.2–3.7	Purchase (1985)
NIGERIA:				
'Major deposit'	L	1	0.61	Hannan et al. (1982)
'Major deposit'	S	6	0.16–0.44	Hannan et al. (1982)
'Major deposit'	BI	1	0.89	Hannan et al. (1982)
Enugu, 3 areas	S	3	0.81–1.2	Ndiokwere, Guinn and Burtner (1983)
POLAND:				
Belchatow mine	BR	3	0.16–4.9	Tomza (1987)
7 mines	BI	31	0.03–4.6	Tomza (1987)
SOUTH AFRICA:				
Witbank coalfield	–	146	0.24–0.38	Hart and Leahy (1983)
SWITZERLAND:				
	BR, L	9	0.07–6.2	Hügi et al. (1989)
	A	9	<0.2–4.4	Hügi et al. (1989)
TURKEY:				
2 'kinds'	L	2	5.8, 8.8	Akcetin, Ayca and Hoste (1973)
5 zones	L	5	0.08–1.1	Ayanoglu and Gunduz (1978)
UK:				
S. Yorkshire	S	8	0.48–1.6	Ward, Kerr and Otsuka (1986)
23 collieries	BI	23	1–10	Hislop et al. (1978)
S. Wales (12 seams)	BI	300	−188 (16)	Chatterjee and Pooley (1977)
USA:				
Northern Great Plains	L	226	0.1–4.5	Zubovic, Hatch and Medlin (1979)
Gulf Province	L	34	0.2–5.2	Swanson et al. (1976)
Texas	L	43	<0.3–1.9	White, Edwards and Du Bose (1983)
Fort Union region	L	80	0.1–3.0	Hatch and Swanson (1976)
Northern Great Plains	S	855	0.06–9.1	Zubovic, Hatch and Medlin (1979)
Rocky Mt Province	S	466	0.04–43	Zubovic, Hatch and Medlin (1979)
Appalachian Region	BI	1478	0.04–35	Zubovic, Hatch and Medlin (1979)
Virginia	BI	83	0.20–5.9	Henderson et al. (1985)
Interior Province	BI	687	0.04–16	Zubovic, Hatch and Medlin (1979)
Rocky Mt Province	BI	366	0.10–7.3	Zubovic, Hatch and Medlin (1979)
Alaska, N. Slope	L-BI	84	<0.05–0.6	Affolter and Stricker (1987)
Alaska, N. Slope	L-BI	–	0.05–10	Merritt (1988)
Pennsylvania	A	53	<0.1–13	Swanson et al. (1976)

emissions contain As (Cadle et al., 1973) and Greenland and Aruscavage (1986) estimated that about 10 t As were emitted during eruptive emissions from Kilauea over a period of one year. Indeed, Walsh, Duce and Fasching (1979) estimated that volcanoes contributed 22% of the total emission of As to the atmosphere.

Although it has been stated categorically that As is present in coal as arsenopyrite (Crossley, 1946; Francis, 1961) and that 'little exists in any other form' (Chapman, 1901), the only good evidence for the nature of the association of As with pyrite has come from a detailed SEM and EMPA study of an eastern US coal (Minkin et al., 1984). They found that the As was most likely to be present in solid solution in the pyrite, and Finkelman et al. (1979) noted that the As was predominantly in fractures in the coal and in microfractures in the pyrite. Arsenic was also detected at isolated points in some pyrite grains, perhaps because 'small arsenic- and selenium-bearing mineral inclusions formed and were incorporated into the pyrite at the time of crystallization' (Minkin et al., 1984). There is evidence for the association of As with pyrite, for example Goldschmidt (1954) found 3000 ppm As in pyrite from a coal and Brown and Swaine (1964) found 1% As in pyrite separated from mudstone in an NSW coal seam. However, not all high-pyrite coals contain high As (Hokr, 1977), an observation that has also been made in respect of many Australian bituminous coals. This probably means that some coals contain As in other forms, for example organically associated, associated with clays, perhaps as arsenate ions or with phosphate minerals, where $(AsO_4)^{3-}$ could replace some $(PO_4)^{3-}$. Studies need to be carried out on a variety of coals, using the SEM and EMPA techniques outlined by Finkelman (1981a), in order to clarify the nature of the mineral association of As in coals. The extent of organically associated As is not clear, although organic bonding has been suggested for some Bulgarian coals (Minchev and Eskenazi, 1966, 1972), for some low-rank Canadian coals (Goodarzi, 1987b) and for some low-S Siberian coals (Kryukova et al., 1985). For most coals, As seems to be mainly associated with the mineral matter, with varying smaller amounts being associated with organic matter.

The determination of As in coal can be achieved by several methods, including AAS, INAA, XRFS and chemical techniques. Several standard methods are based on the molybdenum blue coloration, for example AS (1980a), but such methods are unsatisfactory for concentrations around 1 ppm As. K.W. Riley (1982, pers. commun.) found that determinations of As in 19 Australian bituminous coals showed a range of <0.1–2.7, with a mean of 0.8 ppm As, using an EAAS method, compared with <1–7, mean 1.9 ppm As, using AS (1980a). Perhaps the method of choice will often be an AAS one, for example Aruscavage (1977) and Haynes (1978) used EAAS and Nadkarni (1982b) and Lindahl (1985b) used hydride-generation. Another approach is slurry atomization, in which the coal sample is slurried with nickel nitrate, magnesium nitrate, nitric acid and ethanol (Ebdon and Parry, 1987). At CSIRO, a modification of Haynes (1978) is used, the coal sample being mixed with magnesium nitrate and nickel nitrate prior to ashing at 450–500 °C for 14 h. The ash is treated with a mixture of HNO_3–HF and finally with HNO_3 to give 50 ml of approximately 5% HNO_3 in the solution used for EAAS. Good results have been obtained by INAA and XRFS. Low temperature ashing of coal prior to the determination of As is certainly satisfactory (Walling et al., 1978), but ashing of coal mixed with MgO at 700°C gives 'very little, if any, loss' (Kunstmann and Bodenstein, 1961), if reducing conditions are avoided (Duck and Himus, 1951).

Results for As in coal are given in Table 5.2. The range of values is from <0.1 to several hundred ppm. The values of above ~80 ppm As would seem to be unusual for coals used for power production, although a brown coal with 900–1500 ppm As has been burnt in a Czechoslovakian power station (Bencko and Symon, 1977). In this latter case, bees which had been affected by As moved their colonies to beyond 30 km downwind and some undesirable health effects were noted (Bencko et al., 1980). However, this was an extreme case and there seems to be no evidence for ill-effects from As from coal combustion, where low-As coals are used. An estimate of the ranges for As in most coals is, say, 0.5–80 ppm As and in most Australian coals, 0.2–9 ppm As. Differences in As levels between Gondwana coals and coals from the northern hemisphere are reflected in mean values (as ppm As), namely, 1.5 (Australia), 4 (South Africa), 14 (United States), 15 (United Kingdom), and 11 for coals burnt in nine EC countries (calculated by Sabbioni et al., 1983).

It is interesting to compare the ranges of values for coals with the range of 1–50 ppm As for soils (Swaine, 1955) and the mean value of 7 ppm As for US soils (Shacklette, Boerngen and Keith, 1974). As Chen et al. (1986) noted, stress should not be placed on the high values and in some cases, notably Australian and US sources, information about individual coals is available. Some other publications give mean values. Although As affects the properties of high-strength steels (Kurmanov et al., 1957), the amounts of As in most cokes would not cause problems. Modern methods of combustion based on pulverized coal do not cause the formation of phosphatic deposits on boiler surfaces, which were formerly fouled by deposits containing, *inter alia*, boron arsenate (Swaine and Taylor, 1970) and around 1% As (Brown and Swaine, 1964). However, flyash may contain several hundred ppm As, depending on the coal used, and hence some checks should be made on the As-contents of flyash applied to land and on leachates from flyash dumps. Clearly, no problems are to be expected from low-As coals, for example most Australian coals and other low-S coals. However, PECH (1980) lists As as an element of 'greatest concern', which probably means that As in some coals may be a nuisance under certain conditions.

5.2.3 Barium

Perhaps the earliest detection of Ba in coals was by Binns and Harrow (1897) who reported the presence of barite in a coal seam from Leicestershire, UK. An early determination of Ba in a Japanese coal showed the presence of 95 ppm Ba (Mingaye, 1907). Barite was also found, together with ankerite, in a Leicestershire coal (Crook, 1913) and in a Yorkshire coal (Finn, 1930). Other Ba minerals found in some coals are barytocelestine, witherite and gorceixite. In low-rank coals, there is evidence for the association of Ba with the organic coaly matter via

Table 5.2 Results for arsenic (as ppm As)

Sample details	Rank	Samples	Content	Reference
AUSTRALIA:				
Latrobe Valley, Vic.	BR	10	<1–3	Brown and Swaine (1964)
Latrobe Valley, Vic.	BR	20	<0.03–0.7	Bone and Schaap (1981a)
Latrobe Valley, Vic.	BR	28	<0.01–1.3	Fardy, McOrist and Farrar (1984)
St Vincent Basin, SA	L	5	2.1–3.7	Fardy, McOrist and Farrar (1984)
Leigh Creek, SA	H-B	44	0.7–6	CSIRO (1966)
Collie, WA	S	22	<1–2	Davy and Wilson (1984)
Northern area, NSW	BI	150	<0.1–16	Clark and Swaine (1962a)
Southern area, NSW	BI	50	<0.1–9	Clark and Swaine (1962a)
Western area, NSW	BI	60	<0.6–6	Swaine, Godbeer and Morgan (1984b)
NSW and Qnsld	BI	65	0.1–17	Knott and Warbrooke (1983)
NSW and Qnsld	BI	45	0.2–16	Fardy, McOrist and Farrar (1984)
Bowen Basin, Qnsld	BI	100	<0.3–9	Brown and Swaine (1964)
West Moreton area, Qnsld	BI	90	<1–9	Brown and Swaine (1964)
Callide area, Qnsld	BI	40	<1–4	Brown and Swaine (1964)
Fingal, Tasmania	BI	2	0.8, 1.0	Knott and Warbrooke (1983)
Baralaba, Qnsld	S-A	20	<1–1	CSIRO (unpublished)
BELGIUM:				
From 12 mines	BI, A	40	0.2–30	Block and Dams (1975)
BRAZIL:				
Rio Grande do Sul	BI	13	1.5–218	Bellido and Arezzo (1987)
BULGARIA:				
'Some coals'	–	–	(12)	Kostadinov and Djingova (1980)
CANADA:				
Saskatchewan	L	8	<0.2–5.9	Landheer, Dibbs and Labuda (1982)
Moose River Basin	L	17	3.0–6.0	Van der Flier-Keller and Fyfe (1987b)
Hat Creek No.1, BC	L, S	14	4.0–53	Goodarzi (1987b)
Hat Creek No.2, BC	L, S	32	4.0–30	Goodarzi and van der Flier-Keller (1988)
Alberta and BC	S	23	<0.3–13	Landheer, Dibbs and Labuda (1982)
'Typical'	S, BI	56	0.3–320	Jervis, Ho and Tiefenbach (1982)
Fording mine, BC	BI	22	0.2–2.6	Goodarzi (1988)
Nova Scotia	BI	186	33–270	Hawley (1955)
CHINA:				
'Selection'	–	15	0.54–5.9	Sun and Jervis (1986)
from 110 mines	–	–	0.32–120	Chen et al. (1986)
CZECHOSLOVAKIA:				
	–	6	3.8–31	Flum (1957)
North Bohemia	BR	1100	1–150	Hokr (1977)
FRANCE:				
2 areas	–	2	14, 21	Sabbioni et al. (1983)

Table 5.2 (contd)

Sample details	Rank	Samples	Content	Reference
GERMANY WEST:				
From 1 power station	BR	7	0.81–1.6	Heinrichs (1982)
Mostly Ruhr	BI	27	1.5–50	Kautz, Kirsch and Laufhutte (1975)
For power production	BI	5	3–14	Sabbioni et al. (1983)
NEW ZEALAND:				
Taranaki area	S	41	<1–12	Black (1981)
Waikato area	S	5	1.0–3.0	Lynskey, Gainsford and Hunt (1984)
Westland	S-BI	–	0.3–38	Purchase (1985)
NIGERIA:				
'Major deposit'	L	1	0.80	Hannan et al. (1982)
'Major deposit'	S	6	0.20–3.3	Hannan et al. (1982)
'Major deposit'	BI	1	3.0	Hannan et al. (1982)
POLAND:				
Belchatow mine	BR	3	1.8–8.8	Tomza (1987)
	BI	–	0–40	Widawska-Kusmierska (1981)
For power production	BI	2	(1.7)	Sabbioni et al. (1983)
7 mines	BI	31	0.17–69	Tomza (1987)
SOUTH AFRICA:				
Transvaal, Natal	BI	61	0.27–10	Kunstmann and Bodenstein (1961)
Witbank coalfield	BI	146	1.9–5.7	Hart and Leahy (1983)
	BI	14	0.9–8.2	Willis (1983)
SWITZERLAND:				
	BR, L	9	4.0–42	Hügi et al. (1989)
	A	9	<0.5–17	Hügi et al. (1989)
THAILAND:				
1 deposit	L	3	9–33	CSIRO (unpublished)
TURKEY:				
5 'kinds'	L	5	1–64	Akcetin, Ayca and Hoste (1973)
5 zones	L	5	11–344	Ayanoglu and Gunduz (1978)
USSR:				
South Yakutia	–	–	25–83	Egorov, Laktionova and Borts (1981)
Pechora	–	4	1.2–6.8	Blomqvist (1983)
East Siberia	–	300	<1–32	Kryukova et al. (1985)
UK:				
S. Yorkshire	S	8	0.48–1.2	Ward, Kerr and Otsuka (1986)
23 collieries	BI	23	2–73	Hislop et al. (1978)
USA:				
Northern Great Plains	L	226	0.70–110	Zubovic, Hatch and Medlin (1979)
Gulf Province	L	34	1–16	Swanson et al. (1976)
Texas	L	93	<0.1–31	White, Edwards and Du Bose (1983)
Fort Union region	L	80	1–30	Hatch and Swanson (1976)
Northern Great Plains	S	855	0.20–420	Zubovic, Hatch and Medlin (1979)

Table 5.2 (*contd*)

Sample details	Rank	Samples	Content	Reference
Rocky Mt Province	S	466	0.10–125	Zubovic, Hatch and Medlin (1979)
Appalachian Region	BI	1478	0.12–354	Zubovic, Hatch and Medlin (1979)
Interior Province	BI	687	0.70–240	Zubovic, Hatch and Medlin (1979)
Rocky Mt Province	BI	366	0.19–60	Zubovic, Hatch and Medlin (1979)
Alaska, N. Slope	L-BI	84	<0.07–8	Affolter and Stricker (1987)
Alaska	L-BI	–	0.05–25	Merritt (1988)
Pennsylvania	A	53	<0.7–140	Swanson *et al.* (1976)

carboxyl groups (Morgan, Jenkins and Walker, 1981). Although Ba can be determined by modern instrumental methods, the most convenient method is FAAS, after dissolving the 500 °C ash with aqua regia then HF in a sealed plastic bottle, heated on a steam bath (AS, 1986). A nitrous oxide–acetylene flame is used with the line at 553.5 nm and background correction.

The results shown in Table 5.3 indicate that most coals fall in the range of 20 to about 1000 ppm Ba and mean values of 70 to about 300 ppm Ba have been reported.

There is little interest in Ba in coal, but a mixed vein of barite and witherite in a colliery at Durham, UK, was a major source of these minerals for many years (Francis, 1961). No reports have been found concerning untoward effects from Ba in coal during mining and usage.

5.2.4 Beryllium

Following the initial investigation of Be in some German coals (Goldschmidt and Peters, 1932a), there was a lull until results were published for Canadian (Hawley, 1955) and US coals (Stadnichenko, Zubovic and Sheffey, 1961). Unfortunately there has been some stress on two early results, namely 8000 ppm in a German coal ash (Goldschmidt, 1954) and 1000 ppm in a Soviet coal ash (Zilbermintz and Rusanov, 1936). These results are quite atypical and should not be quoted (even sometimes erroneously as being in coal!) in general references to coal. In some coals, Be is highest in seam-bottoms (Hak and Babcan, 1967). It is generally considered that Be is organically associated in most coals. Singh, Singh and Chandra (1983) suggested that Be appeared to be in quartz and clay minerals in Ghugus coals from India. In some coals, Be may, to a minor degree, be present as beryl or associated with quartz and clays. Modern instrumental techniques are not suitable for determining Be in coal, but there are good methods available, namely OES, AAS and fluorimetry. Two national standard methods use FAAS (ASTM, 1983a; AS, 1986), the procedure being to treat the ash (500 °C) with aqua regia and HF in a plastic bottle (with a screw cap) on a steam bath. The solution is analysed

Table 5.3 **Results for barium** (as ppm Ba)

Sample details	Rank	Samples	Content	Reference
ANTARCTICA:				
Thermally altered	BI	7	<130–500	Brown and Taylor (1961)
ARGENTINA:				
Rio Turbio	–	6	6–8	A. Berset (1982, pers. commun.)
AUSTRALIA:				
Latrobe Valley, Vic.	BR	16	60–800	Brown and Swaine (1964)
Latrobe Valley, Vic.	BR	20	<2–175	Bone and Schaap (1981a)
Latrobe Valley, Vic.	BR	15	7–128	Bolger (1989)
Latrobe Valley, Vic.	BR	28	1.7–172	Fardy, McOrist and Farrar (1984)
St Vincent Basin, SA	L	5	220–440	Fardy, McOrist and Farrar (1984)
Leigh Creek, SA	H-B	80	100–2000	CSIRO (1966)
Collie, WA	S	22	43–519	Davy and Wilson (1984)
Northern area, NSW	BI	1000	<20–1500	Clark and Swaine (1962a)
Southern area, NSW	BI	800	40–1000	Clark and Swaine (1962a)
Western area, NSW	BI	80	<20–300	Swaine, Godbeer and Morgan (1984b)
NSW and Qnsld	BI	65	20–612	Knott and Warbrooke (1983)
NSW and Qnsld	BI	45	33–980	Fardy, McOrist and Farrar (1984)
Bowen Basin, Qnsld	BI	200	15–1500	Brown and Swaine (1964)
West Moreton area, Qnsld	BI	150	70–600	Brown and Swaine (1964)
Callide area, Qnsld	BI	100	50–200	Brown and Swaine (1964)
Fingal, Tasmania	BI	2	78, 208	Knott and Warbrooke (1983)
Baralaba, Qnsld	S-A	20	30–400	CSIRO (unpublished)
BELGIUM:				
From 12 mines	BI, A	48	40–1000	Block and Dams (1975)
BULGARIA:				
'Some coals'	–	–	(16)	Kostadinov and Djingova (1980)
CANADA:				
Saskatchewan	L	8	413–1190	Landheer, Dibbs and Labuda (1982)
Hat Creek No.1, BC	L, S	14	45–345	Goodarzi (1987b)
Hat Creek No.2, BC	L, S	32	32–331	Goodarzi and van der Flier-Keller (1988)
Alberta and BC	S	23	<60–1160	Landheer, Dibbs and Labuda (1982)
'Typical'	S, BI	60	13–3900	Jervis, Ho and Tiefenbach (1982)
Fording mine, BC	BI	22	50–967	Goodarzi (1988)
Nova Scotia	BI	186	2–257	Hawley (1955)
CHINA:				
'Selection'	–	15	25–458	Sun and Jervis (1986)
from 110 mines	–	–	13–1540	Chen et al. (1986)
COLOMBIA:				
3 regions	BI	14	15–95	Rincon et al. (1978)
GERMANY EAST:				
Niederrhein	BR	62	13–151	Pietzner and Wolf (1964)

Table 5.3 (contd)

Sample details	Rank	Samples	Content	Reference
GERMANY WEST:				
From 1 power station	BR	7	19–145	Heinrichs (1982)
Mostly Ruhr	BI	27	45–350	Kautz, Kirsch and Laufhutte (1975)
NEW ZEALAND:				
Taranaki area	S	–	32–230	Purchase (1985)
Waikato area	S	–	7–150	Purchase (1985)
Southland	S	–	22–150	Purchase (1985)
Westland	S-BI	–	6–225	Purchase (1985)
NIGERIA:				
'Major deposit'	L	1	39	Hannan et al. (1982)
'Major deposit'	S	6	30–141	Hannan et al. (1982)
'Major deposit'	BI	1	91	Hannan et al. (1982)
POLAND:				
Belchatow mine	BR	3	76–181	Tomza (1987)
7 mines	BI	31	60–610	Tomza (1987)
SOUTH AFRICA:				
	BI	14	86–474	Willis (1983)
SWITZERLAND:				
	BR, L	9	63–612	Hügi et al. (1989)
	A	9	42–619	Hügi et al. (1989)
THAILAND:				
1 deposit	L	3	30–150	CSIRO (unpublished)
TURKEY:				
5 zones	L	5	77–186	Ayanoglu and Gunduz (1978)
USSR:				
South Yakutia	–	–	160–449	Egorov, Laktionova and Borts (1981)
UK:				
S. Yorkshire	S	8	36–144	Ward, Kerr and Otsuka (1986)
N. England, S. Wales	BI	26	<6–500	D.J. Swaine (unpublished)
23 collieries	BI	23	30–250	Hislop et al. (1978)
USA:				
Gulf Province	L	34	15–700	Swanson et al. (1976)
Texas	L	103	2–1820	White, Edwards and Du Bose (1983)
Fort Union region	L	80	15–2000	Hatch and Swanson (1976)
S. and E. Arkansas	L	53	50–1500	Hildebrand, Clardy and Holbrook (1981)
Eastern USA	BI	23	72–420	Gluskoter et al. (1977)
Western USA	–	28	160–1600	Gluskoter et al. (1977)
Alaska, N. Slope	L-BI	84	150–5000	Affolter and Stricker (1987)

by FAAS, using a nitrous oxide–acetylene flame and background correction on the Be line at 234.9 nm. ASTM (1983a) uses coal ground to $-150\,\mu m$, while AS (1986) requires 98% to pass $-75\,\mu m$. A method has been developed using ICPAES (Mills, Doolan and Knott, 1983).

Results for Be in coal are given in Table 5.4. The range for most coals is estimated as 0.1–15 ppm Be, while most Australian coals are in the range 0.2–8 ppm Be. Mean values of 1.5–2.0 ppm Be indicate that Be is low in most coals. Although there is no evidence for health problems arising from coal power-production, coals with 100 or more ppm in ash, regarded as 'comparatively high', were formerly used in a Czechoslovakian power station and investigations were carried out on local people (Bencko, Vasilieva and Symon, 1980), but no clear ill effects attributable to Be were found. Nevertheless, it should be kept in mind that Be is potentially a toxic element (Darwin and Buddery, 1960). According to Fishbein (1981) 'beryllium in the environment chiefly arises from coal combustion', although a careful study by Gladney and Owens (1976) showed that 'less than 4% of the Be in the coal burned is being emitted to the atmosphere'.

5.2.5 Bismuth

There is little interest in Bi in coal, as this element has no known adverse environmental or health implications. The analysis of a flue dust from a gas works burning a Yorkshire, UK, coal showed the presence of Bi (Ramage, 1927), probably the first evidence of Bi in coal. Goldschmidt (1937) found up to 200 ppm Bi in some German coal ashes and Headlee and Hunter (1955b) found 60 ppm Bi in a Canadian coal ash. Based on the examination of several coals Finkelman (1981a) stated that 'it is probable that bismuth is associated with the sulfides in coal'. Kessler, Sharkey and Friedel (1973) found <0.1–0.2 ppm Bi in some USA bituminous coals. The determination of Bi at sub-ppm levels requires the high sensitivity of SSMS, but EAAS (Heinrichs, 1982) may also be used. The high values found in some of the early ash analyses were from AES, and XRFS can be used for some coals. Heinrichs (1982) found 0.012–0.027 ppm Bi in seven samples of German brown coal collected from a power station and ten German bituminous coals had 0.03–0.53 ppm Bi (Heinrichs, 1975b). Knott and Warbrooke (1983) found <1–2 ppm Bi in 31 Australian bituminous coals and Dale, Fardy and Clayton (1985) found 0.2 ppm Bi in one NSW bituminous coal. In Latrobe Valley, Victoria, brown coals Bi was not detected, the limits of detection being 0.02 ppm Bi (CSIRO, unpublished) and 0.04 ppm Bi (Bone and Schaap, 1981a). During combustion of coal in stoker-fired furnaces, boiler deposits may be formed with up to 2000 ppm Bi (Brown and Swaine, 1964) and Smith (1958) found up to 20 ppm Bi in some pulverized-fuel ash. As a result of measuring inputs of Bi from volcanoes, Patterson and Settle (1987) declared 'it would seem, therefore, that volcanic emissions of Bi to the atmosphere are by far the predominant source of Bi in the atmosphere'. Perhaps this is not surprising when one considers the low levels of Bi in most coals, probably less than 0.05 ppm Bi, based on the scant data available.

5.2.6 Boron

Boron is an interesting element because it is essential for healthy growth, but relatively small excesses may cause harm in some situations, certain plants being the most susceptible. Hence, there is an ongoing interest in

Table 5.4 Results for beryllium (as ppm Be)

Sample details	Rank	Samples	Content	Reference
ANTARCTICA:				
Thermally altered	BI	7	<1–4	Brown and Taylor (1961)
AUSTRALIA:				
Latrobe Valley, Vic.	BR	16	<0.1–0.4	Brown and Swaine (1964)
Latrobe Valley, Vic.	BR	20	<0.05–0.7	Bone and Schaap (1981a)
Latrobe Valley, Vic.	BR	15	<0.2–2.0	Bolger (1989)
St Vincent Basin, SA	L	3	<0.3–1	CSIRO (unpublished)
Leigh Creek, SA	H-B	80	<1–1.5	CSIRO (1966)
Collie, WA	S	22	<1–3	Davy and Wilson (1984)
Northern area, NSW	BI	1000	<0.3–6	Clark and Swaine (1962a)
Southern area, NSW	BI	800	<0.6–6	Clark and Swaine (1962a)
Western area, NSW	BI	150	1–12	Swaine, Godbeer and Morgan (1984b)
NSW and Qnsld	BI	65	0.2–5.3	Knott and Warbrooke (1983)
Bowen Basin, Qnsld	BI	200	<0.2–5	Brown and Swaine (1964)
West Moreton area, Qnsld	BI	150	0.7–3	Brown and Swaine (1964)
Callide area, Qnsld	BI	100	<0.5–2.5	Brown and Swaine (1964)
Fingal, Tasmania	BI	2	2.3, 3.4	Knott and Warbrooke (1983)
Baralaba, Qnsld	S-A	20	<0.4	CSIRO (unpublished)
CANADA:				
'Commercial'	L-BI	14	0.4–1.8	Landheer, Dibbs and Labuda (1982)
Nova Scotia	BI	186	1–2	Hawley (1955)
COLOMBIA:				
3 regions	BI	14	<0.1–1.3	Rincon et al. (1978)
CZECHOSLOVAKIA:				
North Bohemia	BR	11	0.47–9.2	Dubansky (1983)
GERMANY WEST:				
From 1 power station	BR	7	0.21–1.5	Heinrichs (1982)
Mostly Ruhr	BI	27	0.9–3.5	Kautz, Kirsch and Laufhutte (1975)
NEW ZEALAND:				
Mokau area	S	–	0.2–1.5	Gainsford (1985)
Waikato area	S	–	0.08–1.0	Purchase (1984)
Southland	S	–	0.2–0.7	Purchase (1984)
Westland	S-BI	–	0.002–4.8	Purchase (1984)
POLAND:				
	BI	–	0.5–10	Widawska-Kusmierska (1981)
SWITZERLAND:				
	BR, L	9	<0.2–4	Hügi et al. (1989)
	A	9	0.4–8	Hügi et al. (1989)
USSR:				
South Yakutia	–	–	0.9–4	Egorov, Laktionova and Borts (1981)
Pechora	–	3	<5	Blomqvist (1983)
UK:				
N. England, S. Wales	BI	26	<0.6–5	D.J. Swaine (unpublished)
20 areas	BI	232	1–17	Taylor (1973)
S. Wales (12 seams)	BI	300	–33(3)	Chatterjee and Pooley (1977)

Table 5.4 (contd)

Sample details	Rank	Samples	Content	Reference
USA:				
Northern Great Plains	L	226	0.08–14	Zubovic, Hatch and Medlin (1979)
Gulf Province	L	34	0.2–15	Swanson et al. (1976)
Texas	L	137	<0.1–9.9	White, Edwards and Du Bose (1983)
Fort Union region	L	80	<0.2–15	Hatch and Swanson (1976)
Northern Great Plains	S	855	0.05–13	Zubovic, Hatch and Medlin (1979)
Rocky Mt Province	S	466	0.08–32	Zubovic, Hatch and Medlin (1979)
Appalachian Region	BI	1478	0.23–25	Zubovic, Hatch and Medlin (1979)
Interior Province	BI	687	0.05–18	Zubovic, Hatch and Medlin (1979)
Rocky Mt Province	BI	366	0.08–6.2	Zubovic, Hatch and Medlin (1979)
Alaska, N. Slope	L-BI	84	<0.1–15	Affolter and Stricker (1987)
Pennsylvania	A	53	<0.2–5	Swanson et al. (1976)

B in coal, dating from the initial work by Goldschmidt and Peters (1932b). Some high-B coals were found in Spitsbergen (Butler, 1953) and New Zealand (Rafter, 1945), the respective maxima being 2% B and 1.5% B in ash. On the basis of the inverse relationship between ash yield and boron in ash (Butler, 1953; Swaine, 1962a) and on estimates of the order of organic affinity (Table 3.1), it is suggested that B is predominantly organically bound in most coals. However, minor amounts of B are often found in mineral matter, especially clays. Illite takes up B more readily than kaolinite or montmorillonite (Harder, 1961). Fleet (1965) confirmed that kaolinite and montmorillonite take up B from solutions, and Swaine (1971) found up to 27 ppm B in kaolinites separated from New South Wales bituminous coals. Clay minerals are suggested as the prime source of B in some Polish coals (Konieczynski, 1970). Another B-containing mineral, found in some coals, is tourmaline, which is a complex borosilicate with up to about 3% B. Finkelman (1981a) has aptly commented 'it seems that tourmaline may be more abundant in coals than heretofore suspected'. Although the identification of small amounts of tourmaline, probably of very small size, is difficult, it would be an interesting undertaking.

Most of the determinations of B in coal have been carried out by AES or spectrophotometric methods. Direct-current AES methods (Clark and Swaine, 1962a; Swanson and Huffman, 1976; Dale, 1979; Baucells et al., 1984) and ICPAES methods (Mills, Doolan and Knott, 1983; Pougnet and Orren, 1986) are useful. The Australian standard method (AS, 1988) is a spectrophotometric one which measures the colour developed with curcumin. There is a useful compilation of B data for 69 reference materials, including coal and flyash (Gladney and Roelandts, 1987). There are differences of opinion about the volatility of B during the ashing of coal. Ruch, Gluskoter and Shimp (1973) suggest that ashing bituminous coals in covered crucibles at 500–700 °C should prevent losses. Nevertheless, it is advisable to check the retention of B for particular coals and ashing conditions (Swaine, 1985a). There has been some interest in the preparation of graphite from coal as a moderator in nuclear reactors (Hutcheon, 1953; Winter, Leibnitz and Otto, 1963). However, B has a high nuclear cross-section and the B-content of usable graphite must be less than 2 ppm B. Perhaps B has some influence on the mechanical properties of some steels (Borrowdale, Jenkins and Shanahan, 1959). Coal residues, for example flyash and washery wastes, are often disposed of on land or in ponds. Under some conditions, leachates may contain enough B to be an undesirable addition to natural waters, and hence the levels of soluble-B should be checked. Coal ash may be useful in agriculture, for example Mikhailova and Gladkaya (1980) recommended the ash of an East Siberian coal with 210–1620 ppm B. Flyash could be a useful soil amendment material (Swaine, 1981b). Again, plants grown on amended soils should be checked for possible effects on growth, which may be advantageous or not. Plants growing on areas being rehabilitated after coal-mining may show undesirable effects. An interesting method of ascertaining possible B problem areas on rehabilitation sites is the measurement of pH and electrical conductivity, thereby avoiding unnecessary analyses for B (Severson and Gough, 1983).

Results for B in coal are given in Table 5.5, the range for world coals being 0.5–2455 ppm B, but most would probably be between 5 and 400 ppm B, the corresponding range for most Australian coals being 3–150 ppm B. The mean value for most coals is probably in the range 30–60 ppm B.

5.2.7 Bromine

Bromine has no known function in nutrition and the main reason for most of the data on Br in coal is that it can be determined readily by INAA. It is generally agreed that Br is predominantly associated with organic matter in coal, and there is no direct evidence for mineral forms, although perhaps small amounts could be associated with iron oxides and clays. Although INAA seems to be the preferred method of analysis, Br can be determined chemically. A recent method proposed by Rigin (1987) is based on decomposing the coal sample, mixed with K_2CO_3 plus H_2O_2, in a PTFE-lined bomb, passing the resulting solution through a cation-exchange resin and then determining Br^- by an ion chromatographic technique. Another method of removing Br from coal may be by pyrohydrolysis, using a similar approach to that used for F. At temperatures above a few hundred degrees C, there is complete loss of Br, and there are variable losses under LTA conditions (Finkelman, 1981a; Doolan *et al.*, 1984a), so coal samples cannot be prepared for analysis by ashing techniques.

A detailed examination of results for Br in a wide range of coals shows great variability which seems almost haphazard. Results for Br in coal are given in Table 5.6. The spread of results makes it difficult to assess the range for most coals, but 0.5–90 ppm Br seems reasonable, the main

Table 5.5 Results for boron (as ppm B)

Sample details	Rank	Samples	Content	Reference
ANTARCTICA:				
Thermally altered	BI	7	5–39	Brown and Taylor (1961)
ARGENTINA:				
Rio Turbio	–	6	80–100	A. Berset (1982, pers. commun.)
AUSTRALIA:				
Latrobe Valley, Vic.	BR	16	3–70	Brown and Swaine (1964)
Latrobe Valley, Vic.	BR	20	2–30	Bone and Schaap (1981a)
St Vincent Basin, SA	L	3	100–250	CSIRO (unpublished)
Esperance, WA	L	3	100	CSIRO (unpublished)
Leigh Creek, SA	H-B	80	40–300	CSIRO (1966)
Collie, WA	S	22	2–5	Davy and Wilson (1984)
Northern area, NSW	BI	1000	0.9–300	Clark and Swaine (1962a)
Southern area, NSW	BI	800	1.5–150	Clark and Swaine (1962a)
Western area, NSW	BI	150	9–40	Swaine, Godbeer and Morgan (1984b)
NSW and Qnsld	BI	65	8–143	Knott and Warbrooke (1983)
Bowen Basin, Qnsld	BI	200	<1.5–150	Brown and Swaine (1964)
West Moreton area, Qnsld	BI	150	3–90	Brown and Swaine (1964)
Callide area, Qnsld	BI	100	2.5–70	Brown and Swaine (1964)
Fingal, Tasmania	BI	2	17, 23	Knott and Warbrooke (1983)
Fingal, Tasmania	BI	1	10	CSIRO (unpublished)
Baralaba, Qnsld	S-A	20	1.5–8	CSIRO (unpublished)
AUSTRIA:				
East Alpine area	BR	18	5.1–600	Janda and Schroll (1959)
	BI	7	3.3–170	Janda and Schroll (1959)
CANADA:				
Saskatchewan	L	8	741–1858	Landheer, Dibbs and Labuda (1982)
Hat Creek No.1, BC	L, S	14	9–32	Goodarzi (1987b)
Hat Creek No.2, BC	L, S	32	5–18	Goodarzi and van der Flier-Keller (1988)
Alberta and BC	S	20	427–2455	Landheer, Dibbs and Labuda (1982)
Fording mine, BC	BI	22	9–94	Goodarzi (1988)
Nova Scotia	BI	186	6–25	Hawley (1955)
Nova Scota	BI	2	149, 1872	Landheer, Dibbs and Labuda (1982)
COLOMBIA:				
3 regions	BI	14	0.5–10	Rincon et al. (1978)
GERMANY EAST:				
	BR	–	2–236	Winter, Leibnitz and Otto (1963)
Niederrhein	BR	55	6–21	Pietzner and Wolf (1964)
GERMANY WEST:				
Mostly Ruhr	BI	27	9–91	Kautz, Kirsch and Laufhutte (1975)
NEW ZEALAND:				
N. and S. Islands	L-BI	69	3–560	Kear and Ross (1961)
Mokau area	S	–	320–820	Gainsford (1985)

Table 5.5 (contd)

Sample details	Rank	Samples	Content	Reference
Waikato area	S	–	118–329	Mackay and Wilson (1978)
Waikato area	S	–	140–770	Purchase (1984)
Southland	S	–	19–295	Purchase (1984)
Westland	S-BI	–	5–400	Purchase (1984)
POLAND:				
	–	33	6.8–56	Roga et al. (1958)
	BI	–	10–50	Widawska-Kusmierska (1981)
SOUTH AFRICA:				
Transvaal, Natal, OFS	BI	61	11–109	Kunstmann et al. (1963)
SWITZERLAND:				
	BR, L	9	20–150	Hügi et al. (1989)
	A	9	5–60	Hügi et al. (1989)
THAILAND:				
1 deposit	L	3	80–100	CSIRO (unpublished)
USSR:				
Eastern Siberia	BR, BI	7	35–175	Mikhailova and Gladkaya (1980)
Moscow Basin	BR	3	70–80	Laktionova, Egorov and Popinako (1978)
South Yakutia	–	–	5–199	Egorov, Laktionova and Borts (1981)
UK:				
S. Yorkshire	S	8	50–107	Ward, Kerr and Otsuka (1986)
N. England, S. Wales	BI	26	0.5–50	D.J. Swaine (unpublished)
23 collieries	BI	23	<20–160	Hislop et al. (1978)
USA:				
Gulf Province	L	34	<10–200	Swanson et al. (1976)
Texas	L	103	0.8–490	White, Edwards and Du Bose (1983)
Fort Union region	L	80	15–300	Hatch and Swanson (1976)
S. and E. Arkansas	L	53	20–500	Hildebrand, Clardy and Holbrook (1981)
Eastern USA	BI	644	<0.6–160	Zubovic et al. (1980)
Western USA	–	28	16–140	Gluskoter et al. (1977)
Alaska, N. Slope	L-BI	84	20–200	Affolter and Stricker (1987)
Pennsylvania	A	53	2–20	Swanson et al. (1976)

exceptions being some Canadian coals with up to 3550 ppm Br. The range for most Australian coals is probably 0.4–30 ppm Br, the main exceptions being some coals from four seams in the northern area of New South Wales, where values of up to 403 ppm Br were found (Doolan et al., 1984b). It seems that most South African coals are very low, as it has been stated that 'the Br content of uncontaminated coals is generally <2 ppm' (Willis and Hart, 1985). A prime source of Br in the atmosphere is sea spray, but important contributions are also made by coal-burning, volcanoes (Cadle et al., 1973) and the exhausts of cars using leaded fuels. Bromine is listed by PECH (1980) as an element of 'minor concern'.

Table 5.6 Results for bromine (as ppm Br)

Sample details	Rank	Samples	Content	Reference
AUSTRALIA:				
Latrobe Valley, Vic.	BR	28	6.2–19	Fardy, McOrist and Farrar (1984)
Leigh Creek, SA	H-B	2	5, 9	Knott and Warbrooke (1983)
Western area, NSW	BI	39	0.055–0.59	Swaine, Godbeer and Morgan (1984b)
NSW and Qnsld	BI	65	<1–403	Knott and Warbrooke (1983)
NSW and Qnsld	BI	45	0.34–30	Fardy, McOrist and Farrar (1984)
Fingal, Tasmania	BI	2	2, 13	Knott and Warbrooke (1983)
BELGIUM:				
From 12 mines	BI, A	48	9–35	Block and Dams (1975)
BRAZIL:				
Rio Grande do Sul	BI	13	0.93–3.7	Bellido and Arezzo (1987)
CANADA:				
Hat Creek No.1, BC	L, S	14	92–678	Goodarzi (1987b)
Hat Creek No.2, BC	L, S	32	5–60	Goodarzi and van der Flier-Keller (1988)
Alberta and BC	S	23	<0.4–263	Landheer, Dibbs and Labuda (1982)
'Typical'	S, BI	50	0.13–3550	Jervis, Ho and Tiefenbach (1982)
Fording mine, BC	BI	22	24–140	Goodarzi (1988)
Nova Scotia	BI	5	7.6–29	Landheer, Dibbs and Labuda (1982)
CHINA:				
'Selection'	–	15	25–57	Sun and Jervis (1986)
from 110 mines	–	–	0.12–47	Chen et al. (1986)
NEW ZEALAND:				
Taranaki area	S	41	4–30	Black (1981)
Waikato area	S	5	0.8–3	Lynskey, Gainsford and Hunt (1984)
NIGERIA:				
'Major deposit'	L	1	2.6	Hannan et al. (1982)
'Major deposit'	S	6	0.39–1.1	Hannan et al. (1982)
'Major deposit'	BI	1	1.4	Hannan et al. (1982)
Enugu, 3 areas	S	3	6.1–7.3	Ndiokwere, Guinn and Burtner (1983)
POLAND:				
Belchatow mine	BR	3	3.4–7.5	Tomza (1987)
7 mines	BI	31	1.3–58	Tomza (1987)
SOUTH AFRICA:				
Witbank coalfield	BI	146	0.13–2.1	Hart and Leahy (1983)
	BI	14	<0.7–2	Willis (1983)
SWITZERLAND:				
	BR, L	9	0.34–6.9	Hügi et al. (1989)
	A	8	0.5–3.8	Hügi et al. (1989)
TURKEY:				
5 zones	L	5	1.0–14	Ayanoglu and Gunduz (1978)

Table 5.6 (*contd*)

Sample details	Rank	Samples	Content	Reference
UK:				
For power production	–	–	10–80	Hamilton (1974)
S. Yorkshire	S	8	0.79–4.1	Ward, Kerr and Otsuka (1986)
23 collieries	BI	23	7–155	Hislop *et al.* (1978)
USSR:				
Kuzbas area	–	20	3–19	Gulyayeva and Itkina (1962)
USA:				
	S, BI	11	2–45	Prather, Guin and Tarrer (1979)
Illinois Basin	BI	114	0.6–52	Gluskoter *et al.* (1977)
Upper Freeport seam	BI	8	2–90	Finkelman (1981a)
Eastern, USA	BI	23	0.71–26	Gluskoter *et al.* (1977)
Western, USA	–	28	0.50–25	Gluskoter *et al.* (1977)

5.2.8 Cadmium

Although there are no widely recognized beneficial effects of Cd in nutrition, much is known about harmful effects, especially from areas contaminated by some industrial wastes. It is currently an element of prime environmental concern and hence the levels of Cd in coal and coal-related materials are of interest. The first information seems to be that of Jensch (1887) who reported 9–70 ppm Cd in six German coal ashes, seemingly a remarkably good result for the time, when the chemical determination of Cd in the presence of excess Zn was very difficult. Binns and Harrow (1897) found sphalerite in a sample of coal from Leicestershire, UK, and stated that 'a minute yellow precipitate, too small for identification, was considered to be probably cadmium'. The maximum value found by Goldschmidt (1937) was 50 ppm Cd in German coal ashes. In view of the discovery of Gluskoter and Lindahl (1973) that Cd occurs in Illinois, USA, coals 'in solid solution, replacing zinc in the mineral sphalerite', it is generally accepted that this is the main mode of occurrence of Cd in most coals. However, a careful study by Kirsch, Schirmer and Schwarz (1980) of some German coals, showed that Cd is also associated with clay minerals and carbonate minerals, and it is known that pyrite may contain some Cd. In general, the situation seems to be that Cd is predominantly associated with mineral matter in coal, mainly sphalerite, but that the other mineral associations also occur. Perhaps hydrous iron oxides could hold some Cd by adsorption in some coals. In general, only minor amounts of Cd are organically associated in most coals, although there must be the possibility of some reaction with hydroxyl groups in low-rank coals.

Under the correct conditions, ashing at 450 °C is suitable for preparing coals prior to analysis by EAAS, which is the preferred method for sub-ppm concentrations of Cd. In a study of Australian coals Godbeer and Swaine (1979) were able to measure down to 0.01 ppm Cd, coal ash being treated with aqua regia, then HF, prior to the determination by EAAS. The high sensitivity of anodic-stripping voltammetry should enable low-Cd

Table 5.7 Results for cadmium (as ppm Cd)

Sample details	Rank	Samples	Content	Reference
AUSTRALIA:				
Latrobe Valley, Vic.	BR	2	0.01, 0.08	Godbeer and Swaine (1979)
Latrobe Valley, Vic.	BR	20	<0.01–0.10	Bone and Schaap (1981a)
Latrobe Valley, Vic.	BR	15	<0.01–0.05	Bolger (1989)
Leigh Creek, SA	H-B	2	0.10, 0.11	Godbeer and Swaine (1979)
Collie, WA	S	3	0.10–1.4	Godbeer and Swaine (1979)
Northern area, NSW	BI	30	0.042–0.25	Godbeer and Swaine (1979)
Southern area, NSW	BI	20	0.058–0.36	Godbeer and Swaine (1979)
Western area, NSW	BI	70	0.062–0.33	Swaine, Godbeer and Morgan (1984b)
NSW and Qnsld	BI	65	<0.02–0.16	Knott and Warbrooke (1983)
Bowen Basin, Qnsld	BI	15	0.07–0.20	Godbeer and Swaine (1979)
Fingal, Tasmania	BI	2	0.04, 0.06	Knott and Warbrooke (1983)
BULGARIA:				
'Some coals'	–	–	(1.9)	Kostadinov and Djingova (1980)
CANADA:				
Saskatchewan	L	2	7.7, 8.8	Landheer, Dibbs and Labuda (1982)
Moose River Basin	L	17	<0.1–4.8	Van der Flier-Keller and Fyfe (1987b)
Alberta and BC	S	9	6.6–8.6	Landheer, Dibbs and Labuda (1982)
FRANCE:				
2 areas	–	2	<0.01, 0.03	Sabbioni et al. (1983)
GERMANY WEST:				
From 1 power station	BR	7	0.022–0.071	Heinrichs (1982)
Mostly Ruhr	BI	27	0.5–10	Kirsch, Schirmer and Schwarz (1980)
Various	BI	10	0.02–21	Heinrichs (1975b)
NEW ZEALAND:				
Mokau area	S	–	0.011–0.020	Gainsford (1985)
Waikato area	S	–	0.009–0.18	Purchase (1985)
Southland	S	–	0.008–0.015	Purchase (1985)
Westland	S-BI	–	<0.02–0.06	Purchase (1985)
POLAND:				
Belchatow mine	BR	3	3.2–7.9	Tomza (1987)
	BI	–	0–4	Widawska-Kusmierska (1981)
For power production	BI	2	<0.03	Sabbioni et al. (1983)
7 mines	BI	31	1.9–4.9	Tomza (1987)
SOUTH AFRICA:				
	BI	11	0.1–0.8	Watling and Watling (1976)
SWITZERLAND:				
	BR, L	9	0.015–2.4	Hügi et al. (1989)
	A	9	<0.005–0.57	Hügi et al. (1989)
USSR:				
Pechora	–	4	<0.1	Blomqvist (1983)
UK:				
S. Yorkshire	S	8	0.03–0.10	Ward, Kerr and Otsuka (1986)
20 areas	BI	232	0.02–5	Taylor (1973)

Table 5.7 (contd)

Sample details	Rank	Samples	Content	Reference
23 collieries	BI	23	<0.3–3.4	Hislop et al. (1978)
S. Wales (12 seams)	BI	300	−10(1)	Chatterjee and Pooley (1977)
USA:				
Northern Great Plains	L	226	0.06–2.7	Zubovic, Hatch and Medlin (1979)
Gulf Province	L	34	<0.1–5.5	Swanson et al. (1976)
Texas	L	93	0.1–3	White, Edwards and Du Bose (1983)
Fort Union region	L	80	<0.1–0.2	Hatch and Swanson (1976)
Northern Great Plains	S	855	0.04–2.7	Zubovic, Hatch and Medlin (1979)
Rocky Mt Province	S	466	0.03–3.7	Zubovic, Hatch and Medlin (1979)
Appalachian Region	BI	1478	0.01–3.1	Zubovic, Hatch and Medlin (1979)
Virginia	BI	83	0.02–0.21	Henderson et al. (1985)
Interior Province	BI	687	0.01–170	Zubovic, Hatch and Medlin (1979)
Rocky Mt Province	BI	366	0.02–0.99	Zubovic, Hatch and Medlin (1979)
Alaska, N. Slope	L-BI	84	<0.02–0.6	Affolter and Stricker (1987)
Pennsylvania	A	53	<0.02–1.4	Swanson et al. (1976)

coals to be analysed. There is an American standard method for Cd in coal (ASTM, 1983a), which uses the same dissolution technique described for Be (5.2.4) an air–acetylene flame and the line at 228.8 nm with background correction. However, this method is not sensitive enough for sub-ppm levels of Cd, which are characteristic of Australian coals.

Results for Cd in coal are given in Table 5.7. Most coals would be in the range 0.1–3 ppm Cd, while most Australian coals would be covered by 0.01–0.20 ppm Cd. This difference is also shown by the mean of 0.08 ppm Cd for Australian coals, compared with the mean of 0.5 ppm Cd for coals used for power production in nine EC countries (Sabbioni et al., 1983). A graphical illustration of differences between coals from five countries is shown by Godbeer and Swaine (1979), the only very low Cd-coals being from Australia and the Rocky Mountain Province, USA. There are several sources of emission of Cd to the atmosphere, including non-ferrous metal production, waste incineration and volcanoes, together with minor sources, for example tyre-wear and cigarette-smoking. Coal combustion contributes about one-tenth of the amount emitted by volcanoes (Nriagu, 1979), and should be considered as a minor source, especially now that so many low-S coals, and *ipso facto* usually low-Cd coals, are being used for power production. It is salutary to recall the statement by Hutton (1983), namely 'it would appear that the role of coal combustion as a source of atmospheric cadmium in the EC has been exaggerated in the past', surely a comforting statement for coal users.

5.2.9 Caesium

Caesium was discovered by Bunsen and Kirchoff in 1860 during their pioneering studies on emission spectra. There has been scant attention paid to Cs in coal, probably because it has no biological significance and it was analytically difficult until recent times. However, its presence in coal was shown by Ramage (1927) who found Cs in a flue dust from a Yorkshire, UK, coal. It is most unlikely that Cs is organically bound, except perhaps in low-rank coals, although even there it is dubious because another large cation (K^+) is not associated with hydroxyl groups in the same way as Ca, Mg and Na. Uzunov and Karadzhova (1968) suggested that Cs was predominantly related to the mineral matter in coals from the Burgas Basin, Bulgaria, while Palmer and Filby (1984) found that Cs was associated with the clay fraction of a bituminous coal from Ohio, USA. It is geochemically sound to expect Cs to be associated with K-rich minerals, including some clays, micas and feldspars.

The determination of Cs can be carried out by AAS, INAA and XRFS. Most of the results to be quoted are from INAA determinations. The first detection of Cs in an Australian coal was by Sinnatt and Baragwanath (1938) who found roughly 10 ppm Cs in the ash of a brown coal from Yallourn, Victoria, using AES. Results for Cs are given in Table 5.8. It seems likely that most coals would have Cs contents in the range 0.3–5, with a mean of around 1–2 ppm Cs.

5.2.10 Chlorine

The long-term interest in Cl in coal stemmed from the fact that relatively high Cl tended to be associated with deposits on boiler tubes in equipment using mechanical stokers. Wandless (1958) was of the opinion that coals with <3000 ppm Cl are unlikely to produce troublesome deposits. The firing of pulverized coal, as in modern power-station boilers, does not give rise to phosphatic or other deposits. A more important aspect of Cl is that high-Cl coals may give rise to corrosion of furnace walls and superheater tubes, probably because of the formation of HCl. It has been estimated that coal combustion is the largest anthropogenic source of Cl, producing, for example, about 75% of the HCl emitted in western Europe (Lightowlers and Cape, 1988). However, this HCl is only a minor contributor (about 2%) to acidity in the atmosphere. Relevant natural sources of HCl are volcanoes and methylchloride, which reacts with OH radicals to eventually produce HCl in the atmosphere (Lightowlers and Cape, 1988). The behaviour of Cl during combustion has been reviewed by Kear and Menzies (1956) and by Gibb (1983).

Pre-coal vegetation probably took up some Cl, which may be the source of some Cl in coal (Louis, 1927), but it seems more likely that most of the Cl was introduced during diagenesis, probably by Cl-rich groundwater (Caswell, Holmes and Spears, 1984a). There has been much debate, especially in the United Kingdom, about the mode of occurrence of Cl. Inorganic chlorides were favoured by some, while others preferred some form of organic association. Although probably only minor contributors of Cl, halite has been seen in coals from the United Kingdom (Cressey

Table 5.8 Results for caesium (as ppm Cs)

Sample details	Rank	Samples	Content	Reference
AUSTRALIA:				
Latrobe Valley, Vic.	BR	20	<0.01–0.29	Bone and Schaap (1981a)
St Vincent Basin, SA	L	5	<0.7–2.8	Fardy, McOrist and Farrar (1984)
Leigh Creek, SA	H-B	2	<1, 5	Knott and Warbrooke (1983)
NSW	BI	4	0.52–1.1	P.C. Rankin (1977, pers. commun.)
Western area, NSW	BI	66	0.8–2.4	Swaine, Godbeer and Morgan (1984b)
NSW and Qnsld	BI	65	<0.2–7	Knott and Warbrooke (1983)
NSW and Qnsld	BI	45	0.16–4.9	Fardy, McOrist and Farrar (1984)
Fingal, Tasmania	BI	2	2, 3	Knott and Warbrooke (1983)
BELGIUM:				
From 12 mines	BI, A	48	0.05–2.8	Block and Dams (1975)
BRAZIL:				
Rio Grande do Sul	BI	13	1.9–5.5	Bellido and Arezzo (1987)
CANADA:				
Hat Creek No.1, BC	L, S	14	0.3–2.3	Goodarzi (1987b)
Hat Creek No.2, BC	L, S	32	0.3–6.4	Goodarzi and van der Flier-Keller (1988)
'Typical'	S, BI	56	0.02–4.1	Jervis, Ho and Tiefenbach (1982)
Fording mine, BC	BI	22	0.3–1.2	Goodarzi (1988)
CHINA:				
'Selection'	–	15	<0.1–31	Sun and Jervis (1986)
from 110 mines	–	–	0.07–33	Chen et al. (1986)
NIGERIA:				
'Major deposit'	L	1	0.27	Hannan et al. (1982)
'Major deposit'	S	6	0.085–0.26	Hannan et al. (1982)
'Major deposit'	BI	1	3.2	Hannan et al. (1982)
POLAND:				
Belchatow mine	BR	3	0.80–1.3	Tomza (1987)
7 mines	BI	31	0.09–10	Tomza (1987)
SOUTH AFRICA:				
Witbank coalfield	BI	146	0.33–3.1	Hart and Leahy (1983)
SWITZERLAND:				
	BR, L	9	0.11–5.1	Hügi et al. (1989)
	A	9	<0.08–4.9	Hügi et al. (1989)
TURKEY:				
5 'kinds'	L	5	1.1–20	Akcetin, Ayca and Hoste (1973)
5 zones	L	5	0.12–0.33	Ayanoglu and Gunduz (1978)
UK:				
S. Yorkshire	S	8	2.8–6.3	Ward, Kerr and Otsuka (1986)
Durham	BI	10	0.75–4	Smales and Salmon (1955)
23 collieries	BI	23	<0.1–5.3	Hislop et al. (1978)
USA:				
Eastern USA	BI	617	0.02–15	Zubovic, Hatch and Medlin (1979)
Illinois	BI	114	0.5–3.6	Gluskoter et al. (1977)
Western USA	–	28	0.10–0.60	Gluskoter et al. (1977)

and Crossey, 1988) and the United States (Finkelman, 1981a), while Crumley, Fletcher and Wilson (1955) suggested the presence of $CaCl_2$. Replacement of some OH by Cl in, say, apatite can occur, but this could not account for much Cl in coal. It is much more likely that Cl is somehow associated with the organic coaly matter (Saunders, 1980; Martinez-Tarazona, Palacios and Cardin, 1988), a view supported by Finkelman (1981a). Pearce and Hill (1986) suggested that the Cl in UK coals is likely to be present in one form 'uniformly distributed and linked ionically to the coal substance', a view also espoused by Ladner (1984). It is interesting to note that the original suggestion of the predominance of inorganic Cl in Illinois coals (Gluskoter and Rees, 1964) was modified by Gluskoter and Ruch (1971) to favour organically bound Cl. Although modern techniques are helping to unravel the Cl situation in coal, it seems that 'the nature of the chlorine-coal bond is still uncertain' (Hodges, Ladner and Martin, 1983).

The determination of Cl in coal is usually carried out by a chemical method, but INAA and XRFS are also used. There are several standard methods, for example ASTM (1978b) and AS (1980b), which are based on heating a mixture of coal plus Eschka mixture (1:2 anhydrous Na_2CO_3 plus MgO) either in an oxidizing atmosphere or in a bomb. The final measurement of Cl^- ion is by a modified Volhard titrimetric method or by an ion-selective electrode (ASTM, 1983b). Rigin (1987) determines the Cl^- ion by an ion chromatographic method. A method of determining Cl in organic combination in coal (Chakrabarti, 1978, 1982) is not regarded as generally 'suitable for determining organically-bound chlorine in coals' (Ladner, 1984). Another approach uses dimethyl sulfoxide treatment of coal to estimate the ratio of inorganic to organic Cl in coal (Cox, Larson and Carlson, 1984).

Results for Cl in coal are given in Table 5.9. The range for most coals is probably 50–2000 ppm Cl, there being some exceptions, notably in the United Kingdom. It is not considered that Cl in coal would have untoward health effects, but it may cause corrosion, as noted above.

5.2.11 Chromium

'Chromium is an essential trace element required for normal carbohydrate metabolism' (Anderson, 1981), and is not regarded as toxic, except in the hexavalent state, which may occur in some waste waters and industrial situations, but not usually during coal mining or usage. An early value was reported by Mingaye (1907) who found 80 ppm Cr in a Japanese coal, and Goldschmidt and Peters (1933b) reported up to 1000 ppm Cr in the ash of some German coals. Although coal combustion is a source of Cr in the atmosphere there has not been a proper study of this in relation to other sources but, as expected, volcanoes 'emit large amounts' of Cr (Nriagu *et al.*, 1988). The mode of occurrence of Cr is not certain, as is pointed out clearly by Finkelman (1981a), who regards an association with clays as reasonable, together with the possibility of finely divided chromites in some coals. Although Given and Miller (1987) suggested that Cr may be partly complexed with organic matter in some US lignites, the statement by Bouska (1981) that 'Cr in coal and coal ash is almost universally

Table 5.9 Results for chlorine (as ppm Cl)

Sample details	Rank	Samples	Content	Reference
ANTARCTICA:				
Thermally altered	BI	7	<100–300	CSIRO (unpublished)
AUSTRALIA:				
Latrobe Valley, Vic.	BR	25	200–3000	Brown and Swaine (1964)
Latrobe Valley, Vic.	BR	20	400–1000	Bone and Schaap (1981a)
Latrobe Valley, Vic.	BR	15	500–4600	Bolger (1989)
Latrobe Valley, Vic.	BR	28	380–1300	Fardy, McOrist and Farrar (1984)
St Vincent Basin, SA	L	5	2600–4900	Fardy, McOrist and Farrar (1984)
Leigh Creek, SA	H-B	80	500–17000	CSIRO (1966)
Collie, WA	S	22	<50–230	Davy and Wilson (1984)
Northern area, NSW	BI	1000	<100–1200	Clark and Swaine (1962a)
Southern area, NSW	BI	800	<100–400	Clark and Swaine (1962a)
Western area, NSW	BI	60	<100–600	Swaine, Godbeer and Morgan (1984b)
NSW and Qnsld	BI	65	50–3400	Knott and Warbrooke (1983)
NSW and Qnsld	BI	45	24–1900	Fardy, McOrist and Farrar (1984)
Bowen Basin, Qnsld	BI	200	<100–1000	Brown and Swaine (1964)
West Moreton area, Qnsld	BI	150	<100–500	Brown and Swaine (1964)
Callide area, Qnsld	BI	100	<100–200	Brown and Swaine (1964)
Baralaba, Qnsld	S-A	20	200–800	Taylor (1967)
BELGIUM:				
From 12 mines	BI, A	48	250–2500	Block and Dams (1975)
CANADA:				
Hat Creek No.1, BC	L, S	14	99–621	Goodarzi (1987b)
Hat Creek No.2, BC	L, S	32	23–83	Goodarzi and van der Flier-Keller (1988)
'Typical'	S, BI	60	16–3710	Jervis, Ho and Tiefenbach (1982)
Fording mine, BC	BI	22	40–500	Goodarzi (1988)
CHINA:				
'Selection'	–	15	20–1630	Sun and Jervis (1986)
GERMANY WEST:				
From 1 power station	BR	7	410–830	Heinrichs (1982)
	BI	2	1350, 2450	Brumsack, Heinrichs and Lange (1984)
INDIA:	–	10	3220–5510	Das Gupta and Chakrabarti (1951)
NEW ZEALAND:				
Taranaki area	S	41	100–1400	Black (1981)
Waikato area	S	5	74–125	Lynskey, Gainsford and Hunt (1984)
NIGERIA:				
'Major deposit'	L	1	67	Hannan et al. (1982)
'Major deposit'	S	6	20–275	Hannan et al. (1982)
'Major deposit'	BI	1	106	Hannan et al. (1982)
POLAND:				
Belchatow mine	BR	3	24–111	Tomza (1987)
7 mines	BI	31	40–4670	Tomza (1987)

Table 5.9 (*contd*)

Sample details	Rank	Samples	Content	Reference
SOUTH AFRICA:				
Transvaal, Natal, OFS	BI	74	19–341	Kunstmann and van Veijeren (1964)
SPAIN:				
Asturian Basin	–	19	100–3000	Martinez-Tarazona and Cardin (1986)
SWITZERLAND:				
	BR, L	9	<13–125	Hügi *et al.* (1989)
	A	8	12–149	Hügi *et al.* (1989)
UK:				
23 collieries	BI	23	200–9100	Hislop *et al.* (1978)
Midlands	BI	84	350–7100	Thorne *et al.* (1983)
S. Wales	A	21	nil–300	Johnson (1947)
USA:				
Illinois	BI	114	100–5400	Gluskoter *et al.* (1977)
Eastern USA	BI	23	100–8000	Gluskoter *et al.* (1977)
Virginia	BI	83	50–1700	Henderson *et al.* (1985)
Penn. State Data Base	–	444	(580)	Given and Yarzab (1978)
Western USA	–	28	100–1300	Gluskoter *et al.* (1977)

associated with organic matter' is untenable. In addition to the mineral-matter association, it seems reasonable to expect some organically associated Cr in many coals, especially those of low rank.

Most modern instrumental methods are suitable for determining Cr in coal. The US standard method (ASTM, 1983a) and the Australian equivalent (AS, 1986) use FAAS. Dissolution of 500 °C ash is with aqua regia plus HF in a sealed plastic bottle on a steam bath. The FAAS determination uses a nitrous oxide–acetylene flame and the line at 357.9 nm with suitable background correction. The sizes of coal for analysis are −150 μm for ASTM (1983a) and 98% to pass −75 μm for AS (1986). If determinations are required for coals with very low concentrations, say of 1 ppm or less, an EAAS method would be preferred.

Results for Cr in coal are given in Table 5.10. Most coals would have values in the range 0.5–60, with a general mean of around 20 ppm Cr. The range of mean values for coals burnt in nine EC countries (Sabbioni *et al.*, 1983) is 25–60 ppm Cr, which tends to be higher than means for published data.

There is no evidence known to me of untoward effects from Cr in coal, but it is listed as of 'moderate concern' by PECH (1980).

5.2.12 Cobalt

Cobalt has an important role in nutrition and serious deficiencies have been found in grazing animals. It is not regarded as an element of concern in the normal environment. There is a comprehensive review of all aspects

Table 5.10 Results for chromium (as ppm Cr)

Sample details	Rank	Samples	Content	Reference
ANTARCTICA:				
Thermally altered	BI	7	5–26	Brown and Taylor (1961)
AUSTRALIA:				
Latrobe Valley, Vic.	BR	16	<0.2–5	Brown and Swaine (1964)
Latrobe Valley, Vic.	BR	20	<0.2–18	Bone and Schaap (1981a)
Latrobe Valley, Vic.	BR	15	<0.4–10	Bolger (1989)
Latrobe Valley, Vic.	BR	28	0.08–19	Fardy, McOrist and Farrar (1984)
St Vincent Basin, SA	L	5	5.4–41	Fardy, McOrist and Farrar (1984)
Esperance, WA	L	3	<15–40	CSIRO (unpublished)
Leigh Creek, SA	H-B	80	5–80	CSIRO (1966)
Collie, WA	S	22	1–10	Davy and Wilson (1984)
Northern area, NSW	BI	1000	<1.5–80	Clark and Swaine (1962a)
Southern area, NSW	BI	800	3–70	Clark and Swaine (1962a)
Western area, NSW	BI	150	3–30	Swaine, Godbeer and Morgan (1984b)
NSW and Qnsld	BI	31	4.1–20	Knott and Warbrooke (1983)
NSW and Qnsld	BI	45	2–56	Fardy, McOrist and Farrar (1984)
Bowen Basin, Qnsld	BI	200	1.5–70	Brown and Swaine (1964)
West Moreton area, Qnsld	BI	150	<4–25	Brown and Swaine (1964)
Callide area, Qnsld	BI	100	4–20	Brown and Swaine (1964)
Fingal, Tasmania	BI	2	6.9, 17	Knott and Warbrooke (1983
Baralaba, Qnsld	S-A	20	3–15	CSIRO (unpublished)
BELGIUM:				
From 12 mines	BI, A	48	6–90	Block and Dams (1975)
BRAZIL:				
Rio Grande do Sul	BI	13	13.5–76	Bellido and Arezzo (1987)
BULGARIA:				
'Some coals'	–	–	(31)	Kostadinov and Djingova (1980)
CANADA:				
Moose River Basin	L	17	<0.1–498	Van der Flier-Keller and Fyfe (1987b)
Hat Creek No.1, BC	L, S	14	5.7–64	Goodarzi (1987b)
Hat Creek No.2, BC	L, S	32	15–64	Goodarzi and van der Flier-Keller (1988)
Alberta and BC	S	23	<3–37	Landheer, Dibbs and Labuda (1982)
'Typical'	S, BI	59	0.16–35	Jervis, Ho and Tiefenbach (1982)
Fording mine, BC	BI	22	2.2–17	Goodarzi (1988)
Nova Scotia	BI	186	2–9	Hawley (1955)
Nova Scotia and New Brunswick	BI	7	<3–45	Landheer, Dibbs and Labuda (1982)
CHINA:				
'Selection'	–	15	6.7–71	Sun and Jervis (1986)
COLOMBIA:				
3 regions	BI	14	4–80	Rincon et al. (1978)

Table 5.10 (*contd*)

Sample details	Rank	Samples	Content	Reference
FRANCE:				
2 areas	–	2	6.4, 24	Sabbioni *et al.* (1983)
GERMANY WEST:				
From 1 power station	BR	7	1.8–12	Heinrichs (1982)
Mostly Ruhr	BI	27	4–80	Kautz, Kirsch and Laufhutte (1975)
For power production	BI	6	10–29	Sabbioni *et al.* (1983)
JAPAN:				
Various parts	–	27	0.3–35	Iwasaki and Ukimoto (1942)
NEW ZEALAND:				
Taranaki area	S	41	1–15	Black (1981)
Waikato area	S	5	2.7–12	Lynskey, Gainsford and Hunt (1984)
Southland	S	–	0.4–33	Purchase (1985)
Westland	S-BI	–	0.02–71	Purchase (1985)
NIGERIA:				
'Major deposit'	L	1	20	Hannan *et al.* (1982)
'Major deposit'	S	6	6.9–11	Hannan *et al.* (1982)
'Major deposit'	BI	1	43	Hannan *et al.* (1982)
POLAND:				
Belchatow mine	BR	3	13–28	Tomza (1987)
	BI	–	0.6–12	Widawska-Kusmierska (1981)
7 mines	BI	31	1.4–67	Tomza (1987)
SOUTH AFRICA:				
	BI	11	21–70	Watling, Watling and Wardale (1976)
	BI	14	12–63	Willis (1983)
SWITZERLAND:				
	BR, L	9	3.3–89	Hügi *et al.* (1989)
	A	9	4.1–32	Hügi *et al.* (1989)
THAILAND:				
1 deposit	L	3	3–4	CSIRO (unpublished)
TURKEY:				
5 'kinds'	L	5	9–190	Akcetin, Ayca and Hoste (1973)
5 zones	L	5	9.8–544	Ayanoglu and Gunduz (1978)
USSR:				
Moscow Basin	BR	3	4–7	Laktionova *et al.* (1987)
South Yakutia	–	–	5.0–8.5	Egorov, Laktionova and Borts (1981)
Pechora	–	4	8–22	Blomqvist (1983)
UK:				
S. Yorkshire	S	8	0.8–3.2	Ward, Kerr and Otsuka (1986)
N. England, S. Wales	BI	26	1–30	D.J. Swaine (unpublished)
23 collieries	BI	23	3–45	Hislop *et al.* (1978)
S. Wales (12 seams)	BI	300	–384(26)	Chatterjee and Pooley (1977)
USA:				
Northern Great Plains	L	226	0.25–43	Zubovic, Hatch and Medlin (1979)

Table 5.10 (contd)

Sample details	Rank	Samples	Content	Reference
Gulf Province	L	34	3–70	Swanson et al. (1976)
Texas	L	152	2–87	White, Edwards and Du Bose (1983)
Fort Union region	L	80	0.5–10	Hatch and Swanson (1976)
Northern Great Plains	S	855	0.54–66	Zubovic, Hatch and Medlin (1979)
Rocky Mt Province	S	466	0.54–70	Zubovic, Hatch and Medlin (1979)
Appalachian Region	BI	1478	1.5–220	Zubovic, Hatch and Medlin (1979)
Interior Province	BI	687	2.0–200	Zubovic, Hatch and Medlin (1979)
Rocky Mt Province	BI	366	0.81–75	Zubovic, Hatch and Medlin (1979)
Alaska, N. Slope	L-BI	84	<0.1–59	Affolter and Stricker (1987)
Pennsylvania	A	53	5–70	Swanson et al. (1976)

of Co (Smith and Carson, 1981). Although there are data for Co in coals, there are no special studies on Co in coal during mining and usage. The uncommon mineral linnaeite has been reported several times in coals from South Wales (Des Cloizeaux, 1880; Howarth, 1928; North and Howarth, 1928; Goldschmidt, 1954; Firth, 1973) and Finkelman (1981a) found it in some US coals. The suggestion by Finkelman (1981a) that some Co could be associated with clays in some coals is tenable, and there is support for organic affinity in some US coals (Zubovic, Stadnichenko and Sheffey, 1961b; Gluskoter et al., 1977). It seems that Co may occur in coal associated with the mineral matter (linnaeite, other sulfides, clay) and with the organic matter.

Most of the data given in Table 5.11 for Co in coal were obtained by OES, INAA and XRF, but AAS methods are also suitable for many coals. The maximum value of 1500 ppm Co in some German coal ashes (Goldschmidt, 1954) is well above the levels in most coals. The range for most coals is probably 0.5–30 with a mean of 4–8 ppm Co, Australian coals being at the lower end of the range. There are no reports of unwanted effects caused by Co during coal mining and usage.

5.2.13 Copper

Copper is of interest because of biological essentiality and because excesses are undesirable. Perhaps the earliest report of Cu in coal was by Platz (1887) who found 160–540 ppm in the ash of coal from Westphalia, Germany. Chalcopyrite was seen in a coal from Leicestershire, UK (Binns and Harrow, 1897), and in peat from Wales (Percy, 1875), which was mined for Cu in the early nineteenth century (Davidson and Ponsford, 1954). There is no doubt that chalcopyrite is found in many coals (Finkelman, 1981a) and other sulfide minerals of Cu occur occasionally. During the peat stage of coalification, it is surely likely that some Cu was fixed, probably by humic acids (Senesi, Sposito and Martin, 1986), by

Table 5.11 Results for cobalt (as ppm Co)

Sample details	Rank	Samples	Content	Reference
ANTARCTICA:				
Thermally altered	BI	7	<2–64	Brown and Taylor (1961)
AUSTRALIA:				
Latrobe Valley, Vic.	BR	16	<0.2–1.5	Brown and Swaine (1964)
Latrobe Valley, Vic.	BR	20	<0.1–1.8	Bone and Schaap (1981a)
Latrobe Valley, Vic.	BR	15	<0.2–10	Bolger (1989)
Latrobe Valley, Vic.	BR	28	<0.05–2.0	Fardy, McOrist and Farrar (1984)
St Vincent Basin, SA	L	5	2.9–7.1	Fardy, McOrist and Farrar (1984)
Leigh Creek, SA	H-B	80	<2–25	CSIRO (1966)
Collie, WA	S	22	2–18	Davy and Wilson (1984)
Northern area, NSW	BI	1000	0.6–30	Clark and Swaine (1962a)
Southern area, NSW	BI	800	1.5–25	Clark and Swaine (1962a)
Western area, NSW	BI	150	<2–10	Swaine, Godbeer and Morgan (1984b)
NSW and Qnsld	BI	65	1–11	Knott and Warbrooke (1983)
NSW and Qnsld	BI	45	1.6–20	Fardy, McOrist and Farrar (1984)
Bowen Basin, Qnsld	BI	200	<0.6–25	Brown and Swaine (1964)
West Moreton area, Qnsld	BI	150	2–30	Brown and Swaine (1964)
Callide area, Qnsld	BI	100	0.8–8	Brown and Swaine (1964)
Fingal, Tasmania	BI	2	6	Knott and Warbrooke (1983)
Baralaba, Qnsld	S-A	20	0.8–6	CSIRO (unpublished)
BELGIUM:				
From 12 mines	BI, A	48	4–25	Block and Dams (1975)
BRAZIL:				
Rio Grande do Sul	BI	13	5.3–29	Bellido and Arezzo (1987)
BULGARIA:				
'Some coals'	–	–	(3.1)	Kostadinov and Djingova (1980)
CANADA:				
Saskatchewan	L	8	<0.4–5.6	Landheer, Dibbs and Labuda (1982)
Moose River Basin	L	17	5–198	Van der Flier-Keller and Fyfe (1987b)
Hat Creek No.1, BC	L, S	14	1.1–31	Goodarzi (1987b)
Hat Creek No.2, BC	L, S	32	4.0–17	Goodarzi and van der Flier-Keller (1988)
Alberta and BC	S	23	0.42–9.9	Landheer, Dibbs and Labuda (1982)
'Typical'	S, BI	56	0.3–18	Jervis, Ho and Tiefenbach (1982)
Fording mine, BC	BI	22	0.8–4.6	Goodarzi (1988)
Nova Scotia	BI	186	2–34	Hawley (1955)
CHILE:				
Arauco Basin	BI	399	5.6–11	Collao et al. (1987)
CHINA:				
'Selection'	–	15	1.7–15	Sun and Jervis (1986)
COLOMBIA:				
3 regions	BI	14	1–28	Rincon et al. (1978)

Table 5.11 (contd)

Sample details	Rank	Samples	Content	Reference
FRANCE:				
2 areas	–	2	5.7, 9.0	Sabbioni et al. (1983)
GERMANY WEST:				
From 1 power station	BR	7	1.4–3.8	Heinrichs (1982)
Mostly Ruhr	BI	27	7–30	Kautz, Kirsch and Laufhutte (1975)
NEW ZEALAND:				
Waikato area	S	–	0.7–19	Purchase (1985)
Southland	S	–	0.4–8.1	Purchase (1985)
Westland	S-BI	–	0.09–14	Purchase (1985)
NIGERIA:				
'Major deposit'	L	1	3.4	Hannan et al. (1982)
'Major deposit'	S	6	1.8–8.0	Hannan et al. (1982)
'Major deposit'	BI	1	8.5	Hannan et al. (1982)
Enugu, 3 areas	S	3	2.2–3.1	Ndiokwere, Guinn and Burtner (1983)
POLAND:				
Belchatow mine	BR	3	2.6–6	Tomza (1987)
	BI	–	3–50	Widawska-Kusmierska (1981)
7 mines	BI	31	0.46–15	Tomza (1987)
SOUTH AFRICA:				
	BI	11	5–11	Watling and Watling (1976)
Witbank coalfield	BI	146	2.0–7.8	Hart and Leahy (1983)
	BI	14	3.3–14	Willis (1983)
SWITZERLAND:				
	BR, L	9	1.0–5.9	Hügi et al. (1989)
	A	9	0.19–16	Hügi et al. (1989)
THAILAND:				
1 deposit	L	3	1.5–2	CSIRO (unpublished)
TURKEY:				
5 'kinds'	L	5	1.0–17	Akcetin, Ayca and Hoste (1973)
5 zones	L	5	0.26–32	Ayanoglu and Gunduz (1978)
USSR:				
South Yakutia	–	–	3.0–3.5	Egorov, Laktionova and Borts (1981)
	–	5	1.0–9.0	Egorov et al. (1983)
UK:				
S. Yorkshire	S	8	2.2–13	Ward, Kerr and Otsuka (1986)
N. England, S. Wales	BI	26	0.4–60	D.J. Swaine (unpublished)
23 collieries	BI	23	2–20	Hislop et al. (1978)
S. Wales (12 seams)	BI	300	–78(5)	Chatterjee and Pooley (1977)
USA:				
Northern Great Plains	L	226	0.25–43	Zubovic, Hatch and Medlin (1979)
Gulf Province	L	34	1–30	Swanson et al. (1976)
Texas	L	79	0.1–26	White, Edwards and Du Bose (1983)
Fort Union region	L	80	<0.3–3.0	Hatch and Swanson (1976)
Northern Great Plains	S	855	0.25–20	Zubovic, Hatch and Medlin (1979)

Table 5.11 (contd)

Sample details	Rank	Samples	Content	Reference
Rocky Mt Province	S	466	0.06–70	Zubovic, Hatch and Medlin (1979)
Appalachian Region	BI	1478	0.70–930	Zubovic, Hatch and Medlin (1979)
Virginia	BI	83	1.5–18	Henderson et al. (1985)
Interior Province	BI	687	0.20–500	Zubovic, Hatch and Medlin (1979)
Rocky Mt Province	BI	366	0.24–21	Zubovic, Hatch and Medlin (1979)
Alaska, N. Slope	L-BI	84	<0.4–58	Affolter and Stricker (1987)
Pennsylvania	A	53	<0.3–50	Swanson et al. (1976)

adsorption (Ong and Swanson, 1966) and possibly by complexing with hydroxyl groups and forming Cu–porphyrin compounds (Goodman and Cheshire, 1973). Bogs rich in Cu are fairly common, for example in Belorussia, USSR (Manskaya, Drozdova and Emelyanova, 1960), in Canada (Fraser, 1961; Boyle, 1977) and in North Wales (Andrews and Fuge, 1986). There is no apparent reason why some strongly bound organic Cu should not persist during coalification. So the mode of occurrence of Cu in most coals is probably as a sulfide, usually chalcopyrite, and associated with organic matter, especially in low-rank coals.

Modern instrumental methods, except INAA, are suitable for determining Cu in most coals. There are two standard methods (ASTM, 1983a; AS, 1986) which use FAAS, the dissolution technique being the same as for Be (5.2.4), but an air–acetylene flame is used with the line at 324.8 nm and background correction. Langmyhr and Aadalen (1980) proposed a method using solid powdered coal prior to EAAS, but the two standard methods are suitable for most coals.

Results for Cu in coal are given in Table 5.12. The range for most coals is probably 0.5–50 ppm Cu. Means are variable from 8 to 10 ppm Cu for Southern Hemisphere coals to about 15 ppm Cu for US coals and about 30 ppm Cu for European coals. Mean values for the coals burnt in nine EC countries are 8.4–42 ppm Cu (Sabbioni et al., 1983). There is no apparent reason for these regional differences, which should have no environmental significance during the usage of the various coals. However, Cu has been listed as of 'moderate concern' (PECH, 1980), which may mean that some problems could occur with high-Cu coals.

5.2.14 Fluorine

Fluorine is an element of great environmental interest because of its biological role and because it may be harmful under certain conditions. Initially it had been observed that heating some coals could cause corrosion of some ceramic and metal surfaces, could lead to etching of glass and porcelain, and could increase the F-content of malt grains dried by hot gases from coal-burning (Horton, 1950). These effects focussed attention

Table 5.12 Results for copper (as ppm Cu)

Sample details	Rank	Samples	Content	Reference
ANTARCTICA:				
Thermally altered	BI	7	4–50	Brown and Taylor (1961)
ARGENTINA:				
Rio Turbio	–	6	8–10	A. Berset (1982, pers. commun.)
AUSTRALIA:				
Latrobe Valley, Vic.	BR	16	1.5–30	Brown and Swaine (1964)
Latrobe Valley, Vic.	BR	20	0.2–10	Bone and Schaap (1981a)
Latrobe Valley, Vic.	BR	15	0.3–5.0	Bolger (1989)
St Vincent Basin, SA	L	3	2–15	CSIRO (unpublished)
Esperance, WA	L	3	2–10	CSIRO (unpublished)
Leigh Creek, SA	H-B	80	3–60	CSIRO (1966)
Collie, WA	S	22	1–13	Davy and Wilson (1984)
Northern area, NSW	BI	1000	2–100	Clark and Swaine (1962a)
Southern area, NSW	BI	800	2.5–60	Clark and Swaine (1962a)
Western area, NSW	BI	150	2–50	Swaine, Godbeer and Morgan (1984b)
NSW and Qnsld	BI	65	2–25	Knott and Warbrooke (1983)
Bowen Basin, Qnsld	BI	200	4–70	Brown and Swaine (1964)
West Moreton area, Qnsld	BI	150	8–30	Brown and Swaine (1964)
Callide area, Qnsld	BI	100	10–70	Brown and Swaine (1964)
Fingal, Tasmania	BI	2	13, 14	Knott and Warbrooke (1983)
Baralaba, Qnsld	S-A	20	10–25	CSIRO (unpublished)
BELGIUM:				
From 12 mines	BI, A	48	10–160	Block and Dams (1975)
BRAZIL:				
Candiota	–	–	3.4–16	Martins, Braganca de Moraes and Baron (1984)
CANADA:				
Saskatchewan	L	8	5.6–18	Landheer, Dibbs and Labuda (1982)
Moose River Basin	L	17	2.0–44	Van der Flier-Keller and Fyfe (1987b)
Hat Creek No.2	L, S	32	14–80	Goodarzi and van der Flier-Keller (1988)
Alberta and BC	S	23	3.6–61	Landheer, Dibbs and Labuda (1982)
Fording mine, BC	BI	22	9.2–35	Goodarzi (1988)
Nova Scotia and New Brunswick	BI	7	3.6–32	Landheer, Dibbs and Labuda (1982)
CHILE:				
Arauco Basin	BI	399	12–21	Collao *et al.* (1987)
COLOMBIA:				
3 regions	BI	14	9–55	Rincon *et al.* (1978)
FRANCE:				
2 areas	–	2	5.5, 15	Sabbioni *et al.* (1983)
GERMANY WEST:				
From 1 power station	BR	7	0.60–5.5	Heinrichs (1982)
Mostly Ruhr	BI	27	10–60	Kautz, Kirsch and Laufhutte (1975)

Table 5.12 (*contd*)

Sample details	Rank	Samples	Content	Reference
NEW ZEALAND:				
Taranaki area	S	41	4–25	Black (1981)
Waikato area	S	–	0.8–8.1	Purchase (1985)
Southland	S	–	4–20	Purchase (1985)
Westland	S-BI	–	0.2–5.3	Purchase (1985)
POLAND:				
Belchatow mine	BR	3	19–90	Tomza (1987)
	BI	–	8–150	Widawska-Kusmierska (1981)
For power production	BI	2	(5.3)	Sabbioni et al. (1983)
7 mines	BI	31	22–150	Tomza (1987)
ROMANIA:				
'Several types'	–	–	1.9–15	Savul and Ababi (1958)
SOUTH AFRICA:				
	BI	11	5–29	Watling and Watling (1976)
	BI	14	4.2–16	Willis (1983)
SWITZERLAND:				
	BR, L	9	4–31	Hügi et al. (1989)
	A	9	0.5–66	Hugi et al. (1989)
THAILAND:				
1 deposit	L	3	5–8	CSIRO (unpublished)
TURKEY:				
5 'kinds'	L	5	6.8–27	Akcetin, Ayca and Hoste (1973)
USSR:				
Moscow Basin	BR	3	5–8	Laktionova et al. (1987)
South Yukutia	–	–	7–37	Egorov, Laktionova and Borts (1981)
Pechora	–	4	6–24	Blomqvist (1983)
UK:				
	–	200	10–240	Aubrey (1954)
S. Yorkshire	S	8	2.7–7.6	Ward, Kerr and Otsuka (1986)
N. England, S. Wales	BI	26	7–70	D.J. Swaine (unpublished)
20 areas	BI	232	7–144	Taylor (1973)
23 collieries	BI	23	12–50	Hislop et al. (1978)
S. Wales (12 seams)	BI	300	–441 (36)	Chatterjee and Pooley (1977)
USA:				
Northern Great Plains	L	226	2.2–78	Zubovic, Hatch and Medlin (1979)
Gulf Province	L	34	3.3–289	Swanson et al. (1976)
Texas	L	150	3–433	White, Edwards and Du Bose (1983)
Fort Union region	L	80	0.3–15	Hatch and Swanson (1976)
Northern Great Plains	S	855	1.2–80	Zubovic, Hatch and Medlin (1979)
Rocky Mt Province	S	466	0.16–120	Zubovic, Hatch and Medlin (1979)
Appalachian Region	BI	1478	0.13–280	Zubovic, Hatch and Medlin (1979)
Virginia	BI	83	7.4–40	Henderson et al. (1985)
Interior Province	BI	687	2.7–170	Zubovic, Hatch and Medlin (1979)

Table 5.12 (contd)

Sample details	Rank	Samples	Content	Reference
Rocky Mt Province	BI	366	1.5–68	Zubovic, Hatch and Medlin (1979)
Alaska, N. Slope	L-BI	84	1–32	Affolter and Stricker (1987)
Alaska	L-BI	–	2–90	Merritt (1988)
Pennsylvania	A	53	5.9–274	Swanson et al. (1976)
YUGOSLAVIA:				
Kosova Basin	BR	2	35, 50	Daci, Berisha and Gashi (1983)

on F in coal, which seems to have been publicized first by Lessing (1934) who examined some UK coals.

The mode of occurrence of F in coal is by no means settled. Fluorapatite has been seen (Crossley, 1944a; Durie and Schafer, 1964), but others have pointed out that this is not always found, and that sometimes there is more F than needed for fluorapatite (Kunstmann et al., 1963). This has led to the suggestion that fluorite may be present. It is well known that clays can accommodate varying amounts of F, the F^- replacing some OH^-. Kaolinites with 125–5500 ppm F and montmorillonites with 870–1350 ppm F were reported by Thomas et al. (1977). According to Beising and Kirsch (1974) F in German coals is mainly associated with illite. Finkelman (1981a) has suggested amphiboles and micas as other F minerals in coal. Perhaps tourmaline and topaz could be present in some coals (Godbeer and Swaine, 1987). The identification of several of these minerals, especially when fine-grained, is difficult. Some suggestions about the association of F with organic matter have been made but it seems that this could only be a minor allocation. A study of coal with SIMS and XPS showed that fusinite had enhanced concentrations of F (McIntyre et al., 1985), but the evidence for F directly bonded to C does not seem to be conclusive.

The determination of F in coal was not an easy task, contamination being troublesome (Schultz, Hattman and Booher, 1973) and modern instrumental methods, except SSMS, have not provided the answer. Information on analytical aspects of F is given in Section 4.5. The key to a successful method is stated well by Campbell (1987), namely 'the conversion of the fluoride in the sample into a form suitable for analysis is a very critical operation'. In recent years, the final measurement of the F^- set free has been achieved successfully by an ion-selective electrode. The combustion of coal plus oxygen in a bomb is the basis of the US standard method (ASTM, 1979). However, after several years of investigations at three Australian laboratories (Godbeer and Swaine, 1987; Doolan, 1987; Rice, 1988) the pyrohydrolysis method has been shown to be suitable for Australian coals, and it may well be the method of choice for all coals. Coal mixed with an equal amount of finely ground silica is combusted for 15 min at 1200 °C in a silica tube furnace in a stream of moist oxygen. The evolved gases are absorbed in 0.025 M NaOH and the F^- eventually measured by an ion-selective electrode. This approach is the basis of the Australian standard method (AS, 1989). There are some variations, namely the use of catalysts and the use of an induction furnace. Although the induction furnace is favoured by Bettinelli (1983a) and Rice (1988), it is not recommended when a silica tube furnace is available because of uncertainty about sustaining the temperature during the determination. Perhaps the use of catalysts ensures completion of decomposition, even if the temperature fluctuates. The silica tube furnace is used by Gao, Yan and Yang (1984), Godbeer and Swaine (1987) and Doolan (1987). It is clear that the ASTM (1979) method gives lower results than the Godbeer and Swaine (1987) method. The amount of F not released is variable, but may often be 20–50%. Distinct differences in results by the two methods were shown in the round-robin (Ring and Hansen, 1984) for the three SARM reference materials, this being the reason for not recommending any values for F, certified or uncertified. However, there are some reliable values for several reference standards (Godbeer, 1987) analysed by the pyrohydrolysis method (Godbeer and Swaine, 1987). The final F^- measurement can also be made by an ion-chromatographic method (Hall, MacLaurin and Vaive, 1986; Conrad and Brownlee, 1988). The method of fusion with NaOH, used in the past, is the method of choice of Sager (1987).

Results for F in coal are given in Table 5.13. The range for most Australian coals is 20–300, with a mean of 110 ppm F, except for brown coals which have very low values in most cases. Due to the differences between pyrohydrolysis and other methods, it is difficult to estimate the range for most other coals, but 20–500, with a mean of about 150 ppm F is probably of the right order. Results for coals from six countries (Millancourt et al., 1986) were within this range. It is interesting to note that the range for US soils is 10–3680, with a mean of 400 ppm F (Shacklette, Boerngen and Keith, 1974). It has been known for a long time that F is emitted by volcanoes (Clarke, 1916), but Barnard and Nordstrom (1982) suggest that fluorides in wet fall-out are mainly anthropogenic; the relative contribution from coal burning seems to be unknown. An unusual case of fluorosis caused by coal burning was reported from a village in China, where coal with 170–1026 ppm F was used for cooking and heating with open stoves. This gave high values of F in indoor air (Dai et al., 1986). More research is needed to ascertain the fate of F during the combustion of coal in modern power stations, thereby permitting a proper assessment to be made of the emission of F to the atmosphere.

5.2.15 Gallium

The main interest in Ga in coal is the possible use of coal ash as an industrial source of Ga. There are no deleterious effects from Ga in coal during mining and usage. Perhaps the first indication of the presence of Ga in coal was its detection in a flue dust from burning a Yorkshire, UK, coal (Ramage, 1927). Goldschmidt and Peters (1931) found up to 400 ppm Ga in some coal ashes. Cooke (1938) found up to 600 ppm Ga in flue dusts from NSW bituminous coals. Sorption studies of Ga on peat suggested that adsorption could account for a close association between Ga and peat,

Table 5.13 Results for fluorine (as ppm F)

Sample details	Rank	Samples	Content	Reference
AUSTRALIA:				
Latrobe Valley, Vic.	BR	1	8	Godbeer and Swaine (1987)
Latrobe Valley, Vic.	BR	20	4–79	Bone and Schaap (1981a)
Latrobe Valley, Vic.	BR	15	0.05–9.4	Bolger (1989)
St Vincent Basin, SA	L	1	180	Godbeer and Swaine (1987)
Esperance, WA	L	1	52	Godbeer and Swaine (1987)
Collie, WA	S	5	16–55	Godbeer and Swaine (1987)
Leigh Creek, SA	H-B	3	191–367	Godbeer and Swaine (1987)
Northern area, NSW	BI	15	39–266	Godbeer and Swaine (1987)
Southern area, NSW	BI	13	15–236	Godbeer and Swaine (1987)
Western area, NSW	BI	59	90–147	Swaine, Godbeer and Morgan (1984b)
NSW and Qnsld	BI	65	15–425	Knott and Warbrooke (1983)
Bowen Basin, Qnsld	BI	8	22–196	Godbeer and Swaine (1987)
West Moreton area, Qnsld	BI	3	66–102	Godbeer and Swaine (1987)
Callide area, Qnsld	BI	6	50–458	Godbeer and Swaine (1987)
Fingal, Tasmania	BI	2	152, 284	Knott and Warbrooke (1983)
Fingal, Tasmania	BI	4	42–130	Godbeer and Swaine (1987)
CANADA:				
'Commercial'	L-BI	3	47–89	Evans et al. (1985)
CZECHOSLOVAKIA:				
From power stations	BR, BI	25	(273)	Lustigova and Kubant (1983)
GERMANY WEST:				
From 1 power station	BR	7	80–106	Heinrichs (1982)
Mostly Ruhr	BI	27	20–370	Kautz, Kirsch and Laufhutte (1975)
INDIA:				
Jiharia area	–	–	316–1208	Rao et al. (1951)
NEW ZEALAND:				
From 7 mines	L-BI	25	<3–24	Soong, Godbeer and Swaine (1984)
Taranaki area	S	–	5–70	Purchase (1985)
Waikato area	S	–	2–64	Purchase (1985)
Southland	S	–	<4–25	Purchase (1985)
Westland	S-BI	–	5.5–13	Purchase (1985)
POLAND:				
	BI	–	50–400	Widawska-Kusmierska (1981)
SOUTH AFRICA:				
Transvaal, Natal, OFS	BI	55	50–465	Kunstmann et al. (1963)
SWITZERLAND:				
	BR, L	6	45–1420	W.C. Godbeer (1988, pers. commun.)
	A	6	40–345	W.C. Godbeer (1988, pers. commun.)
USSR:				
Pechora	–	4	40–83	Blomqvist (1983)
Kuzbas area	–	15	60–270	Gulyayeva and Itkina (1962)
UK:				
	BI	–	27–202	Lim (1979)

Table 5.13 (contd)

Sample details	Rank	Samples	Content	Reference
USA:				
Northern Great Plains	L	226	15–1300	Zubovic, Hatch and Medlin (1979)
Gulf Province	L	34	24–350	Swanson et al. (1976)
Fort Union region	L	80	<20–120	Hatch and Swanson (1976)
Northern Great Plains	S	855	20–1400	Zubovic, Hatch and Medlin (1979)
Rocky Mt Province	S	466	2.0–900	Zubovic, Hatch and Medlin (1979)
Appalachian Region	BI	1478	20–1900	Zubovic, Hatch and Medlin (1979)
Interior Province	BI	687	18–630	Zubovic, Hatch and Medlin (1979)
Rocky Mt Province	BI	366	15–940	Zubovic, Hatch and Medlin (1979)
Alaska, N. Slope	L-BI	84	<20–310	Affolter and Stricker (1987)
Pennsylvania	A	53	<20–290	Swanson et al. (1976)

probably through functional groups of humic acids (Eskenazi, 1967). Ratynskii and Zharov (1979) regard organic and inorganic associations as probable, but seem to favour clays as the site of Ga in some coals (Ratynskii and Glushnev, 1967a). For US lignites, Given and Miller (1987) suggest that Ga appeared to be partly complexed with organic matter. Other authors favour an association with mineral matter (Inagaki, 1967; Widawska-Kusmierska, 1975). Cahill and Shiley (1981) found Ga in sphalerite. Careful studies of several coals confirmed the presence of a Ga-porphyrin, albeit in very small amounts (Bonnett and Czechowski, 1987). The only other report of a Ga-porphyrin was an occurrence in a feldspar (Haberlandt, 1944). It seems that Ga occurs in coals associated with clays, sulfide minerals and, to a lesser extent, with organic matter.

The determination of Ga can be done by modern instrumental techniques, especially INAA and XRFS, and AES and spectrophotometric methods are also used. Results for Ga in coal are given in Table 5.14. The range of values for most coals is 1–20, with a mean of about 5 ppm Ga. The Victorian brown coals with <0.006–0.2 ppm Ga are much lower than other coals.

5.2.16 Germanium

There has been a great interest in Ge in coal as a source of Ge for semiconductor materials and this has led to a spate of publications, probably more than for any other trace element in coal. It is not feasible to cover such a vast array of information, and in any case this is not necessary because much has been dealt with and reviewed by Clark and Swaine (1962a), Abernethy and Gibson (1963), Bethell (1963), Swaine (1975), Bouska (1981) and Valkovic (1983). In particular, the last two authors

Table 5.14 Results for gallium (as ppm Ga)

Sample details	Rank	Samples	Content	Reference
ANTARCTICA:				
Thermally altered	BI	7	4–10	Brown and Taylor (1961)
AUSTRALIA:				
Latrobe Valley, Vic.	BR	16	<0.006–0.2	Brown and Swaine (1964)
St Vincent Basin, SA	L	5	9.9–12.0	Fardy, McOrist and Farrar (1984)
St Vincent Basin, SA	L	3	<0.04–0.3	CSIRO (unpublished)
Esperance, WA	L	3	<0.08	CSIRO (unpublished)
Collie, WA	S	5	0.1–3	CSIRO (unpublished)
Leigh Creek, SA	H-B	80	1–20	CSIRO (1966)
Northern area, NSW	BI	1000	0.5–20	Clark and Swaine (1962a)
Southern area, NSW	BI	800	1.5–20	Clark and Swaine (1962a)
Western area, NSW	BI	150	2–20	Swaine, Godbeer and Morgan (1984b)
NSW and Qnsld	BI	65	1.9–16.0	Knott and Warbrooke (1983)
NSW and Qnsld	BI	45	2.2–12.0	Fardy, McOrist and Farrar (1984)
Bowen Basin, Qnsld	BI	200	1–9	Brown and Swaine (1964)
West Moreton area, Qnsld	BI	150	3–30	Brown and Swaine (1964)
Callide area, Qnsld	BI	100	2–15	Brown and Swaine (1964)
Fingal, Tasmania	BI	2	12–14	Knott and Warbrooke (1983)
Fingal, Tasmania	BI	1	5	CSIRO (unpublished)
Baralaba, Qnsld	S-A	20	1.5–6	CSIRO (unpublished
BELGIUM:				
From 12 mines	BI, A	48	1–20	Block and Dams (1975)
CANADA:				
Hat Creek No.1, BC	L, S	14	<8–13	Goodarzi (1987b)
Fording mine, BC	BI	22	2.8–9	Goodarzi (1987b)
CHILE:				
Arauco Basin	BI	399	62–88	Collao et al. (1987)
CHINA:				
5 areas, S-W China	BI	250	3.0–29	Zhou and Ren (1981)
COLOMBIA:				
3 regions	BI	14	2–7	Rincon et al. (1978)
GERMANY EAST:				
4 areas	BR	100	(2.2–9.5)	Malberg (1961)
HUNGARY:				
	BR, BI	7	7–80	Takacs and Horvath (1959)
INDIA:				
Different areas	BI, S-A	27	<1–48	Ghosh, Biswas and Banerjee (1979)
NEW ZEALAND:				
Mokau area	S	–	0.5–6.7	Gainsford (1985)
Waikato area	S	–	0.06–5.8	Purchase (1985)
Southland	S	–	0.4–2.2	Purchase (1985)
Westland	S-BI	–	0.06–7.6	Purchase (1985)
NIGERIA:				
Enugu, 3 areas	S	3	7.4–8.0	Ndiokwere, Guinn and Burtner (1983)

Table 5.14 (*contd*)

Sample details	Rank	Samples	Content	Reference
POLAND:				
Belchatow mine	BR	3	0.61–7.0	Tomza (1987)
7 mines	BI	31	0.76–17	Tomza (1987)
SOUTH AFRICA:				
	BI	11	43–66	Watling and Watling (1976)
	BI	14	7.1–18	Willis (1983)
SWITZERLAND:				
	BR, L	9	0.2–9.6	Hügi *et al.* (1989)
	A	9	1.0–9.5	Hügi *et al.* (1989)
THAILAND:				
1 deposit	L	3	0.3–0.8	CSIRO (unpublished)
TURKEY:				
5 'kinds'	L	5	2.2–7.9	Akcetin, Ayca and Hoste (1973)
USSR:				
1 deposit	L	6	2.7–12	Borisova *et al.* (1974a)
Moscow Basin	BR	3	2–4	Laktionova *et al.* (1987)
South Yakutia	–	–	3.1–10	Egorov, Laktionova and Borts (1981)
UK:				
W. Midlands, 26 seams	BI	26	2–14	Dalton and Pringle (1962)
N. England, S. Wales	BI	26	0.6–7	D.J. Swaine (unpublished)
23 collieries	BI	23	1.5–9	Hislop *et al.* (1978)
USA:				
Gulf Province	L	34	2–30	Swanson *et al.* (1976)
Texas	L	83	1–28	White, Edwards and Du Bose (1983)
Fort Union region	L	80	0.5–7	Hatch and Swanson (1976)
S. and E. Arkansas	L	53	3–30	Hildebrand, Clardy and Holbrook (1981)
Eastern USA	BI	644	0.71–35	Zubovic *et al.* (1980)
Western USA	–	28	0.80–6.5	Gluskoter *et al.* (1977)
Alaska, N. Slope	L-BI	84	<0.3–20	Affolter and Stricker (1987)
Pennsylvania	A	53	1.5–20	Swanson *et al.* (1976)

have covered Eastern Bloc papers, not readily available elsewhere. Earlier chapters have discussed the origin (Chapter 2), the occurrence (Chapter 3) and analysis (Chapter 4).

The presence of Ge in coal was established by Goldschmidt (1930) using the optical and X-ray spectroscopic methods, which enabled him to find out so much about a host of trace elements in coal. Then, Goldschmidt and Peters (1933a) found 1.1% Ge in the ash of a bituminous coal from the, now famous, Yard seam, Durham, UK, which established coal as a possible source of Ge. Lower, yet enhanced, levels of Ge were found in some Australian coal ashes (Cooke, 1938) and very high concentrations were found in a lignite from the United States (Stadnichenko, Murata and Axelrod, 1950). Investigations were soon being carried on Japanese coals

Table 5.15 Results for germanium (as ppm Ge)

Sample details	Rank	Samples	Content	Reference
ANTARCTICA:				
Thermally altered	BI	7	1.4–58	Brown and Taylor (1961)
AUSTRALIA:				
Coolungoolun, Vic.	BR	4	10–17	Pilkington (1957)
Latrobe Valley, Vic.	BR	16	<0.02–0.03	Brown and Swaine (1964)
St Vincent Basin, SA	L	3	0.1–3	CSIRO (unpublished)
Esperance, WA	L	3	<0.08	CSIRO (unpublished)
Collie, WA	S	22	<0.1–2	Davy and Wilson (1984)
Leigh Creek, SA	H-B	80	<1–6	CSIRO (1966)
Northern area, NSW	BI	1000	0.2–50	Clark and Swaine (1962a)
Southern area, NSW	BI	800	1–30	Clark and Swaine (1962a)
Western area, NSW	BI	150	0.06–30	Swaine, Godbeer and Morgan (1984b)
NSW and Qnsld	BI	65	<0.5–66	Knott and Warbrooke (1983)
Bowen Basin, Qnsld	BI	200	0.7–15	Brown and Swaine (1964)
West Moreton area, Qnsld	BI	150	<0.2–8	Brown and Swaine (1964)
Callide area, Qnsld	BI	100	0.3–3	Brown and Swaine (1964)
Fingal, Tasmania	BI	2	1.8, 2.3	Knott and Warbrooke (1983)
Fingal, Tasmania	BI	1	9	CSIRO (unpublished)
Baralaba, Qnsld	S-A	20	<0.5–0.9	CSIRO (unpublished)
CANADA:				
Nova Scotia	BI	186	1–8	Hawley (1955)
HUNGARY:				
	BR, BI	7	1–55	Takacs and Horvath (1959)
INDIA:				
Several areas	–	29	<10–78	Banerjee, Rao and Lahiri (1974)
NEW ZEALAND:				
Mokau area	S	–	0.09–15	Gainsford (1985)
Waikato	S	–	<0.06–0.7	Lynskey, Gainsford and Hunt (1984)
Westland	S-BI	–	0.08–18	Purchase (1985)
POLAND:				
4 mines	BR	73	0.1–4.8	Gregorowicz (1957)
	BI	–	0.1–50	Olczak (1985)
SOUTH AFRICA:				
Transvaal, Natal	BI	49	1.0–16	Kunstmann and Hamersma (1955)
	BI	14	0.7–10	Willis (1983)
SWITZERLAND:				
	BR, L	9	0.8–40	Hügi et al. (1989)
	A	9	0.3–20	Hügi et al. (1989)
SPAIN:				
7 areas	L	52	0.2–21	Martin Perez (1963)
11 areas	BI	142	0.2–80	Martin Perez (1963)
10 areas	A	29	0.2–6	Martin Perez (1963)
UK:				
N. England, S. Wales	BI	26	0.07–25	D.J. Swaine (unpublished)
20 areas	BI	232	1–25	Taylor (1973)

Table 5.15 (contd)

Sample details	Rank	Samples	Content	Reference
USSR:				
	–	–	<2–210	Ratynskii (1946)
Kizelovsk	–	–	1–9	Ershov (1958)
USA:				
Texas	L	60	0.4–25	White, Edwards and Du Bose (1983)
Eastern USA	BI	644	<0.07–36	Zubovic et al. (1980)
West Virginia	BI	238	<0.7–215	Headlee and Hunter (1951)
Western USA	–	28	0.10–3.0	Gluskoter et al. (1977)
Alaska	L-BI	–	0–10	Merritt (1988)

(Inagaki, 1951b, 1952b), on UK coals (Aubrey, 1952), on US coals (Stadnichenko et al., 1953) and on Australian coals (Adcock and Muir, 1953; Pilkington, 1957; Durie and Schafer, 1958). There is general agreement that organically bound Ge is common in most coals. The presence of large amounts of Ge in coalified wood (Breger and Schopf, 1955) and in lignitic inclusions (Stadnichenko et al., 1953; Hallam and Payne, 1958) establishes that there can be marked enrichment in Ge during the early stages of coalification. According to Manskaya and Drozdova (1968) Ge could form bonds with OH groups in humic acid during peat formation. Support for organic bonding in coal is given by Bernstein (1985) who stated that 'during coalification, highly condensed aromatic organogermanium compounds will form; these will have increased stability due to a greater number of Ge–C bonds'. Pyrite is an unlikely source of Ge (Swaine, 1975), but sphalerite is well known for enhanced contents of Ge. There is no reason why some Ge should not be associated with clays. I agree with Finkelman (1981a) that Ge is organically bound in most coals, but some coals may have much of their Ge associated with sphalerite or clays.

The determination of Ge has been carried out mainly by AES and chemical methods, but XRFS and AAS have also been used. Nadkarni (1984) has used ICPAES for USA coals.

Results for Ge in coals are given in Table 5.15. The overall range for most coals is around 0.5–50, with a mean of about 6 ppm Ge. There are many results for Ge in coal ash, but these cannot be converted to a coal basis because ash yields are not given.

Germanium has no known biological function and in view of present knowledge of Ge compounds 'no environmental nor human health hazards are apparent' (Furst, 1987).

5.2.17 Gold

Gold is an ultratrace constituent of coal and is rarely determined. There are early reports of Au in coal, for example it was detected in a coal from

Wyoming, USA (Chance, 1899), about 1 ppm Au was found in some coals from Utah and Wyoming, USA (Jenny, 1903) and up to 3 ppm Au in coal from Cambria, Wyoming-North Dakota, USA (Stone 1912). Goldschmidt and Peters (1933b) found 0.2–1 ppm Au in some German coal ashes, and Fersman (1958) referred to some coal ashes with up to 1 ppm Au. Leutwein (1956) mentioned some peats with 'rather high' amounts of Au. Finkelman (1981a, 1988) detected fine particles of Au and a particle of gold telluride during his studies of different coals. These are the most likely modes of occurrence of Au in coal, although auriferous pyrite (Bethell, 1963) and a relationship with quartz (Ratynskii et al., 1982) have been suggested. It is known that some plants have small particles of Au in them (Lakin, Curtin and Hubert, 1974), so perhaps such plants could have been the source of Au in some coals. However, a detrital origin seems more plausible (Finkelman, 1981a). The determination of Au in coal is usually INAA, but other methods may be applicable after preconcentration.

There are some results (as ppb Au) for Au in coal, namely <1–10 for Australia (Fardy, McOrist and Farrar, 1984), 1.0–2.5 for the Lithgow seam, NSW, Australia (1985, R.H. Filby, pers. commun.), 10–140 for Belgium (Block and Dams, 1975), mean value 3 for Bulgaria (Kostadinov and Djingova, 1980), 1–96 for Canada (van der Flier-Keller and Fyfe, 1987b), 1–7 for Swiss brown, lignite and <1–11 for Swiss anthracitic coals (Hügi et al., 1989), 1.2–10 for the United Kingdom (Hislop et al., 1978), 2–150 for Donets Basin, USSR (Dzyuba and Lepkii, 1981), 3.5–44 for brown coals from one deposit, USSR (Ratynskii et al., 1982), and <0.02–1.8 for 65 Kentucky, US coals (Chyi, 1982). Perhaps up to about 10 ppb Au could be expected in most coals, except for special coals, such as those from Wyoming, Germany and the Soviet Union, mentioned above. However, there is a need for much more data to clarify the situation.

5.2.18 Hafnium

Goldschmidt (1937) noted that some coal ashes showed enrichments in Hf, but not to any marked extent. There is only one publication dealing with Hf in detail (Eskenazi, 1987a). She found that Hf was mainly associated with mineral matter in Bulgarian coals, but organically bound Hf dominated in some low-ash coals. It was suggested that Hf, leached from minerals, was removed by organic matter at the peat stage. In view of the close geochemical association between Hf and Zr it is not surprising to find that zircons in coal contain small amounts of Hf, as confirmed by Finkelman (1981a). Indeed, the suggestion that zircon is the site of most of the Hf in coal (Finkelman, 1981a) seems reasonable.

There is more information on the concentrations of Hf in coal than could have been anticipated. This is because INAA has been used for most coals and Hf is sensitive and readily determined, as part of an analysis for other elements. X-ray fluorescence spectrometry is also a suitable method. Results for Hf in coal are given in Table 5.16. There is an overall agreement between the results, which indicates that most coals would fall in the range 0.4–5 ppm Hf, with low-rank coals tending to be at the lower end of the range.

Table 5.16 Results for hafnium (as ppm Hf)

Sample details	Rank	Samples	Content	Reference
AUSTRALIA:				
Latrobe Valley, Vic.	BR	28	0.001–1.1	Fardy, McOrist and Farrar (1984)
St Vincent Basin, SA	L	5	0.62–5.2	Fardy, McOrist and Farrar (1984)
Leigh Creek, SA	H-B	2	<3, 7	Knott and Warbrooke (1983)
NSW	BI	4	1.0–4.7	P.C. Rankin (1977, pers. commun.)
Western area, NSW	BI	39	1.9–5.1	Swaine, Godbeer and Morgan (1984b)
NSW and Qnsld	BI	65	1.3–7.0	Knott and Warbrooke (1983)
NSW and Qnsld	BI	45	0.63–5.6	Fardy, McOrist and Farrar (1984)
Fingal, Tasmania	BI	2	<3, 6	Knott and Warbrooke (1983)
BELGIUM:				
From 12 mines	BI, A	48	0.1–3.0	Block and Dams (1975)
BRAZIL:				
Rio Grande do Sul	BI	13	1.5–4.3	Bellido and Arezzo (1987)
BULGARIA:				
7 areas	BR	67	0.01–8.1	Eskenazi (1987a)
2 areas	L	48	0.03–4.2	Eskenazi (1987a)
3 areas	BI	38	0.004–2.2	Eskenazi (1987a)
CANADA:				
Saskatchewan	L	7	0.66–1.4	Landheer, Dibbs and Labuda (1982)
Hat Creek No.1, BC	L, S	14	<0.2–7.0	Goodarzi (1987b)
Hat Creek No.2, BC	L, S	32	1.5–4.8	Goodarzi and van der Flier-Keller (1988)
Alberta and BC	S	23	0.62–3.2	Landheer, Dibbs and Labuda (1982)
'Typical'	S, BI	55	0.11–4.9	Jervis, Ho and Tiefenbach (1982)
Fording mine, BC	BI	22	0.4–2.3	Goodarzi (1988)
CHINA:				
'Selection'	–	15	0.92–4.9	Sun and Jervis (1986)
from 110 mines	–	–	0.31–16	Chen et al. (1986)
NIGERIA:				
'Major deposit'	L	1	2.1	Hannan et al. (1982)
'Major deposit'	S	6	0.46–1.7	Hannan et al. (1982)
'Major deposit'	BI	1	0.67	Hannan et al. (1982)
SOUTH AFRICA:				
Witbank coalfield	BI	146	1.2–3.0	Hart and Leahy (1983)
SWITZERLAND:				
	BR, L	8	0.055–1.1	Hügi et al. (1989)
	A	9	0.11–3.3	Hügi et al. (1989)
TURKEY:				
5 zones	L	5	0.08–1.8	Ayanoglu and Gunduz (1978)
USA:				
Eastern USA	BI	644	<0.1–4.0	Zubovic et al. (1980)
Illinois	BI	114	0.13–1.5	Gluskoter et al. (1977)
Western USA	–	28	0.26–1.3	Gluskoter et al. (1977)

There are no known biological or environmental effects of Hf during coal mining and usage.

5.2.19 Indium

Goldschmidt (1937) found up to 2 ppm In in coal ashes, while Ramage (1927) had detected In spectroscopically in a flue dust from a Yorkshire, UK, coal. The presence of In in coal had been inferred from its presence in pulverized-fuel flyash with <10–20 ppb (Smith, 1958) and in superheater deposits from a stoker-fired boiler, which had up to 40 ppm In (Brown and Swaine, 1964). Inagaki (1951a) detected In in some Japanese coal ashes. Based on its geochemical associations in minerals generally, Finkelman (1981a) has proposed that In is probably associated with sulfide or carbonate minerals in most coals. Cahill and Shiley (1981) found appreciable concentrations of In in some sphalerites from Illinois, USA, coals. In Bulgarian coals, which were examined in detail by Eskenazi (1980), In was found to be mainly associated with the mineral matter, including clays, and to a lesser extent with organic matter. It would seem that In in most coals would be mainly associated with sulfide and carbonate minerals and probably clays, but that organic bonding would be only minor. Although INAA is most commonly used to determine In, AES and AAS can be used, usually after preconcentration of In. The EAAS method of Zhou, Chao and Meier (1984) should be adaptable for coal ash.

There are some data for In in coal (as ppb In), which will be listed without details which can be found in other tables for the particular references given. Australian coals had <50 ppb In (CSIRO, unpublished) and 1 NSW bituminous coal (Dale, Fardy and Clayton, 1985) had 50 ppb In. Coals from Belgium had 10–70 (Block and Dams, 1975), from Bulgaria 2–167 (Eskenazi, 1980), from East Marica, Bulgaria, 16–102 (Eskenazi, 1982a), from Hat Creek, Canada, <8–56 (Goodarzi, 1987b), from New Zealand <40–250 (Purchase, 1985), from Pakistan 127 (Chaudhary et al., 1984), from Poland 8–167 (Tomza, 1987), from Illinois Basin, USA, 100–630, from eastern United States 130–370, and from western United States 10–250 (Gluskoter et al., 1977). It would seem that most coals would have In contents in the range 10–200 ppb In. It is surprising to see that Merritt (1988) lists In among elements of 'chief environmental concern', as there are no known untoward effects of low concentrations of In on health or the environment. There is an extensive survey of all aspects of In (Smith, Carson and Hoffmeister, 1978).

5.2.20 Iodine

Iodine is an important element in human nutrition and much is known about its distribution and biological activity. There is a good review of the geochemistry of I (Fuge and Johnson, 1986), which indicates that earlier data may have analytical shortcomings. The earliest report concerning I in coal deals with I in combustion products from coal (Bussy, 1839), while I was detected in several coals from different parts of Europe by Duflos (1847), MacAdam (1852), and Graf (1852). Von Fellenberg (1923) found 0.09 ppm I in another coal and Stoklasa (1927) found 0.9 ppm I in a coal.

Peat contains I, as shown by McClendon (1939) who found 12–40 ppm I in some, presumably US, peats, and the range of values for European peats has been given as 1.2–32 ppm I (CIEB, 1956). Goldschmidt (1954) classifies I as a biophile element, which infers that it is concentrated organically. It seems likely that most of the I in coals is organically bound, while a varying proportion, mainly in low-I coals, may be inorganically associated perhaps with clays (Finkelman, 1981a).

Data for coals will be given (as ppm I) without details, which can be found under relevant references in other tables. For seven Australian bituminous coals <0.3–2.8 (Cahill and Mills, 1983), for Belgian coals 1.3–4.1 (Block and Dams, 1975), for Canadian bituminous and subbituminous coals <0.4–0.5 (Jervis, Ho and Tiefenbach, 1982), for 15 Chinese coals <0.2–9.2 (Sun and Jervis, 1986), for two Czechoslovakian coals 0.1, 3 (Behringer, 1925; Wald, 1953), for German brown coals 1.4–5.7 (Wache, 1931) and for hard coals <1–4.2 (Kautz, Kirsch and Laufhutte, 1975), for mid-European coals 0.85–11 (Wilke-Dörfurt and Römersperger, 1930), for Nigerian coals 0.32–1.9 (Hannan et al., 1982), for Polish coals 0.4–20 (Tomza, 1987), for South African coals 1.5–3.1 (Köhler, 1931), for Yorkshire, UK, coals 0.51–1.2 (Ward, Kerr and Otsuka, 1986), for UK coals <1–22 (Hislop et al., 1978), for Soviet coals 3–23 (Gulyayeva and Itkina, 1962), for Illinois, USA, coals 0.2–14, for eastern US coals 0.3–5.0 and for western US coals 0.2–1.0 (Gluskoter et al., 1977). The determination of I may be carried out by INAA or chemically, care being taken to avoid losses during handling. Ashing, even LTA, would give losses of I.

Most coals would probably fall in the range 0.5–15 ppm I. Although there is no emphasis on emissions from coal burning as a source of I to the environment, there is no doubt that most of the I would be released from stacks (Stolper, 1956), some perhaps as elemental I, but most probably attached to flyash particles. According to Fuge and Johnson (1986) 'smoke from burning fossil fuels' has 1–200 ppm I ('in soot'). On the basis of calculations made by Wilke-Dörfurt and Römersperger (1930), the annual emission of I to the atmosphere from coal burning could be about 10 million kg. In view of its essentiality this can only be regarded favourably. Other sources of I for the atmosphere include volcanoes, volatilization from vegetation and soils and transfers from the hydrosphere, for example as sea spray. Some I is released from vehicle exhausts and can be detected in roadside vegetation (Fuge, 1988). The first detection of I in volcanic emissions was by St-Claire Deville and Leblanc (1857) who found an ammonium salt of iodine in exhalations from fumaroles of Vulcano in the Lipari Islands. No references have been found to damage to vegetation near coal-burning power stations. It is interesting that the burning of seaweed in the past was a major contributor of I to the local environment. It will be recalled that the discovery of I was made by J.B. Courtois in 1811 who found it in seaweed (Weeks, 1945).

5.2.21 Lead

There is much interest in Pb in relation to health and the environment because it can give rise to illness in certain situations. There is no well-established role for Pb as an essential trace element in nutrition. It is not surprising that some attention has been given to Pb in coal. The earliest reference to Pb in coal concerns Upper Silesian coals which had 200–760 ppm Pb in ash (Jensch, 1887). There are several early references to galena in coal (Phillips, 1896; Binns and Harrow, 1897; Crook, 1913). Goldschmidt and Peters (1933b) found up to 1000 ppm Pb in some German coal ashes. There is little dispute about the occurrence of Pb in the mineral matter, although Marczak and Parzentny (1985) found that 23% of the Pb in some Polish coals was in the organic matter. Finkelman (1981a) discovered grains of PbSe, probably clausthalite, in Appalachian, US, and some other coals. After an assessment of the mode of occurrence of Pb in coals, Finkelman (1981a) regarded the situation as varying for different areas, the main associations being with galena, PbSe, and some Ba minerals, where Pb can replace Ba (sulfates, carbonates, phosphates, silicates). It seems reasonable to add pyrite (Finkelman, 1988) and some organic association (probably in low-rank coals).

There are several methods for the determination of Pb, namely AES, AAS and XRFS. The two standard methods (ASTM, 1983a; AS, 1986) use FAAS after dissolution (see Be, Section 5.2.4); an air–acetylene flame is used with the line at 283.3 nm and background correction. For low concentrations, say less than 10 ppm Pb, an EAAS method is required. It has been known for some time that Pb is not lost by ashing at 500 °C or even higher (Cuttitta and Warr, 1958). Block (1975) found that excess acid, especially HNO_3, should be avoided in order to attain maximum absorbance. Other methods that have been used include hydride generation–AAS (Nadkarni, 1982b) and IDSSMS (Machlan et al., 1976).

Results for Pb in coal are given in Table 5.17. Most coals would fall in a range of about 2–80 ppm Pb, Australian, US and South African coals being lower, namely 2–40 ppm Pb. The mean value for most Australian, South African and US coals is probably around 10–15 ppm Pb, whereas European coals have mean values of 30–60 ppm Pb, confirmed by the mean value of 54 ppm Pb for the nine coals used for power production in the EC countries (Sabbioni et al., 1983). According to Patterson and Settle (1987) the emission of Pb from volcanoes is relatively small, but is still very similar to that from coal burning (Bennett, 1981). The emissions of Pb from petrol and oil combustion far outweigh those from all other sources combined. According to Lovering (1976) up to about 6% of Pb in coal may reach the atmosphere from coal burning, but there are no sound reasons for coal to be regarded as a source of Pb which is detrimental to the environment. There is a review of the geochemistry of Pb which puts its role in the environment into perspective (Swaine, 1978b).

5.2.22 Lithium

Goldschmidt (1937) reported finding a German coal ash with 500 ppm Li, but Butler (1953) found up to 5000 ppm Li in ash from some Spitsbergen coals and Headlee and Hunter (1953) found up to 3000 ppm Li in ash from some coals from West Virginia, USA. The only reference to the association of Li, at least partly, with the organic coaly matter is to some coals from the Burgas Basin, Bulgaria (Uzunov and Karadzhova, 1968).

Table 5.17 Results for lead (as ppm Pb)

Sample details	Rank	Samples	Content	Reference
ANTARCTICA:				
Thermally altered	BI	7	8–96	Brown and Taylor (1961)
ARGENTINA:				
Rio Turbio	–	6	<70–80	A. Berset (1982, pers. commun.)
AUSTRALIA:				
Latrobe Valley, Vic.	BR	16	<0.4–15	Brown and Swaine (1964)
Latrobe Valley, Vic.	BR	20	0.1–9	Bone and Schaap (1981a)
Latrobe Valley, Vic.	BR	15	<0.2–7.1	Bolger (1989)
St Vincent Basin, SA	L	3	3–25	CSIRO (unpublished)
Esperance, WA	L	3	6–9	CSIRO (unpublished)
Leigh Creek, SA	H-B	80	<2–50	CSIRO (unpublished)
Collie, WA	S	22	<1–10	Davy and Wilson (1984)
Northern area, NSW	BI	1000	1–40	Clark and Swaine (1962a)
Southern area, NSW	BI	800	4–60	Clark and Swaine (1962a)
Western area, NSW	BI	150	2–24	Swaine, Godbeer and Morgan (1984b)
NSW and Qnsld	BI	65	5–22	Knott and Warbrooke (1983)
Bowen Basin, Qnsld	BI	200	1.5–25	Brown and Swaine (1964)
West Moreton area, Qnsld	BI	150	5–40	Brown and Swaine (1964)
Callide area, Qnsld	BI	100	2.5–30	Brown and Swaine (1964)
Fingal, Tasmania	BI	2	13, 21	Knott and Warbrooke (1983)
Baralaba, Qnsld	S-A	20	4–20	CSIRO (unpublished)
BELGIUM:				
Different mines	–	8	8–110	Block (1975)
CANADA:				
Saskatchewan	L	8	5.6–18	Landheer, Dibbs and Labuda (1982)
Alberta and BC	S	23	1.9–19	Landheer, Dibbs and Labuda (1982)
Nova Scotia and New Brunswick	BI	7	1.8–53	Landheer, Dibbs and Labuda (1982)
COLOMBIA:				
3 regions	BI	14	<2–20	Rincon et al. (1978)
GERMANY WEST:				
From 1 power station	BR	7	0.9–4.2	Heinrichs (1982)
Mostly Ruhr	BI	27	20–270	Kautz, Kirsch and Laufhutte (1975)
Various	BI	10	0.1–112	Heinrichs (1975b)
34 mines	BI	204	0–390	Riepe (1986)
KOREA:				
	–	–	7.5–23	Bae and Kim (1980)
NEW ZEALAND:				
Taranaki area	S	–	0.5–15	Purchase (1985)
Waikato area	S	–	0.2–16	Purchase (1985)
Southland	S	–	1.2–17	Purchase (1985)
Westland	S-BI	–	0.1–25	Purchase (1985)
POLAND:				
Chelm deposit	BI	219	4–150	Widawska-Kusmierska (1981)

Table 5.17 (contd)

Sample details	Rank	Samples	Content	Reference
ROMANIA:				
'Several types'	–	–	0.10–13	Savul and Ababi (1958)
SOUTH AFRICA:				
	BI	11	3–17	Watling and Watling (1976)
	BI	14	1.9–25	Willis (1983)
SWITZERLAND:				
	BR, L	9	0.1–7.0	Hügi et al. (1989)
	A	9	0.4–51	Hügi et al. (1989)
THAILAND:				
1 deposit	L	3	4–300	CSIRO (unpublished)
USSR:				
Moscow Basin	BR	3	8–15	Laktionova et al. (1987)
South Yakutia	–	–	5–28	Egorov, Laktionova and Borts (1981)
UK:				
N. England, S. Wales	BI	26	2–70	D.J. Swaine (unpublished)
20 areas	BI	232	1–900	Taylor (1973)
23 collieries	BI	23	8–63	Hislop et al. (1978)
S. Wales (12 seams)	BI	300	–237(32)	Chatterjee and Pooley (1977)
USA:				
Northern Great Plains	L	226	1.4–17	Zubovic, Hatch and Medlin (1979)
Gulf Province	L	34	<2.8–129	Swanson et al. (1976)
Texas	L	108	0.3–32	White, Edwards and Du Bose (1983)
Fort Union region	L	80	<1.4–11	Hatch and Swanson (1976)
Northern Great Plains	S	855	0.72–58	Zubovic, Hatch and Medlin (1979)
Rocky Mt Province	S	466	0.95–76	Zubovic, Hatch and Medlin (1979)
Appalachian Region	BI	1478	0.37–86	Zubovic, Hatch and Medlin (1979)
Interior Province	BI	687	0.78–590	Zubovic, Hatch and Medlin (1979)
Rocky Mt Province	BI	366	0.76–137	Zubovic, Hatch and Medlin (1979)
Alaska, N. Slope	L-BI	84	<0.8–21	Affolter and Stricker (1987)
Pennsylvania	A	53	1.0–24	Swanson et al. (1976)

In the absence of any direct evidence, one must resort to knowledge of the general geochemistry of Li in the sedimentary environment. It is reasonable to expect Li in most coals to be associated mainly with clays and possibly, to a lesser extent, with micas and tourmaline, as surmised by Finkelman (1981a).

The determination of Li is usually carried out by AES and AAS. Nowadays ICPAES is considered useful for determining Li in coal (Mills, Doolan and Knott, 1983). There is a standard method for Li in coal (AS, 1986), which uses FAAS, the dissolution being done as for Be (Section

5.2.4); an air-acetylene flame is used with the line at 670.8 nm and background correction. The sample should be ground so that 98% is less than 75 μm. Results for Li in coal are given in Table 5.18. It is difficult to assess the relatively few results available, apart from the United States, but most coals would probably be within the range 1–80, with a mean of about 20 ppm Li. References have not been found to ill-effects from the ingestion of Li, but Li compounds have been used as depressants in medicine. Perhaps there may be beneficial effects from Li in a normal diet. Indeed, Cannon, Harms and Hamilton (1975) suggested a positive outlook for the role of Li in health.

It seems most unlikely that Li in coal would have adverse environmental or health effects.

Table 5.18 Results for lithium (as ppm Li)

Sample details	Rank	Samples	Content	Reference
AUSTRALIA:				
Latrobe Valley, Vic.	BR	20	<0.06–1.3	Bone and Schaap (1981a)
Latrobe Valley, Vic.	BR	15	<0.2–0.5	Bolger (1989)
St Vincent Basin, SA	L	3	<3–25	CSIRO (unpublished)
Northern area, NSW	BI	20	4–19	CSIRO (unpublished)
Southern area, NSW	BI	10	14–50	CSIRO (unpublished)
Western area, NSW	BI	140	7–37	Swaine, Godbeer and Morgan (1984b)
COLOMBIA:				
3 regions	BI	14	3–30	Rincon et al. (1978)
NEW ZEALAND:				
Mokau area	S	4	1.4–3.9	Gray (1986)
Waikato	S	–	0.4–15	Purchase (1985)
Southland	S	–	0.2–10	Purchase (1985)
Westland	S-BI	–	2.3–5.3	Purchase (1985)
POLAND:				
	BI	–	5–50	Widawska-Kusmierska (1981)
USA:				
Northern Great Plains	L	226	0.37–33	Zubovic, Hatch and Medlin (1979)
Gulf Province	L	34	0.9–145	Swanson et al. (1976)
Texas	L	60	1.6–180	White, Edwards and Du Bose (1983)
Fort Union region	L	80	<0.3–23	Hatch and Swanson (1976)
Northern Great Plains	S	855	0.22–61	Zubovic, Hatch and Medlin (1979)
Rocky Mt Province	S	466	0.52–87	Zubovic, Hatch and Medlin (1979)
Appalachian Region	BI	1478	0.70–350	Zubovic, Hatch and Medlin (1979)
Interior Province	BI	687	0.44–210	Zubovic, Hatch and Medlin (1979)
Rocky Mt Province	BI	366	0.58–70	Zubovic, Hatch and Medlin (1979)
Alaska, N. Slope	L-BI	84	0.5–84	Affolter and Stricker (1987)
Pennsylvania	A	53	4–162	Swanson et al. (1976)

5.2.23 Manganese

There is interest in Mn because of its biological essentiality and because excess can cause problems in the environment. Jensch (1887) found 1400–20 400 ppm Mn in six coal ashes from Silesia, and ankerite with about 9000 ppm Mn was found in two coal seams from Lancashire, UK (Crook, 1913; Sinnatt, Grounds and Bayley, 1921). In low-rank coals, Mn is probably largely organically bound through carboxylic acid groups (Brown and Swaine, 1964; Benson and Holm, 1985). Organic bonding for Mn in some Hungarian brown coals is favoured by Odor (1967), while Idzikowski (1960) indicates that Mn is not limited to inorganic associations in some Polish coals. An interesting finding by Bonnett and Czechowski (1981) was that of a Mn-porphyrin, but this only accounted for a very small amount of the total Mn in the coal. There is evidence for Mn in carbonate minerals (Brown and Swaine, 1964; Swaine, 1986b) and an association with clays is possible. Finkelman (1981a) rightly suggests 'that in the absence of sufficient carbonates, manganese may be associated with the clays'. Also, some Mn can occur with pyrite. For most low-rank coals, the predominant mode of occurrence is organic. For higher rank coals there are several modes, namely in carbonate minerals, probably predominant in many coals, in clays, also a major form for some coals; organically bound Mn and Mn in pyrite are considered to be minor sources in most coals.

There are plenty of methods for determining Mn in coals, namely modern instrumental methods and spectrophotometric methods. There are two standard methods (ASTM, 1983a; AS, 1986), using the same dissolution technique described earlier (Section 5.2.4). The ensuing FAAS determination uses an air–acetylene flame and the line at 279.5 nm with background correction.

Results for Mn in coals are given in Table 5.19. Despite some high values, it is probable that the range for most coals is 5–300, with a mean of 50–70 ppm Mn, while the corresponding range for Australian coals is about 2–400, with a mean of about 130 ppm Mn. There are no reports of serious health or environmental effects during coal mining and usage. However, before using flyash as a soil amendment to correct any trace-element deficiencies, including Mn, it is advisable to do some plant experiments designed to represent the conditions in the flyash-amended soils for a particular plant species.

5.2.24 Mercury

There is continued interest in Hg in coal because of emissions to the atmosphere during coal burning. There are no known beneficial effects of Hg in the environment, but it is detrimental to biological systems if present in excessive amounts. Perhaps the first determination of Hg in coal was done by Stock and Cucuel (1934) who found up to 0.022 ppm Hg in German coals and 28 ppm Hg in chimney soot. Cinnabar was reported occasionally from German coals (Stutzer, 1940). Extensive studies of Donbas and other Soviet coals led Dvornikov (1981a,b) to propose that Hg occurs in three forms, namely HgS, metallic Hg and organomercury compounds. Cahill and Shiley (1981) found Hg in some sphalerite samples

Table 5.19 Results for manganese (as ppm Mn)

Sample details	Rank	Samples	Content	Reference
ANTARCTICA:				
Thermally altered	BI	7	19–78	Brown and Taylor (1961)
ARGENTINA:				
Rio Turbio	–	6	30–80	A. Berset (1982, pers. commun.)
AUSTRALIA:				
Yallourn, Vic.	BR	16	10–60	Baragwanath (1962)
Latrobe Valley, Vic.	BR	16	1–200	Brown and Swaine (1964)
Latrobe Valley, Vic.	BR	20	0.7–61	Bone and Schaap (1981a)
Latrobe Valley, Vic.	BR	15	3.0–154	Bolger (1989)
Latrobe Valley, Vic.	BR	28	0.45–55	Fardy, McOrist and Farrar (1984)
St Vincent Basin, SA	L	5	31–240	Fardy, McOrist and Farrar (1984)
Esperance, WA	L	3	25	CSIRO (unpublished)
Leigh Creek, SA	H-B	80	<2–500	CSIRO (1966)
Collie, WA	S	22	<1–43	Davy and Wilson (1984)
Northern area, NSW	BI	1000	2.5–640	Clark and Swaine (1962a)
Southern area, NSW	BI	800	2.5–900	Clark and Swaine (1962a)
Western area, NSW	BI	150	2–800	Swaine, Godbeer and Morgan (1984b)
NSW and Qnsld	BI	65	1–455	Knott and Warbrooke (1983)
NSW and Qnsld	BI	45	<5–410	Fardy, McOrist and Farrar (1984)
Bowen Basin, Qnsld	BI	200	1.5–700	Brown and Swaine (1964)
West Moreton area, Qnsld	BI	150	2–300	Brown and Swaine (1964)
Callide area, Qnsld	BI	100	2.5–2000	Brown and Swaine (1964)
Fingal, Tasmania	BI	2	324, 425	Knott and Warbrooke (1983)
Baralaba, Qnsld	S-A	20	9–800	CSIRO (unpublished)
BELGIUM:				
From 12 mines	BI, A	48	10–400	Block and Dams (1975)
CANADA:				
Saskatchewan	L	8	24–127	Landheer, Dibbs and Labuda (1982)
Moose River Basin	L	17	<80–700	Van der Flier-Keller and Fyfe (1987b)
Hat Creek No.1, BC	L, S	14	19–122	Goodarzi (1987b)
Hat Creek No.2, BC	L, S	32	1.6–1000	Goodarzi and van der Flier-Keller (1988)
Alberta and BC	S	23	10–172	Landheer, Dibbs and Labuda (1982)
'Typical'	S, BI	60	6–285	Jervis, Ho and Tiefenbach (1982)
Nova Scotia	BI	186	9–254	Hawley (1955)
Nova Scotia and New Brunswick	BI	7	51–526	Landheer, Dibbs and Labuda (1982)
Fording Mine, BC	BI	22	2–293	Goodarzi (1988)
CHINA:				
'Selection'	–	15	5.0–177	Sun and Jervis (1986)
COLOMBIA:				
3 regions	BI	14	2–20	Rincon et al. (1978)

Table 5.19 (*contd*)

Sample details	Rank	Samples	Content	Reference
CZECHOSLOVAKIA:				
	–	9	70–970	Valeska and Havlova (1959)
GERMANY WEST:				
From 1 power station	BR	7	42–68	Heinrichs (1982)
	BI	2	55, 68	Brumsack, Heinrichs and Lange (1984)
For power production	BI	3	82–125	Sabbioni *et al.* (1983)
NEW ZEALAND:				
Mokau area	S	–	31–133	Gainsford (1985)
Waikato area	S	–	11–146	Lynskey, Gainsford and Hunt (1984)
NIGERIA:				
'Major deposit'	L	1	2.2	Hannan *et al.* (1982)
'Major deposit'	S	6	6.6–193	Hannan *et al.* (1982)
'Major deposit'	BI	1	25	Hannan *et al.* (1982)
Enugu, 3 areas	S	3	8.0–8.8	Ndiokwere, Guinn and Burtner (1983)
POLAND:				
Belchatow mine	BR	3	62–143	Tomza (1987)
7 mines	BI	31	13–650	Tomza (1987)
SOUTH AFRICA:				
Transvaal, Natal, OFS	BI	33	11–129	Kunstmann and van Rensburg (1967)
	BI	11	24–272	Watling, Watling and Wardale (1976)
SWITZERLAND:				
	BR, L	9	15–144	Hügi *et al.* (1989)
	A	9	3–234	Hügi *et al.* (1989)
THAILAND:				
1 deposit	L	3	70–100	CSIRO (unpublished)
TURKEY:				
5 'kinds'	L	5	24–85	Akcetin, Ayca and Hoste (1973)
USSR:				
Moscow Basin	BR	3	10–13	Laktionova *et al.* (1987)
South Yakutia	–	–	94–298	Egorov, Laktionova and Borts (1981)
Pechora	–	4	36–151	Blomqvist (1983)
UK:				
Midlands	BI	84	10–230	Thorne *et al.* (1983)
N. England, S. Wales	BI	26	1–600	D.J. Swaine (unpublished)
23 collieries	BI	23	11–250	Hislop *et al.* (1978)
S. Wales (12 seams)	BI	300	–2790(62)	Chatterjee and Pooley (1977)
USA:				
Northern Great Plains	L	226	7.3–660	Zubovic, Hatch and Medlin (1979)
Gulf Province	L	34	7.4–690	Swanson *et al.* (1976)
Texas	L	132	1–1075	White, Edwards and Du Bose (1983)

Table 5.19 (*contd*)

Sample details	Rank	Samples	Content	Reference
Fort Union region	L	80	<20–118	Hatch and Swanson (1976)
Northern Great Plains	S	855	1.4–450	Zubovic, Hatch and Medlin (1979)
Rocky Mt Province	S	466	1.4–3500	Zubovic, Hatch and Medlin (1979)
Appalachian Region	BI	1478	0.75–1400	Zubovic, Hatch and Medlin (1979)
Interior Province	BI	687	1.4–1100	Zubovic, Hatch and Medlin (1979)
Rocky Mt Province	BI	366	0.90–590	Zubovic, Hatch and Medlin (1979)
Alaska, N. Slope	L-BI	84	<5–170	Affolter and Stricker (1987)
Alaska	L-BI	–	1–290	Merritt (1988)
Pennsylvania	A	53	<5–210	Swanson et al. (1976)

from the Illinois Basin and a sample of sphalerite from the same area was analysed at CSIRO (North Ryde) and found to contain 2.6 ppm Hg (W.R. Ryall, 1979, pers. commun.). Tkach (1975) found Hg associated with epigenetic pyrite in Donbas coals and Finkelman et al. (1979) found a good correlation between Hg and pyritic S. Porritt and Swaine (1976) and Ruch, Gluskoter and Kennedy (1971) found incomplete evidence for an Hg–pyrite association. It seems likely that Hg and HgS are mainly found in the Soviet coals near Hg-mineralization areas (Dvornikov, 1981b), where coals have up to 20 ppm Hg (Dvornikov, 1967d). This means that Hg in other coals is probably associated with pyrite and sometimes sphalerite, with organically bound Hg still an uncertainty.

Although Hg is lost during normal ashing procedures, it may be retained by the LTA method using radiofrequency heating, the so-called oxygen–plasma method (Doolan et al., 1984a). It should not be assumed that LTA will suit all coals and that every set-up will give conditions that favour the retention of Hg. Special care must be taken because of the volatility of Hg and the low concentrations in most coals demand the avoidance of contamination. Determinations of Hg can be done by NAA, AAS and spectrophotometry, but the usual method is by a flameless cold-vapour AAS technique. Although very precise results have been obtained by special NAA methods, for example by Ruch, Gluskoter and Shimp (1974) and by Porritt and Swaine (1976), these methods are not recommended for general use as they require special equipment and skill only found in certain laboratories. There is a standard method (ASTM, 1983c) in which the sample of coal is combusted in an oxygen bomb containing dilute HNO_3 to absorb Hg. The bomb washings are treated with potassium permanganate solution, followed by hydroxylamine hydrochloride solution and then stannous chloride solution as the reducing agent. The measurement is made by a flameless cold-vapour AAS technique, the line at 253.7 nm being commonly used. Safety and other precautions are given in ASTM (1983c). Among the many variants of this approach, two use

Table 5.20 Results for mercury (as ppm Hg)

Sample details	Rank	Samples	Content	Reference
AUSTRALIA:				
Latrobe Valley, Vic.	BR	2	0.048, 0.071	Porritt and Swaine (1976)
Latrobe Valley, Vic.	BR	20	<0.01–1.3	Bone and Schaap (1981a)
Latrobe Valley, Vic.	BR	15	<0.01–0.07	Bolger (1989)
Leigh Creek, SA	H-B	2	0.12, 0.27	Knott and Warbrooke (1983)
Northern area, NSW	BI	15	0.026–0.26	Porritt and Swaine (1976)
Southern area, NSW	BI	5	0.073–0.17	Porritt and Swaine (1976)
Western area, NSW	BI	55	0.015–0.07	Swaine, Godbeer and Morgan (1984b)
NSW and Qnsld	BI	65	0.01–0.34	Knott and Warbrooke (1983)
Bowen Basin, Qnsld	BI	4	0.043–0.40	Porritt and Swaine (1976)
Fingal, Tasmania	BI	2	0.02, 0.06	Knott and Warbrooke (1983)
BELGIUM:				
From 12 mines	BI, A	48	0.2–2.0	Block and Dams (1975)
BULGARIA:				
'Some coals'	–	–	(0.25)	Kostadinov and Djingova (1980)
CANADA:				
Saskatchewan	L	9	0.04–0.08	Faurschou, Bonnell and Janke (1982)
Alberta	S	39	<0.02–0.44	Faurschou, Bonnell and Janke (1982)
New Brunswick	BI	56	0.022–1.3	Faurschou, Bonnell and Janke (1982)
Nova Scotia	BI	182	0.03–0.90	Faurschou, Bonnell and Janke (1982)
FRANCE:				
2 areas	–	2	0.19, 0.28	Sabbioni *et al.* (1983)
GERMANY EAST:				
	L	–	0.16–1.5	Rösler *et al.* (1977)
GERMANY WEST:				
From 1 power station	BR	–	0.071–0.16	Heinrichs (1982)
Mostly Ruhr	BI	27	<0.7–1.4	Kautz, Kirsch and Laufhutte (1975)
Various	BI	10	0.10–0.98	Heinrichs (1975b)
For power production	BI	4	0.1–1.0	Sabbioni *et al.* (1983)
NEW ZEALAND:				
Waikato area	S	–	0.02–0.24	Purchase (1985)
Southland	S	–	0.19–0.56	Purchase (1985)
Westland	S-BI	–	0.2–0.24	Purchase (1985)
POLAND:				
For power production	BI	2	(0.15)	Sabbioni *et al.* (1983)
SOUTH AFRICA:				
	BI	11	0.08–7.0	Watling and Watling (1976)
USSR:				
Donbas	BI	3000	0.70–1.3	Karasik *et al.* (1967)
UK:				
S. Yorkshire	S	8	0.29–0.46	Ward, Kerr and Otsuka (1986)
20 areas	BI	232	0.03–2.0	Taylor (1973)

Table 5.20 (contd)

Sample details	Rank	Samples	Content	Reference
USA:				
Northern Great Plains	L	226	0.01–12	Zubovic, Hatch and Medlin (1979)
Gulf Province	L	34	0.03–1.0	Swanson et al. (1976)
Texas	L	67	<0.1–1.5	White, Edwards and Du Bose (1983)
Fort Union region	L	80	<0.01–0.60	Hatch and Swanson (1976)
Northern Great Plains	S	855	0.01–3.8	Zubovic, Hatch and Medlin (1979)
Rocky Mt Province	S	466	0.01–8.0	Zubovic, Hatch and Medlin (1979)
Appalachian Region	BI	1478	0.01–3.2	Zubovic, Hatch and Medlin (1979)
Virginia	BI	83	0.01–0.50	Henderson et al. (1985)
Interior Province	BI	687	0.01–1.5	Zubovic, Hatch and Medlin (1979)
Rocky Mt Province	BI	366	0.01–0.90	Zubovic, Hatch and Medlin (1979)
Alaska, N. Slope	L-BI	84	0.02–0.40	Affolter and Stricker (1987)
Pennsylvania	A	53	0.03–1.3	Swanson et al. (1976)

different methods of releasing Hg from the coal sample, namely, high-temperature combustion at 1350 °C in a tube furnace (Doolan, 1982) and non-oxidative pyrolysis in nitrogen at 800 °C (Ebdon, Wilkinson and Jackson, 1982).

Results for Hg in coal are given in Table 5.20. Although there are results above 1 ppm Hg, these can be regarded as unusual, except perhaps for certain areas. Most coals should be in the range of 0.02–1.0 ppm Hg, while most Australian coals are in the range 0.01–0.25, with a mean of about 0.10 ppm Hg. The mean of coals burnt in power stations in nine EC countries is 0.30 ppm Hg. The high results and mean value of 3.3 ppm Hg obtained for Hg in some US coals (Joensuu, 1971) are not regarded as representative and even the mean value of 1 ppm Hg used to calculate the global release of Hg from coal burning is much too high. It is a pity that this high estimate is still quoted in papers dealing with the cycling of Hg. There are important natural sources of Hg which contribute to the atmosphere, for example volcanoes (Eshleman, Siegel and Siegel, 1971; Cadle et al., 1973; Varekamp and Buseck, 1986), volatilization from vascular plants (Kama and Siegel, 1980) and degassing from land and water surfaces. There is a general account of Hg in the environment (USGS, 1970), but the coal values should be treated as exceptional because they refer to those areas in the Donets Basin, USSR, where mineralization has caused unusually high contents in some coals. Apart from coal, there are several industrial sources of Hg (Airey, 1982), including cement-making (Fukuzaki et al., 1986) and cremation (Swaine, 1984a). Takizawa, Minagawa and Fujii (1981) investigated the regional distribution of Hg in Japan and found that 'total mercury in air was considerably greater in the volcanic and hot spring regions'. The notion that all the Hg in coal is emitted to atmosphere during coal burning was dispelled by Diehl et al. (1972) and Swaine (1977a), who found varying amounts of Hg in flyash. Since modern power stations aim at removing at least 99% of flyash, it is clear that any calculation of Hg release from coal must take into consideration Hg in flyash. The often naïve assumptions about releases of trace elements from coal burning are usually based on the incorrect use of data and a poor understanding of operational matters. A prime example of such a situation is given by Hg.

5.2.25 Molybdenum

The interest in Mo stems from its importance biologically and environmentally, for example it is essential for plant growth, but excess may harm animals. Jorissen (1896, 1913) detected small amounts of Mo in ash and soot of coal from Liege, Belgium, Thilo (1934) found about 50 ppm Mo in two German coal ashes and ter Meulen (1932) found 0.21 ppm Mo in a coal sample. Goldschmidt (1937) reported a maximum of 500 ppm Mo in some coal ashes, but there seemed to be little interest in Mo in coal during the following 20 years. The mode of occurrence of Mo in coals ranges from mostly inorganic to mostly organic, with varying proportions of each. An association of Mo with pyrite has been found by Chou (1984) and Finkelman (1981a), while Petrov (1963) found Mo in sulfide minerals in coal from Uzbekistan, USSR, and Ebersbach (1960) concluded that Mo was mainly associated with sulfides in coals from the Menselevitz district, East Germany. Swaine (1975) suggested that under reducing conditions in coal swamps 'MoS_2 might have been precipitated'. Almassy and Szalay (1956) and Szilagyi (1971) favoured organically bound Mo for Hungarian coals. For coals from the Soviet Union, two forms are favoured, namely organically bound and associated with sulfide minerals (Kuznetsova and Saukov, 1961), a view which seems to be in keeping with the position of Mo in the affinity series of Ratynskii and Glushnev (1967a). A similar situation was found by Leventhal, Briggs and Baker (1983) for Chatanooga shale, Central Tennessee, USA, where Mo concentrated in organic- and sulfide-rich units. Korolev (1957) made the interesting suggestion that 'in coals in which molybdenum is not directly connected with organic substances, its accumulation depends to a considerable extent on the intensity factor of the reducing process'. The determination of Mo in coal has been done by AES, AAS, INAA and XRFS, but INAA has limitations and XRFS has a detection limit of 1–2 ppm Mo. For low concentrations it is necessary to use EAAS.

Results for Mo in coal are given in Table 5.21. Most coals would probably have Mo contents in the range 0.1–10 ppm Mo with the exception of some coals from West Germany and Illinois, USA. This is reflected in the mean values of 14 and 8 ppm Mo for West German and Illinois coals, compared with 1–2 ppm Mo for most other coals. Some Canadian coals with very high Mo contents have been omitted from these considerations. Untoward effects from Mo during coal mining and usage are not likely. However, vegetation on areas where flyash is disposed of and on coal mine spoil heaps may produce some problems in grazing

Table 5.21 Results for molybdenum (as ppm Mo)

Sample details	Rank	Samples	Content	Reference
ANTARCTICA:				
Thermally altered	BI	7	<0.4–4	Brown and Taylor (1961)
AUSTRALIA:				
Latrobe Valley, Vic.	BR	16	<0.02–0.9	Brown and Swaine (1964)
Latrobe Valley, Vic.	BR	20	<0.2–0.4	Bone and Schaap (1981a)
Latrobe Valley, Vic.	BR	15	0.1–0.3	Bolger (1989)
St Vincent Basin, SA	L	3	0.03–0.4	CSIRO (unpublished)
Esperance, WA	L	3	0.07–0.5	CSIRO (unpublished)
Leigh Creek, SA	H-B	80	<0.6–20	CSIRO (1966)
Collie, WA	S	22	<1–2	Davy and Wilson (1984)
Northern area, NSW	BI	1000	0.2–10	Clark and Swaine (1962a)
Southern area, NSW	BI	800	0.2–5	Clark and Swaine (1962a)
Western area, NSW	BI	150	0.4–7	Swaine, Godbeer and Morgan (1984b)
NSW and Qnsld	BI	65	0.5–7	Knott and Warbrooke (1983)
Bowen Basin, Qnsld	BI	200	0.1–15	Brown and Swaine (1964)
West Moreton area, Qnsld	BI	150	0.1–10	Brown and Swaine (1964)
Callide area, Qnsld	BI	100	0.1–2.5	Brown and Swaine (1964)
Baralaba, Qnsld	S-A	20	0.2–1.5	CSIRO (unpublished)
BELGIUM:				
From 12 mines	BI, A	48	0.3–4.0	Block and Dams (1975)
BULGARIA:				
Different mines	L-BI	7	1.1–10	Bekyarova and Rouschev (1971)
CANADA:				
Moose River Basin	–	17	5.0–15	Van der Flier-Keller and Fyfe (1987b)
Hat Creek No.1, BC	L, S	14	<0.08–5.2	Goodarzi (1987b)
Hat Creek No.2, BC	L, S	32	1.4–8.8	Goodarzi and van der Flier-Keller (1988)
Alberta and BC	S	–	1.9–106	Van Voris et al. (1985)
Nova Scotia	BI	186	2–12	Hawley (1955)
Nova Scotia and New Brunswick	BI	–	99–210	Van Voris et al. (1985)
Fording Mine, BC	BI	22	0.4–6.0	Goodarzi (1988)
COLOMBIA:				
3 regions	BI	14	<1–7	Rincon et al. (1978)
FRANCE:				
2 areas	–	2	1.8, 2.9	Sabbioni et al. (1983)
GERMANY WEST:				
From 1 power station	BR	–	0.7–1.8	Heinrichs (1982)
Mostly Ruhr	BI	27	6–30	Kautz, Kirsch and Laufhutte (1975)
NEW ZEALAND:				
Waikato area	S	–	<0.01–0.7	Purchase (1985)
Southland	S	–	0.1–0.3	Purchase (1985)
Westland	S-BI	–	0.02–4.0	Purchase (1985)
POLAND:				
Belchatow mine	BR	3	1.0–2.9	Tomza (1987)
7 mines	BI	31	1–11	Tomza (1987)

Table 5.21 (*contd*)

Sample details	Rank	Samples	Content	Reference
SOUTH AFRICA:				
	BI	14	<1.0–2.7	Willis (1983)
SWITZERLAND:				
	BR, L	9	2–200	Hügi *et al.* (1989)
	A	9	0.9–20	Hügi *et al.* (1989)
THAILAND:				
1 deposit	L	3	0.5–0.8	CSIRO (unpublished)
USSR:				
Unspecified deposit	L	–	88–985	Ratynskii, Shpirt and Krasnobaeva (1980)
UK:				
S. Yorkshire	S	8	0.49–1.0	Ward, Kerr and Otsuka (1986)
N. England, S. Wales	BI	26	0.1–20	D.J. Swaine (unpublished)
23 collieries	BI	23	<1.0–4.3	Hislop *et al.* (1978)
'Main fields'	BI	15	1.1–6.1	Sabbioni *et al.* (1983)
S. Wales (12 seams)	BI	300	–409(13)	Chatterjee and Pooley (1977)
USA:				
Northern Great Plains	L	226	0.35–280	Zubovic, Hatch and Medlin (1979)
Gulf Province	L	34	0.5–10.0	Swanson *et al.* (1976)
Texas	L	79	0.5–23	White, Edwards and Du Bose (1983)
Fort Union region	L	80	<0.3–10.0	Hatch and Swanson (1976)
Northern Great Plains	S	855	0.13–41	Zubovic, Hatch and Medlin (1979)
Rocky Mt Province	S	466	0.19–19	Zubovic, Hatch and Medlin (1979)
Appalachian Region	BI	1478	0.18–29	Zubovic, Hatch and Medlin (1979)
Interior Province	BI	687	0.37–128	Zubovic, Hatch and Medlin (1979)
Rocky Mt Province	BI	366	0.24–57	Zubovic, Hatch and Medlin (1979)
Alaska, N. Slope	L-BI	84	<0.1–3.0	Affolter and Stricker (1987)
Alaska	L-BI	–	0.2–15	Merritt (1988)
Pennsylvania	A	53	0.5–15	Swanson *et al.* (1976)

animals. A survey of eight coal mines in the Northern Great Plains province, USA, showed that some vegetation could cause molybdenosis in sheep and cattle (Erdman, Ebens and Case, 1978).

5.2.26 Nickel

There are good biological and environmental reasons for knowing about Ni generally and in coal in particular. It may be useful in nutrition and it may have some undesirable health effects if present in excess, although Underwood (1977) classed it as 'a relatively non-toxic element'. The early discovery of millerite in coal seams (Miller, 1842; Des Cloizeaux, 1880, and others) showed the presence of Ni in coals. There is a comprehensive review of Ni in coals and flyash (Swaine, 1980b) which covers much more than can be dealt with here. Some reports of very high Ni in certain coal ashes are interesting, namely Goldschmidt (1937), Jones and Miller (1939) and Vorobev (1940), who found up to 8000, 80 000 and 10 000 ppm Ni, respectively. The mode of occurrence of Ni in coal is summarized well by Finkelman (1981a) who concluded that the main modes of occurrence are probably associated with sulfides and organically associated. He later added clays (Finkelman, 1988). As well as Ni derived from vegetation and percolating waters from nearby rocks (Swaine, 1961), there could have been some accretion from volcanic emanations and meteoritic particles. Finkelman and Stanton (1978) identified meteoritic ablation products or cosmic spheres in some US coals. This is an interesting, albeit minor, source of Ni in coal.

The determination of Ni in coal has been done by AES, AAS, XRFS and INAA, although the latter requires special techniques. There are two standard methods (ASTM, 1983a; AS, 1986), which use FAAS with an air–acetylene flame with background correction for the line at 232.0 nm (ASTM, 1983a) and 352.5 nm (AS, 1986). The pretreatment of the coal has been described under Be (Section 5.2.4). Langmyhr and Aadalen (1980) described an EAAS method using powdered samples of coal.

Results for Ni in coal are given in Table 5.22. Apart from exceptional very high values, the range for most coals is around 0.5–50, with a mean of 15–20 ppm Ni. Higher mean values are reported for West German coals (43 ppm Ni) and for the coals used for power production in nine EC countries (41 ppm Ni).

Among the natural sources of Ni for the atmosphere are volcanoes (Cadle *et al.*, 1973) and windblown dusts, each supplying several times that emitted by coal combustion, which constitutes about 1.5% of the global emission of Ni to the atmosphere (Bennett, 1982). Under some conditions, certain plants may be harmed by relatively high levels of Ni, but this is unlikely to occur on coal mine spoil heaps. In general, it seems reasonable to assume that there would not be untoward effects from Ni in coal during mining and usage.

5.2.27 Niobium

There is little interest in Nb, as it has no known effects in the environment, and apart from data on US coals there are few references to Nb in coal. The earliest reference seems to be Headlee and Hunter (1955b) who found <50–85 ppm Nb (called Cb in their paper) in ashes of coals from West Virginia, USA. Ratynskii and Glushnev (1967a) placed Nb in the middle of their affinity series, thereby indicating that it is associated organically and with mineral matter in Soviet coals. Finkelman (1981a) suggested that most of the Nb in coal from Waynesburg, USA, was probably associated with rutile. It is not possible to suggest a mode of occurrence for Nb in coals generally.

Most of the results for Nb in coal have been done by the US Geological Survey, using AES, while some other results were from XRFS. The few results that are available will be given below (as ppm Nb), main details of

Table 5.22 Results for nickel (as ppm Ni)

Sample details	Rank	Samples	Content	Reference
ANTARCTICA:				
Thermally altered	BI	7	4–96	Brown and Taylor (1961)
AUSTRALIA:				
Latrobe Valley, Vic.	BR	16	0.2–8	Brown and Swaine (1964)
Latrobe Valley, Vic.	BR	20	0.1–6.8	Bone and Schaap (1981a)
Latrobe Valley, Vic.	BR	15	0.5–18	Bolger (1989)
St Vincent Basin, SA	L	3	2–4	CSIRO (unpublished)
Esperance, WA	L	3	<0.8	CSIRO (unpublished)
Leigh Creek, SA	H-B	80	4–60	CSIRO (1966)
Collie, WA	S	22	2–22	Davy and Wilson (1984)
Northern area, NSW	BI	1000	0.6–100	Clark and Swaine (1962a)
Southern area, NSW	BI	800	<0.4–60	Clark and Swaine (1962a)
Western area, NSW	BI	150	<0.5–50	Swaine, Godbeer and Morgan (1984b)
NSW and Qnsld	BI	65	<1–29	Knott and Warbrooke (1983)
Bowen Basin, Qnsld	BI	200	0.3–90	Brown and Swaine (1964)
West Moreton area, Qnsld	BI	150	<0.8–30	Brown and Swaine (1964)
Callide area, Qnsld	BI	100	0.7–20	Brown and Swaine (1964)
Fingal, Tasmania	BI	2	6, 9	Knott and Warbrooke (1983)
Baralaba, Qnsld	S-A	20	1.5–6	CSIRO (unpublished)
BELGIUM:				
From 12 mines	BI, A	48	10–160	Block and Dams (1975)
BRAZIL:				
Rio Grande do Sul	BI	13	21–124	Bellido and Arezzo (1987)
CANADA:				
Saskatchewan	L	6	4.8–22	Landheer, Dibbs and Labuda (1982)
Moose River Basin	L	17	4.2–191	Van der Flier-Keller and Fyfe (1987b)
Crowsnest and Hat Creek, BC	L, S	9	1.8–64	Nichols and D'Auria (1981)
Alberta and BC	S	15	4.7–342	Landheer, Dibbs and Labuda (1982)
Nova Scotia	BI	186	6–74	Hawley (1955)
CHILE:				
Arauco Basin	BI	399	12–28	Collao et al. (1987)
CHINA:				
From 110 mines	–	–	4–94	Chen et al. (1986)
COLOMBIA:				
3 regions	BI	14	3–20	Rincon et al. (1978)
GERMANY WEST:				
From 1 power station	BR	7	1.2–3.8	Heinrichs (1982)
Mostly Ruhr	BI	27	15–95	Kautz, Kirsch and Laufhutte (1975)
NEW ZEALAND:				
Taranaki area	S	41	1–20	Black (1981)

Table 5.22 (contd)

Sample details	Rank	Samples	Content	Reference
Waikato area	S	–	2.6–38	Purchase (1985)
Southland	S	–	2.5–8.6	Purchase (1985)
Westland	S-BI	–	0.06–28	Purchase (1985)
POLAND:				
Belchatow mine	BR	3	32–50	Tomza (1987)
	BI	–	6–30	Widawska-Kusmierska (1981)
7 mines	BI	31	1.4–108	Tomza (1987)
SOUTH AFRICA:				
	BI	11	11–27	Watling and Watling (1976)
	BI	14	6.9–32	Willis (1983)
SWITZERLAND:				
	BR, L	9	2–40	Hügi et al. (1989)
	A	9	2–30	Hügi et al. (1989)
THAILAND:				
1 deposit	L	3	3–10	CSIRO (unpublished)
USSR:				
Moscow Basin	BR	3	4–5	Laktionova et al. (1987)
South Yakutia	–	–	6.4–36	Egorov, Laktionova and Borts (1981)
UK:				
S. Yorkshire	S	8	0.38–1.1	Ward, Kerr and Otsuka (1986)
N. England, S. Wales	BI	26	3–60	D.J. Swaine (unpublished)
23 collieries	BI	23	8–35	Hislop et al. (1978)
S. Wales (12 seams)	BI	300	−147(12)	Chatterjee and Pooley (1977)
USA:				
Northern Great Plains	L	226	0.52–84	Zubovic, Hatch and Medlin (1979)
Gulf Province	L	34	3–70	Swanson et al. (1976)
Texas	L	136	0.8–79	White, Edwards and Du Bose (1983)
Fort Union region	L	80	<0.7–7.0	Hatch and Swanson (1976)
Northern Great Plains	S	855	0.32–67	Zubovic, Hatch and Medlin (1979)
Rocky Mt Province	S	466	0.45–69	Zubovic, Hatch and Medlin (1979)
Appalachian Region	BI	1478	1.1–220	Zubovic, Hatch and Medlin (1979)
Virginia	BI	83	4.7–31	Henderson et al. (1985)
Interior Province	BI	687	0.87–580	Zubovic, Hatch and Medlin (1979)
Rocky Mt Province	BI	366	0.35–340	Zubovic, Hatch and Medlin (1979)
Alaska, N. Slope	L-BI	84	1–100	Affolter and Stricker (1987)
Pennsylvania	A	53	3–70	Swanson et al. (1976)
YUGOSLAVIA:				
Kosova Basin	BR	2	35, 50	Daci, Berisha and Gashi (1983)

the samples being given in other tables under the relevant references. The coals are bituminous, unless stated otherwise. The results are 1–9 for subbituminous coals from Collie, Western Australia (Davy and Wilson, 1984), <1–10 for NSW and Queensland coals and 6, 16 for Fingal, Tasmania, coals (Knott and Warbrooke, 1983), <1–1 for Colombian coals (Rincon et al., 1978), 0.65 for a mixture of subbituminous coals from New Zealand (Purchase, 1987) and 3.8–18 for South African coals (Willis, 1983). The results for the United States are 1–70 for Gulf Coast lignites (Swanson et al., 1976), 1–80 for Texas lignites (White, Edwards and Du Bose, 1983), <1.0–1.5 for Fort Union lignites (Hatch and Swanson, 1976), <0.1–18 for coals from eastern United States (Zubovic et al., 1980), 0.32–9.2 for Virginia coals (Henderson et al., 1985), a mean of 3.0 for New Mexico coals (Campbell and McCord, 1988), <0.5–15 for Alaskan coals (Affolter and Stricker, 1987) and <0.5–15 for Pennsylvanian anthracites (Swanson et al., 1976). The range for Nb in most coals would probably be 1–20 ppm Nb. Niobium in coal is not expected to produce any harmful effects.

5.2.28 Phosphorus

Because P is often below 1000 ppm in coal it is being regarded as a trace element. There are thousands of determinations of P in coal and coal ash which are carried out because of the importance of P in coke for steel-making. However, most of these results are not published and hence this account is somewhat limited. For some NZ coals, Black (1981) stated that 'phosphorus appears to be associated with the coal substance rather than with the mineral fraction'. Smith (1941a) felt that organic P could only be present in very small amounts in South African coals. Finkelman (1981a) has assessed the situation well and concluded that P in most coals is likely to be present in phosphates, with an uncertain proportion of organic P.

The usual method of determining P is by a spectrophotometric technique utilizing the molybdivanadate complex (ASTM, 1984; AS, 1985), but the molybdenum-blue complex method is also used (BS, 1971), especially for low-P coals where its better sensitivity is an advantage (Durie, Schafer and Swaine, 1965). XRFS, AES and, later, ICPAES are also used.

Results for P in coal are given in Table 5.23. In the absence of sufficient data, it is difficult to assess the range for most coals, but 10–3000 ppm P is perhaps a reasonable estimate. Environmental problems from P in coal are not to be expected, but the levels of P in coke are of prime importance in steel-making.

5.2.29 Platinum group elements

The platinum group elements (PGE) are Ru, Rh, Pd, Os, Ir, Pt. There is an early reference (USGS, 1896) to two Australian coal ashes reported to contain 0.23 and 3.6% Pt. The identity of these coals remains unknown and no further reference to them has been found. They would surely constitute a valuable source of Pt! Inagaki (1951a) detected Pt and Pd in some Japanese coal ashes. There are only two special publications, one on PGE by Finkelman and Aruscavage (1981) and the other on Pt by Chyi (1982).

Table 5.23 Results for phosphorus (as ppm P)

Sample details	Rank	Samples	Content	Reference
ANTARCTICA:				
Thermally altered	BI	7	<1000–11 000	CSIRO (unpublished)
AUSTRALIA:				
Latrobe Valley, Vic.	BR	12	2–20	Brown and Swaine (1964)
Latrobe Valley, Vic.	BR	20	<3–77	Bone and Schaap (1981a)
Leigh Creek, SA	H-B	80	30–3100	CSIRO (1966)
Collie, WA	S	22	45–750	Davy and Wilson (1984)
Northern area, NSW	BI	1000	20–3900	Clark and Swaine (1962a)
Southern area, NSW	BI	800	30–1000	Clark and Swaine (1962a)
Western area, NSW	BI	50	40–550	CSIRO (unpublished)
NSW and Qnsld	BI	65	<10–3040	Knott and Warbrooke (1983)
Bowen Basin, Qnsld	BI	200	10–3500	Brown and Swaine (1964)
West Moreton area, Qnsld	BI	150	20–1800	Brown and Swaine (1964)
Callide area, Qnsld	BI	100	50–130	Brown and Swaine (1964)
Fingal, Tasmania	BI	2	40, 80	Knott and Warbrooke (1983)
Baralaba, Qnsld	S-A	20	360–6000	Taylor (1967)
CANADA:				
Saskatchewan	L	26	60–290	Faurschou, Bonnell and Janke (1982)
Moose River Basin	L	17	<50–1100	Van der Flier-Keller and Fyfe (1987b)
Alberta	S	39	50–880	Faurschou, Bonnell and Janke (1982)
Nova Scotia	BI	140	<20–1410	Faurschou, Bonnell and Janke (1982)
New Brunswick	BI	46	380–2910	Faurschou, Bonnell and Janke (1982)
GERMANY WEST:				
From 1 power station	BR	7	27–52	Heinrichs (1982)
Mostly Ruhr	BI	27	40–1240	Kautz, Kirsch and Laufhutte (1975)
NEW ZEALAND:				
Mokau area	S	–	<15–490	Gainsford (1985)
Waikato area	S	5	2–112	Purchase (1985)
SOUTH AFRICA:				
	–	62	20–2490	FRISA (1960)
SWITZERLAND:				
	BR, L	9	30–3000	Hügi et al. (1989)
	A	6	80–300	Hügi et al. (1989)
UK:				
Midlands	–	84	30–600	Thorne et al. (1983)
N. England, S. Wales	BI	26	<10–400	D.J. Swaine (unpublished)
	BI	–	10–1000	Monkhouse (1950)
USA:				
Illinois Basin	BI	114	10–340	Gluskoter et al. (1977)
	BI	30	20–1430	Abernethy and Gibson (1963)
Virginia	BI	83	<4–6000	Henderson et al. (1985)
Western USA	–	28	10–510	Gluskoter et al. (1977)
Alaska, N. Slope	L-BI	84	<40–2300	Affolter and Stricker (1987)
	L, A	4078	(120)	Zubovic, Hatch and Medlin (1979)

Finkelman and Aruscavage (1981) suggested that PGE in coal probably have a detrital origin. On the basis of a strong correlation between Pt and Ga, Chyi (1982) feels that Pt may occur 'as colloidal inorganic compounds and trapped in coal macerals'. Finkelman and Aruscavage (1981) used an AAS method after preconcentrating the PGE with Au, the PGE being collected in an Au bead by a classical fire-assay technique, whereas Chyi (1982) used a RNAA method with a final removal of Pt on an ion-exchange column, the limit of detection being 1 ppb Pt, which is very similar to Finkelman's limits for Pt, Pd and Rh.

Table 5.24 Results for Platinum group elements (as ppb)

Location	Rank	No.	Rh	Pd	Ir	Pt	Reference
Belgium	BI, A	4			7–9		Block and Dams (1975)
Canada	L	18		0–250	0–0.05	0–80	Van der Flier-Keller and Fyfe (1987b)
Germany	–	–	100–1000	50–1000		100–500	Goldschmidt and Peters (1933b)
Switzerland	BR, L	9			<1–56		Hügi et al. (1989)
	A	9			<2		Hügi et al. (1989)
UK	BI	23		<400	<2–11		Hislop et al. (1978)
USA	–	7	<0.5	<1–3		<2–6	Finkelman and Aruscavage (1981)
USA	BI	25				1.5–210	Chyi (1982)

The few results that have been found are set out in Table 5.24, values being in ppb. No results were found for Ru or Os. In a survey of information on Pd and Os there are no data for coal (Smith, Carson and Ferguson, 1978). Perhaps the only interest in PGE in coal will be geochemical and possibly as indicators of mineralization in nearby areas. There are no anticipated problems from PGE in coal.

5.2.30 Radium

The main interest in Ra in coal is its contribution to the total radioactivity (Chapter 8). By α-particle emission, ^{226}Ra gives ^{222}Rn which is part of the total radon in rocks, soils, water and the atmosphere. The statement by Szalay and Almassy (1956) that radioactivity from Hungarian coal is mostly derived from U is of general application. Hence, the levels of Ra in coal must be related to U and Th from which Ra is formed, and may be expected to be present in U and Th minerals, for example carnotite (Wilson, 1923).

The determination of Ra is carried out by an emanation method. Contents of Ra in coal are very small and are expressed as parts per trillion (ppt), i.e. 1 in 10^{12}. There are a few published results for Ra (as ppt Ra) namely, 0.8–1.3 for three NSW coals (Bayliss and Whaite, 1966), 0.01–0.97

for French coals (Moureu and Lepape, 1914), 0.01–0.1 for Donets Basin, USSR, coals (Burkser, Shapiro and Kondoguri, 1931), 0.5 (average) for 20 Kuznetsk, USSR, coals (Burkser et al., 1934), 0.001–0.067 for UK coals (Drakeley and Smith, 1922), 0.05–0.33 for ten UK coals (Anderson and Turner, 1956), 0.03–0.37 for Alabama, USA, coals (Lloyd and Cunningham, 1913) and 0.26 for a Denver, USA, coal (Wilson, 1923). A range of around 0.01–1 ppt Ra may well represent many coals. Morris and Bobrowski (1979) found that flyash from western US coals was lower in Ra than flyash from eastern coals, an unexpected result, because of the generally higher U in some western coals, but information of the U contents of the relevant coals is needed for a proper explanation.

The very low levels of Ra in coal are unlikely to be harmful. It is interesting to recall the concept of hormesis, that is, the 'phenomenon in living organisms in which exposure to traces or low levels of hazardous physical or chemical agents stimulates the natural physiological defence mechanisms in a manner which benefits health and survival' (Brown, 1987). It is known that small doses of radiation can stimulate plant growth (Ananyan, 1962). Again, concentration must be considered in order to keep a proper perspective.

5.2.31 Rare earth elements

The rare earth elements (REE) are a group of 15 elements with atomic numbers 57–71, i.e. La–Lu. The data are of some interest to geochemists, although they are often just recorded as part of multielement determinations. Rare earth elements in coal have no environmental importance. It seems reasonable to discuss this group together and rather than giving details of all the data, a selection will be made from publications with data for seven or more REE. No data were found for promethium (Pm) or for thulium (Tm). The first determinations of REE in coal were carried out on ash by Goldschmidt and Peters (1933b) and by Thilo (1934), using AES and X-ray methods. It is worth noting that Goldschmidt's studies on trace elements in coal depended on these two methods which had to be developed in his laboratories to suit the analysis of coal. The modern counterparts of these methods are ICPAES and XRFS. The mode of occurrence of REE in most coals seems to be predominantly an association with mineral matter (Palmer and Filby, 1984; Shpirt et al., 1984; Eskenazi, 1987b, 1987c; Goodarzi and van der Flier-Keller, 1988). Ershov (1961) favoured organic bonding for coals from the Kizelovsk Basin, USSR, while Finkelman (1982a) estimated that no more than 10% of the total REE in some US lignites has an organic association. He found monazite and xenotime in lignites, in keeping with their presence in bituminous coals (Finkelman, 1981a). Clays are also likely sites in some coals as suggested by Hart and Leahy (1983) and inferred by Fardy, Hügi and Swaine (1987) and carbonate minerals should not be ruled out. Goldschmidt and Peters (1933b) and others felt that some biological accumulation of REE had occurred during the early stages of coal formation, but this can only be an indication of a mechanism, which does not necessarily mean that the enhanced REE levels remain organically bound during diagenesis.

Table 5.25 Results for rare earth elements (as ppm)

	Australia (1)	Australia (2)	Australia (3)	Belgium (4)	Bulgaria (5)
La	6.0–12	3–18	4.2–24	3–35	2.9–53
Ce	14–28	4–23	9–46	5–55	7.4–82
Pr	1.7–5.3				
Nd	6.3–24		<3.5–23		
Sm	1.7–6.7	0.8–3.4	0.82–4.6	1–6	0.5–19
Eu	0.32–1.9	0.17–0.55	0.17–0.79	0.2–2.0	0.1–1.5
Gd	1.6–6.5				
Tb	0.26–1.4	0.13–0.40	0.11–0.78		0.1–2.1
Dy	1.8–9.6	0.9–2.5	<1.6–3.5	1.0–2.6	
Ho	0.47–2.5		0.07–0.97		
Er	2.3–6.9				
Yb	1.4–7.8	0.6–1.5	0.42–2.5	0.2–3.0	0.3–3.6
Lu		0.12–0.35	0.07–0.50	0.02–3.0	0.04–1.0

	Bulgaria (6)	Canada (7)	Canada (8)	China (9)	Germany (10)
La	0.6–15	3.0–13	2.5–19	0.58–92	0.94–7.2
Ce	1.4–27	1.4–25	6.0–43	3.4–183	3.2–8.2
Pr					
Nd		3.6–13	1.7–22	3.8–85	1.7–14
Sm	0.3–8.1	0.6–2.4	0.4–4.8	0.08–14.4	0.21–0.81
Eu	0.02–0.8	0.1–0.6	0.1–0.9	0.021–2.0	
Gd					0.72–2.7
Tb	0.1–2.1	0.1–0.5		0.07–2.1	
Dy		0.6–16	0.9–2.7		0.21–0.66
Ho		0.2–11	0.1–0.7		
Er					
Yb	0.08–1.7	0.3–1.3	0.6–10	0.046–6.2	0.18–0.38
Lu	0.002–2.0	0.2–0.7	0.1–0.3	0.014–1.0	

Modern instrumental methods, INAA, ICPAES, EAAS, XRFS and SSMS are used for determining REE in coals, and, in particular, SSMS covers the REE well. For example, 12 REE were determined by SSMS in four NSW, Australia, bituminous coals by P.C. Rankin (1977, pers. commun.). Another early method was RNAA which was used by Schofield and Haskin (1964) to determine 13 REE in two US bituminous coals.

Results for REE in coal are given in Table 5.25. There are also data for some REE in other references, namely for eight REE (mean values; Bulgaria: Kostadinov and Djingova, 1980; China: Sun and Jervis, 1986); for seven REE (New South Wales, Australia: Swaine, Godbeer and Morgan, 1984b; Turkey: Akcetin, Ayca and Hoste, 1973); for six REE (Canada: Jervis, Ho and Tiefenbach, 1982; Nigeria: Hannan et al., 1982; Turkey: Ayanoglu and Gunduz, 1978); for five REE (Australia: Knott and Warbrooke, 1983; Poland: Tomza, 1987); for four REE (Brazil: Bellido and Arezzo, 1987; Nigeria: Ndiokwere, Guinn and Burtner, 1983); for

Table 5.25 (contd)

	Switzerland (11)	UK (12)	USA (13)	USA (14)	USA (15)
La	0.7–24	2–38	2.7–20	<2–58	1.0–27
Ce	2.2–38	8–68	4.4–46	0.44–110	5–49
Pr					3.6–9.9
Nd	1.3–15			<1.4–60	2.7–22
Sm	0.29–2.8	0.5–2.5	0.4–3.8	0.4–14	
Eu	0.024–0.70	0.2–2.0	0.1–0.87	<0.1–3.1	<0.1–3.8
Gd					0.66–7.6
Tb	<0.02–0.71	0.17–0.97	0.04–0.65	<0.1–2.0	<0.1–1.0
Dy	<0.07–3.4	0.5–3.1	0.5–3.3		2.4–3.4
Ho					
Er					0.5–2.2
Yb	0.047–2.2	0.5–1.5	0.27–1.5	<0.2–4.8	<0.1–2.2
Lu	0.011–0.42		0.02–0.44	<0.1–0.8	0.05–0.40

(1) NSW, BI, 4 (P.C. Rankin, 1977, pers. commun.).
(2) NSW and Queensland, BI, 7 (Cahill and Mills, 1983).
(3) NSW and Queensland, BI, 40 (Fardy, McOrist and Farrar, 1984).
(4) From 12 mines, BI, A, 48 (Block and Dams, 1975).
(5) 4 deposits, BR, L, 11 (Eskenazi, 1987b).
(6) Pirin deposit, S, 34 (Eskenazi, 1987c).
(7) Fording Mine, BI, 22 (Goodarzi, 1988).
(8) Hat Creek No.2, B.C. L, S, 32 (Goodarzi and van der Flier-Keller, 1988).
(9) From 110 mines (Chen et al., 1986).
(10) From 1 power station, BR, 7 (Heinrichs, 1982).
(11) BR, L, A, 18 (Fardy, Hügi and Swaine, 1987).
(12) 23 collieries, BI, 23 (Hislop et al., 1978).
(13) Illinois, BI, 114 (Gluskoter et al., 1977).
(14) Eastern USA, BI, 644 (Zubovic et al., 1980).
(15) Virginia, BI, 83 (Henderson et al., 1985).

three REE (New Zealand: Purchase, 1985; Texas lignite: White, Edwards and Du Bose, 1983); for two REE (South Africa: Hart and Leahy, 1983; Alaska, USA: Affolter and Stricker, 1987) and for Yb only (US lignite: Hatch and Swanson, 1976; United States: Swanson et al., 1976). Most of the above references have results for La, but there are additional results for La in Brown and Taylor (1961, Antarctica), Brown and Swaine (1964, Australia), Davy and Wilson (1984, Western Australia), Clark and Swaine (1962a, Australia), Bellido and Arezzo (1987, Brazil), Rincon et al. (1978, Colombia), Egorov, Laktionova and Borts (1981, the Soviet Union), Ward, Kerr and Otsuka (1986, the United Kingdom) and Zubovic et al. (1979, the United States). Approximate values (as ppm) for ranges for most coals are 1–40 La, 2–70 Ce, 1–10 Pr, 3–30 Nd, 0.5–6 Sm, 0.1–2 Eu, 0.4–4 Gd, 0.1–1 Tb, 0.5–4 Dy, 0.1–2 Ho, 0.5–3 Er, 0.3–3 Yb and 0.03–1 Lu. A brown coal from Loy Yang, Victoria, had low contents of 12 REE, ranging from <0.005 for Ho to 0.42 for Ce, the total REE content being 0.95 ppm (P.C. Rankin, 1977, pers. commun.).

Geochemical information on the REE is collated by Henderson (1984). Health and environmental effects of REE have not been sufficiently well studied to make definitive statements. Sabbioni, Pietra and Gaglione (1982) refer to a case of pneumoconiosis in a person working with REE dusts, and mention that 'releases of rare-earths to the environment' may be emitted by coal-fired power stations. Of course there will be emissions of REE during coal firing, but the concentrations in deposition from the atmosphere would be too low to be regarded as harmful.

5.2.32 Rhenium

Apart from some investigations in the Soviet Union, there has been a dearth of information on Re in coal. This is in keeping with no known good or bad effects of Re in the environment and with the difficulty of its determination. Shpirt *et al.* (1984) found that Re is mainly in mineral matter, whereas Kuznetsova and Saukov (1961) favoured a Re-organic association through active groups of humic acids and as a finely dispersed sulfide phase (colloidal Re-bearing molybdenum sulfide). This dual mode of occurrence was also put forward by Ratynskii, Shpirt and Krasnobaeva (1980) who found Re related to vitrinite and sulfide, while Maksimova and Shmariovich (1982) suggested a sulfide phase ReS_2 or organic bonding.

There are details of two analytical methods for determining Re in coal, one spectrophotometric (Kuznetsova, 1961) and the other INAA (Burmistrov and Chekanov, 1984). The only results for Re in coal seem to be 0.28–3.0 ppm Re in five Spanish lignites (Gracia and Martin, 1968), <0.03–2.4 ppm Re in 69 Spanish lignites from five locations (Martin and Garcia-Rosell, 1971) and 0.77–20 ppm Re in samples from a lignite deposit in the Soviet Union (Ratynskii, Shpirt and Krasnobaeva, 1980).

5.2.33 Rubidium

There are no reasons to regard Rb as having environmental or health significance, but it has been determined together with other elements of more interest and hence there are some data. The first evidence that Rb is in coal came from the detection of Rb in a flue dust from a Yorkshire coal by Ramage (1927) and it was also detected in ash from Victorian brown coals (Sinnatt and Baragwanath, 1938). Headlee and Hunter (1955b) found 270–500 ppm Rb in ash from coals from West Virginia, USA, and Butler (1953) found 50–500 ppm Rb in ash from Spitsbergen coals. Finkelman (1981a) favours the mode of occurrence for Rb in coal to be illite and mixed-layer clays and Palmer and Filby (1984) stated that Rb is 'clearly associated with the clays'.

The determination of Rb in coal can be done effectively by AES, INAA and XRFS. Results for Rb in coal are given in Table 5.26. The range for most coals is around 2–50 ppm Rb. There are no indications that Rb in coal has any undesirable effects during mining and usage.

5.2.34 Scandium

There is some interest in Sc for making special alloys and solid state devices, but it is not considered to be essential or harmful in biological systems. The first determination of Sc was carried out by Goldschmidt and Peters (1933b) who found 0.1–8 ppm Sc in some brown and bituminous coals. There is only one publication dealing with a special investigation of

Table 5.26 Results for rubidium (as ppm Rb)

Sample details	Rank	Samples	Content	Reference
AUSTRALIA:				
Leigh Creek, SA	H-B	2	2, 15	Knott and Warbrooke (1983)
Collie, WA	S	22	<0.2–3.0	Davy and Wilson (1984)
NSW	BI	5	10–21	K.W. Riley (1980, pers. commun.)
Western area, NSW	BI	11	17–31	R.H. Filby (1984, pers. commun.)
NSW and Qnsld	BI	65	<1–52	Knott and Warbrooke (1983)
Fingal, Tasmania	BI	2	17, 25	Knott and Warbrooke (1983)
BELGIUM:				
From 12 mines	BI, A	48	1.5–110	Block and Dams (1975)
BRAZIL:				
Rio Grande do Sul	BI	13	10–42	Bellido and Arezzo (1987)
BULGARIA:				
'Some coals'	–	–	(20.6)	Kostadinov and Djingova (1980)
CANADA:				
Moose River Basin	L	17	<0.1–20	Van der Flier-Keller and Fyfe (1987b)
Hat Creek No.1, BC	L, S	14	5.5–22	Goodarzi (1987b)
Hat Creek No.2, BC	L, S	32	3.6–15	Goodarzi and van der Flier-Keller (1988)
Crowsnest and Hat Creek BC	L, S	10	4.7–25	Nichols and D'Auria (1981)
'Typical'	S, BI	50	1.3–30	Jervis, Ho and Tiefenbach (1982)
Fording Mine, BC	BI	22	2.3–18	Goodarzi (1988)
CHINA:				
From 110 mines	–	–	1.4–94	Chen et al. (1986)
GERMANY WEST:				
From 1 power station	BR	7	4.0–14	Heinrichs (1982)
	BI	2	14, 17	Brumsack, Heinrichs and Lange (1984)
NEW ZEALAND:				
Waikato area	S	–	0.1–2.4	Purchase (1985)
Southland	S	–	0.07–0.3	Purchase (1985)
Westland	S-BI	–	1.1–5.2	Purchase (1985)
NIGERIA:				
'Major deposit'	L	1	4.5	Hannan et al. (1982)
'Major deposit'	S	6	1.1–870	Hannan et al. (1982)
'Major deposit'	BI	1	41	Hannan et al. (1982)
POLAND:				
Belchatow mine	BR	3	5.2–8.6	Tomza (1987)
SOUTH AFRICA:				
	BI	14	5.0–20	Willis (1983)
UK:				
S. Yorkshire	S	8	13–46	Ward, Kerr and Otsuka (1986)
Durham	BI	12	5.6–33	Smales and Salmon (1955)

Table 5.26 (contd)

Sample details	Rank	Samples	Content	Reference
USA:				
Texas	L	62	<0.1–82	White, Edwards and Du Bose (1983)
Virginia	BI	83	6–115	Henderson et al. (1985)
Illinois Basin	BI	114	2–46	Gluskoter et al. (1977)
Eastern USA	BI	23	9.0–63	Gluskoter et al. (1977)
Kentucky, Tennessee	–	44	25–99	Lyon et al. (1978)
Western USA	–	28	0.30–29	Gluskoter et al. (1977)

Sc and related materials (Swaine, 1964). As a result of studying Australian bituminous coals Swaine (1964) decided that between 40 and 95% of the Sc was organically bound. Singh, Singh and Chandra (1983) felt that Sc had more affinity with the silicates in Ghugus coals from India. Association with mineral matter was the predominant mode of occurrence of Sc in Soviet coals, based on the affinity series of Ratynskii and Glushnev (1967a). For lignites from the Soviet Union, Borisova et al. (1974b) found Sc to be mainly associated with mineral matter, and suggested later (Borisova et al., 1976) that there is more organically bound Sc in brown than in bituminous coals from the Soviet Union. From studies of some US lignites, Given and Miller (1987) concluded that Sc was partly complexed with organic matter. There is agreement that Sc occurs associated with the mineral matter and with the organic matter in varying proportions. Perhaps clays and phosphate minerals are sites for Sc in some coals (Finkelman, 1981a).

The AES, AAS, INAA, XRFS and spectrophotometric methods have been used for determining Sc in coal, but the latter were used mainly in the Soviet Union several years ago (Guseva, 1959; Kaplan and Olshevskaya, 1963; Belopolskii and Popov, 1964). Kulskaya and Vdovenko (1957) and Swaine (1964) used AES methods, and the latter also developed a microspectrographic technique using 1–3 mg samples. Sen Gupta (1982) put forward an EAAS method which depended on prior separation of Sc. Bettinelli, Baroni and Pastorelli (1987) used EAAS and ICPAES techniques for determining Sc in coal ash. Results for Sc in coal are given in Table 5.27. The range for most coals is probably 1–10, with a mean of about 4 ppm Sc. In addition to emissions from coal burning, volcanoes emit some Sc to the atmosphere, but there are no reports of ill-effects on the environment from Sc.

5.2.35 Selenium

From a nutritional and environmental point of view Se is a very important element which has been investigated widely for many years. There are also uses for Se in several industries (Zingaro and Cooper, 1974). The cycling of Se in the environment has been discussed by Swaine (1978c). It seems that Se may have been observed in the fourteenth century when Villanova

Table 5.27 Results for scandium (as ppm Sc)

Sample details	Rank	Samples	Content	Reference
AUSTRALIA:				
Latrobe Valley, Vic.	BR	25	<0.2–2	Swaine (1964)
Latrobe Valley, Vic.	BR	28	0.01–5.5	Fardy, McOrist and Farrar (1984)
St Vincent Basin, SA	L	5	1–7	Fardy, McOrist and Farrar (1984)
Leigh Creek, SA	H-B	80	1.5–9	CSIRO (1966)
Collie, WA	S	5	<0.4–1	CSIRO (unpublished)
Northern area, NSW	BI	1000	<0.3–15	Clark and Swaine (1962a)
Southern area, NSW	BI	800	<0.3–8	Clark and Swaine (1962a)
Western area, NSW	BI	60	1.1–10	Swaine, Godbeer and Morgan (1984b)
NSW and Qnsld	BI	65	1–19	Knott and Warbrooke (1983)
NSW and Qnsld	BI	45	2.4–14	Fardy, McOrist and Farrar (1984)
Bowen Basin, Qnsld	BI	200	1.5–9	Brown and Swaine (1964)
West Moreton area, Qnsld	BI	150	3–30	Brown and Swaine (1964)
Callide area, Qnsld	BI	100	1.5–10	Brown and Swaine (1964)
Fingal, Tasmania	BI	2	3, 10	Knott and Warbrooke (1983)
Baralaba, Qnsld	S-A	20	1.5–5	CSIRO (unpublished)
BELGIUM:				
From 12 mines	BI, A	48	2–12	Block and Dams (1975)
BRAZIL:				
Rio Grande do Sul	BI	13	7.8–20	Bellido and Arezzo (1987)
BULGARIA:				
'Some coals'	–	–	(23)	Kostadinov and Djingova (1980)
CANADA:				
Saskatchewan	L	8	1.3–4.3	Landheer, Dibbs and Labuda (1982)
Hat Creek No.1, BC	L, S	14	1.1–16	Goodarzi (1987b)
Hat Creek No.2, BC	L, S	32	3.4–14	Goodarzi and van der Flier-Keller (1988)
Alberta and BC	S	23	0.92–12	Landheer, Dibbs and Labuda (1982)
'Typical'	S, BI	60	0.5–8.6	Jervis, Ho and Tiefenbach (1982)
Nova Scotia and New Brunswick	BI	7	0.1–11	Landheer, Dibbs and Labuda (1982)
Fording Mine, BC	BI	22	1.8–6.2	Goodarzi (1988)
CHINA:				
'Selection'	–	15	2.3–10	Sun and Jervis (1986)
From 110 mines	–	–	0.12–18	Chen et al. (1986)
COLOMBIA:				
3 regions	BI	14	0.5–20	Rincon et al. (1978)
GERMANY WEST:				
From 1 power station	BR	7	0.30–0.91	Heinrichs (1982)
	BI	2	1.9, 3.6	Brumsack, Heinrichs and Lange (1984)
For power production	BI	5	2.6–5.8	Sabbioni et al. (1983)

Table 5.27 (contd)

Sample details	Rank	Samples	Content	Reference
NEW ZEALAND:				
Mokau area	S	–	0.7–5.1	Gainsford (1985)
Waikato area	S	–	0.2–2.9	Purchase (1985)
Southland	S	–	0.4–1.6	Purchase (1985)
Westland	S-BI	–	<0.09–2.5	Purchase (1985)
NIGERIA:				
'Major deposit'	L	1	4.6	Hannan et al. (1982)
'Major deposit'	S	6	2.0–3.8	Hannan et al. (1982)
'Major deposit'	BI	1	6.5	Hannan et al. (1982)
Enugu, 3 areas	S	3	1.1–1.4	Ndiokwere, Guinn and Burtner (1983)
POLAND:				
Belchatow mine	BR	3	1.7–2.7	Tomza (1987)
7 mines	BI	31	0.14–15	Tomza (1987)
SOUTH AFRICA:				
Witbank coalfield	BI	146	2.7–6.8	Hart and Leahy (1983)
	BI	14	4.8–13	Willis (1983)
SWITZERLAND:				
	BR, L	9	0.24–5.8	Hügi et al. (1989)
	A	9	0.70–8.7	Hügi et al. (1989)
TURKEY:				
5 'kinds'	L	5	0.7–5.8	Akcetin, Ayca and Hoste (1973)
5 zones	L	5	1.2–16.2	Ayanoglu and Gunduz (1978)
UK:				
S. Yorkshire	S	8	3.6–42	Ward, Kerr and Otsuka (1986)
N. England, S. Wales	BI	26	<0.2–15	D.J. Swaine (unpublished)
23 collieries	BI	23	1.3–8	Hislop et al. (1978)
USA:				
Gulf Province	L	34	1–15	Swanson et al. (1976)
Texas	L	85	0.5–12	White, Edwards and Du Bose (1983)
Fort Union region	L	80	<0.5–7.0	Hatch and Swanson (1976)
Western USA	–	28	0.50–4.5	Gluskoter et al. (1977)
Eastern USA	BI	644	0.6–27	Zubovic et al. (1980)
Illinois	BI	114	1.2–7.7	Gluskoter et al. (1977)
Alaska, N. Slope	L-BI	84	<0.3–20	Affolter and Stricker (1987)
Pennsylvania	A	53	0.7–20	Swanson et al. (1976)

noticed a coloured deposit, described as sulphur–rubeum, during the vaporization of S, but it was not discovered until 1818 when Berzelius found Se as an impurity in S (Weeks, 1945). Goldschmidt and Hefter (1933) found Se in an anthracite from Yorkshire, UK. There is a recent report of coal from China which contains 8.4% Se (Yang et al., 1982), surely an extraordinary enrichment of a trace element usually found at levels of 1 ppm Se or less. There are several modes of occurrence of Se in coal. Pyrite is a host for Se, as suggested by several people, and detected at isolated points in grains of pyrite by SEM (Minkin et al., 1984). Finkelman (1981a) discovered PbSe, probably clausthalite, in coals from the Appalachian Basin, USA, and Savelev and Timofeev (1977) found native Se and ferroselite adsorbed on coal (about 50% of the total Se). It has been reported recently (Hayes et al., 1987) that selenite ions can be bonded directly to goethite surfaces. This prompts the suggestion that such a complex could have formed in coal swamps, and that the breakdown of these complexes could have yielded ferroselite or the Se could have remained with the more stable iron oxides found in mature coal. Oman et al. (1988), after studying Se in coal from the Powder River Basin, USA, found less organically bound Se in coals from the southern and central parts of the Basin than in those from the northern part. Organically bound Se is clearly important in many coals. It is concluded that Se occurs in coal organically associated, in pyrite (probably in solid solution), in galena, as PbSe, at least in some coals and possibly in clays. Information on the modes of occurrence of Se in coal is important for a better appreciation of the behaviour of Se in the environment (Coleman, Finkelman and Bragg, 1987).

Early methods for determining Se in coal included RNAA, which was used by Porritt and Swaine (1976). Several other methods have been used, namely XRFS, INAA and AAS (see Chapter 4). Ashing is not advised, although Doolan et al. (1984a) only lost appreciable Se from one of the six coals tested with LTA. At CSIRO (North Ryde) the method of Sanzolone and Chao (1981) was adapted for coals. The coal sample is digested with HNO_3 + $HClO_4$ and then HF. After forming Se(IV), extraction with toluene gives a solution to which nickel nitrate is added prior to the final EAAS measurement. INAA is used at CSIRO (Lucas Heights). Lindahl (1985b) gives details of a hydride generation–FAAS method for determining Se in coal and EAAS methods have been proposed after distillation of Se (Woo et al., 1985) and after coprecipitation with As (Woo et al., 1987).

Results for Se are given in Table 5.28. It is difficult to suggest a range for all coals, but coals from Australia, New Zealand, South Africa, Alaska and the Fort Union region, USA, are covered by the range 0.2–1.6 ppm Se. Mean values (as ppm Se) are 0.5 (South Africa, but too few samples), 0.9 (Australia), 2 (West Germany), 3 (United Kingdom) and 0.5–4 (United States). For the nine countries in the EC, coals used in power production had a mean of 1.7 ppm Se (Sabbioni et al., 1983).

Although coal burning contributes a high proportion of the Se in the atmosphere (Andren, Klein and Talmi, 1975), there are several important natural sources including volcanoes (Cadle et al., 1973; Greenland and Aruscavage, 1986), volatilization from soils and plants (Zieve and Peterson, 1984) and from ocean surfaces (Mosher and Duce, 1983).

Biomethylation of Se is an important mechanism for cycling Se (Cooke and Bruland, 1987). Lag and Steinnes (1978) concluded that 'anthropogenic selenium was not the major source of the atmospheric input and that natural mechanisms must be responsible for this process' in Norway. A feature of the environmental behaviour of Se is that the difference between a concentration causing deficiency and what may be harmful is relatively small. In the past, stress has been on the toxic effects on grazing animals

Table 5.28 Results for selenium (as ppm Se)

Sample details	Rank	Samples	Content	Reference
AUSTRALIA:				
Latrobe Valley, Vic.	BR	2	0.28–0.36	Porritt and Swaine (1976)
Latrobe Valley, Vic.	BR	20	0.2–1.2	Bone and Schaap (1981a)
Latrobe Valley, Vic.	BR	15	<0.02–0.6	Bolger (1989)
Latrobe Valley, Vic.	BR	28	0.34–1.5	Fardy, McOrist and Farrar (1984)
Leigh Creek, SA	H-B	2	<0.5, 1.0	Knott and Warbrooke (1983)
Northern area, NSW	BI	15	0.25–2.5	Porritt and Swaine (1976)
Southern area, NSW	BI	5	0.21–0.63	Porritt and Swaine (1976)
Western area, NSW	BI	46	0.9–2.2	Swaine, Godbeer and Morgan (1984b)
NSW and Qnsld	BI	65	0.4–3.0	Knott and Warbrooke (1983)
NSW and Qnsld	BI	45	0.18–2.6	Fardy, McOrist and Farrar (1984)
Bowen Basin, Qnsld	BI	4	0.25–1.6	Porritt and Swaine (1976)
BELGIUM:				
From 12 mines	BI, A	48	0.4–4.0	Block and Dams (1975)
BULGARIA:				
'Some coals'	–	–	(0.63)	Kostadinov and Djingova (1980)
CANADA:				
Saskatchewan	L	–	3.0–4.0	Van Voris et al. (1985)
Hat Creek No.2, BC	L, S	32	0.5–2.0	Goodarzi and van der Flier-Keller (1988)
Alberta and BC	S	–	2.0–5.0	Van Voris et al. (1985)
Nova Scotia and New Brunswick	BI	–	3.0–11	Van Voris et al. (1985)
Fording Mine, BC	BI	22	0.6–3.2	Goodarzi (1988)
CHILE:				
Arauco Basin	BI	399	0.03–0.13	Collao et al. (1987)
CHINA:				
High-Se area	–	–	5.7–32	Mei (1985)
High-Se area, Enshi	–	10	128–834	Yang et al. (1983)
From '110 mines'	–	–	0.5–13	Chen et al. (1986)
'Selection'	–	15	<0.3–25	Sun and Jervis (1986)
GERMANY WEST:				
From 1 power station	BR	7	0.13–0.38	Heinrichs (1982)
Mostly Ruhr	BI	27	0.6–5.5	Kautz, Kirsch and Laufhutte (1975)
NEW ZEALAND:				
Taranaki area	S	–	0.2–2.0	Purchase (1985)
Waikato area	S	–	<0.2–0.6	Purchase (1985)
Southland	S	–	0.59–0.84	Purchase (1985)
POLAND:				
Belchatow mine	BR	3	1.5–65	Tomza (1987)
7 mines	BI	31	0.44–24	Tomza (1987)
SOUTH AFRICA:				
	BI	14	<0.4–0.9	Willis (1983)
SWITZERLAND:				
	BR, L	8	0.85–7.8	Hügi et al. (1989)
	A	7	0.65–6.3	Hügi et al. (1989)

Table 5.28 (contd)

Sample details	Rank	Samples	Content	Reference
TURKEY:				
5 zones	L	5	0.82–3.1	Ayanoglu and Gunduz (1978)
UK:				
S. Yorkshire	S	8	0.90–9.9	Ward, Kerr and Otsuka (1986)
'Main fields'	BI	15	0.3–5.1	Sabbioni et al. (1983)
USA:				
Northern Great Plains	L	226	0.10–3.4	Zubovic, Hatch and Medlin (1979)
Gulf Province	L	34	1.8–17	Swanson et al. (1976)
Texas	L	47	0.5–18	White, Edwards and Du Bose (1983)
Fort Union region	L	80	0.1–2.0	Hatch and Swanson (1976)
Northern Great Plains	S	855	0.10–16	Zubovic, Hatch and Medlin (1979)
Rocky Mt Province	S	466	0.10–9.6	Zubovic, Hatch and Medlin (1979)
Appalachian Region	BI	1478	0.12–150	Zubovic, Hatch and Medlin (1979)
Interior Province	BI	687	0.02–75	Zubovic, Hatch and Medlin (1979)
Rocky Mt Province	BI	366	0.10–13	Zubovic, Hatch and Medlin (1979)
Alaska, N. Slope	L-BI	84	<0.1–1.2	Affolter and Stricker (1987)
Pennsylvania	A	53	0.6–13	Swanson et al. (1976)

consuming Se-enriched plants in certain limited areas. The endemic Se-poisoning in Enshi county of Hubei province, China, is the outstanding case of severe effects on humans (Yang et al., 1983). The local high-Se coal was regarded as the source of Se for the soil and plants, and the outbreak of selenosis was caused by a failure of the rice crop that led to the eating of high-Se plants and few protein-rich foods (Yang et al., 1983). However, the more important aspect of Se in health is the prevalence of Se deficiency in several animals and possibly in humans. There is a good review of human Se nutrition by Levander (1987) and medical and biological effects are dealt with in another publication with 845 references (NAS, 1976). In China, there are widespread areas where Se-deficiency is considered to be a factor in two diseases, namely Keshan disease (an endemic cardiomyopathy) and Kashin-Beck disease (an endemic osteoarthropathy). These diseases may affect millions of people (Levander, 1987). There is evidence that small amounts of Se are essential for good nutrition in humans. It has been suggested that 'Se reaching soils from coal and oil burning may be beneficial rather than detrimental' (Swaine, 1978c). It is clear that knowledge of Se in coal is relevant to a proper understanding of the geochemical cycle of Se and to avoiding problems that could arise from the disposal of flyash or the use of flyash as a soil amendment. Environmental considerations of Se during coal mining and usage are important.

5.2.36 Silver

Although Ag has no known role in nutrition and there is no environmental interest, there are plenty of results for Ag in coal, always determined incidentally with other elements. The detection of Ag in a flue dust from a Yorkshire coal (Ramage, 1927) provided the first indication that Ag was present in coal. Then Goldschmidt and Peters (1933b) found 0.5–10 ppm Ag in coal ashes. The detection of Ag in the ash of a Yallourn brown coal (Sinnatt and Baragwanath, 1938) was the first finding of Ag in an Australian coal. The first extensive survey of Ag in coal was carried out by Clark and Swaine (1962a) on a wide range of coals. There is an extensive review of all aspects of Ag (Smith and Carson, 1977b). Finkelman (1981a) has discussed Ag thoroughly and proposed that the modes of occurrence are silver sulfide (probably argentite), together with organically bound, adsorbed on coal, in pyrite and other minerals (calcite, siderite, barite, hematite) and native Ag, as suggested by Boyle (1968). The only positive identification was silver sulfide.

The determination of Ag has been carried out by AES, AAS and INAA. Results for Ag in coal are given in Table 5.29 and Krecji-Graf (1983) quotes several results for Ag in coal ash. The range for most coals is probably 0.02–2, with a mean of less than about 0.1 ppm Ag.

5.2.37 Strontium

There is surprisingly little mention of Sr prior to about 1960. Perhaps the earliest report is by Mingaye (1907) who found 10 ppm Sr in a Japanese coal. Some results for Sr in coal ashes are 85–170 (Noll, 1933), 1400 (Thilo, 1934), up to 11 000 (Rafter, 1945) and 500–10 000 ppm Sr (Butler, 1953). The mode of occurrence of Sr in low-rank coals is organically bound, probably through carboxylic acid groups (Brown and Swaine, 1964; Morgan, Jenkins and Walker, 1981; Benson and Holm, 1985). Odor (1967) regards organic bonding as most important. Brown and Swaine (1964) found 0.5% Sr in calcites from Australian bituminous coals. The rare mineral goyazite occurs in some Australian coals (Ward, 1978) and this and other Sr-bearing minerals of the crandallite group have been found in US coals (Finkelman, 1981a; Palmer and Filby, 1983). It is reasonable to suggest that organically bound Sr predominates in most low-rank coals, whereas other coals also contain Sr associated with phosphate minerals and, in some cases, calcite. The determination of Sr can be achieved by AES, AAS, XRFS and a special NAA technique, developed by Leonhardt et al. (1982). One of the earliest applications of FAAS in coal analysis was the determination of Sr in some NSW coal ashes (Belcher and Brooks, 1963). It seems fitting that there is now an Australian standard method (AS, 1986) which uses coal ground so that 98% passes a $-75\,\mu m$ sieve. The 500 °C ash is treated with aqua regia plus HF in a plastic bottle (with a screw cap) on a steam bath. The solution is sprayed into a nitrous oxide–acetylene flame and the absorption is measured at the 460.7 nm line with background correction.

Results for Sr in coal are given in Table 5.30. Most coals would probably be in the range of 15–500, with a mean of about 100 ppm Sr. Higher mean

Table 5.29 Results for silver (as ppm Ag)

Sample details	Rank	Samples	Content	Reference
ANTARCTICA:				
Thermally altered	BI	7	<0.5	Brown and Taylor (1961)
AUSTRALIA:				
Latrobe Valley, Vic.	BR	16	<0.03–0.03	Brown and Swaine (1964)
Latrobe Valley, Vic.	BR	15	<0.01–0.13	Bolger (1989)
St Vincent Basin, SA	L	3	<0.01–1	CSIRO (unpublished)
Leigh Creek, SA	H-B	80	<0.1–0.3	CSIRO (1966)
Northern area, NSW	BI	1000	<0.01–0.6	Clark and Swaine (1962a)
Southern area, NSW	BI	800	<0.03–1	Clark and Swaine (1962a)
Western area, NSW	BI	150	<0.02–0.4	Swaine, Godbeer and Morgan (1984b)
NSW and Qnsld	BI	65	0.02–0.09	Knott and Warbrooke (1983)
Bowen Basin, Qnsld	BI	200	<0.1–0.9	Brown and Swaine (1964)
West Moreton area, Qnsld	BI	150	<0.04–0.6	Brown and Swaine (1964)
Callide area, Qnsld	BI	100	<0.03	Brown and Swaine (1964)
Fingal, Tasmania	BI	2	0.04	Knott and Warbrooke (1983)
Baralaba, Qnsld	S-A	20	<0.1–2.5	CSIRO (unpublished)
BELGIUM:				
From 12 mines	BI, A	48	0.4–2.0	Block and Dams (1975)
CANADA:				
Saskatchewan	L	4	0.2–1.9	Landheer, Dibbs and Labuda (1982)
Hat Creek No.1, BC	L, S	14	<1.3–2.1	Goodarzi (1987b)
Alberta and BC	S	7	0.8–3.3	Landheer, Gibbs and Labuda (1982)
Nova Scotia and New Brunswick	BI	5	0.7–2.3	Landheer, Dibbs and Labuda (1982)
COLOMBIA:				
3 regions	BI	14	<0.1–1.7	Rincon et al. (1978)
GERMANY WEST:				
Mostly Ruhr	BI	27	0.2–1.2	Kautz, Kirsch and Laufhutte (1975)
NEW ZEALAND:				
Mokau area	S	–	<0.02–0.3	Gainsford ((1985)
Waikato area	S	–	0.02–0.1	Purchase (1985)
Southland	S	–	0.009–0.9	Purchase (1985)
Westland	S-BI	–	0.004–0.1	Purchase (1985)
POLAND:				
Belchatow mine	BR	3	0.60–1.5	Tomza (1987)
7 mines	BI	31	1.2–4.9	Tomza (1987)
SWITZERLAND:				
	BR, L	9	<0.005–0.6	Hügi et al. (1989)
	A	9	<0.03–0.6	Hügi et al. (1989)
TURKEY:				
	L	1	5.9	Akcetin, Ayca and Hoste (1973)
5 zones	L	5	0.004–0.068	Ayanoglu and Gunduz (1978)
UK:				
S. Yorkshire	S	8	0.26–0.94	Ward, Kerr and Otsuka (1986)
N. England, S. Wales	BI	26	<0.01–0.09	D.J. Swaine (unpublished)

Table 5.29 (contd)

Sample details	Rank	Samples	Content	Reference
USA:				
Eastern USA	BI	617	0.01–1.4	Zubovic et al. (1979)
Illinois	BI	114	0.02–0.08	Gluskoter et al. (1977)
Western USA	–	28	0.01–0.07	Gluskoter et al. (1977)
Alaska	L-BI	–	0.05–0.5	Merritt (1988)

values have been reported for some low-rank coals and for South African coals. There is no apparent reason to expect any deleterious effects from Sr in coal.

5.2.38 Tantalum

Although there are no known effects of Ta, biologically or environmentally, there are a surprising number of results, mostly from analyses carried out for other elements which were required. Palmer and Filby (1984) suggested an inorganic association for Ta, perhaps with Ti, Zr or phosphate, which is broadly in line with Finkelman (1981a). Determinations of Ta have been carried out by INAA and XRFS, the higher sensitivity of INAA being often advantageous.

Results for Ta are given in Table 5.31. Most coals would be expected to have Ta contents in the range of about 0.1–1 ppm Ta. There is some uniformity about the results, with no outstanding high values. An overall mean value of about 0.3 ppm Ta seems reasonable.

5.2.39 Tellurium

In view of the difficulty of determining Te at sub-ppm levels, which would be expected in coal, it is not surprising that there is a dearth of information. The usual concentration effect that can be gained for many elements by ashing is not possible because of the losses of Te by volatilization. Although Te is one of the few trace elements in coal about which virtually nothing is known, Goldschmidt (1954) made the bland statement that 'tellurium has been reported in small amounts in certain coals'. Using SSMS, Kessler, Sharkey and Friedel (1973) found <0.1–1 ppm Te in 11 bituminous coals from the United States, Chiou and Manuel (1984) found a mean of 0.80 ppm Te in five US coals, using an ion exchange technique to remove interfering elements, prior to the final FAAS measurement. A hydride–FAAS method was used by Nadkarni (1982b) to analyse NBS 1632 (coal reference standard) in which he found 0.50 ppm Te. A single μm-sized Au–Te particle was detected in an Upper Freeport, USA, coal by Finkelman (1981a). It seems that 1 ppm Te is the upper limit for Te in coal and this is below the limit of detection by XRFS.

Volcanic emissions from Kilauea, Hawaii, contain some Te, mostly in the particulate fraction (Greenland and Aruscavage, 1986). In the absence

Table 5.30 Results for strontium (as ppm Sr)

Sample details	Rank	Samples	Content	Reference
ANTARCTICA:				
Thermally altered	BI	7	80–260	Brown and Taylor (1961)
AUSTRALIA:				
Latrobe Valley, Vic.	BR	16	10–800	Brown and Swaine (1964)
Latrobe Valley, Vic.	BR	15	<1–58	Bolger (1989)
Latrobe Valley, Vic.	BR	20	0.5–192	Bone and Schaap (1981a)
Latrobe Valley, Vic.	BR	28	<6–250	Fardy, McOrist, and Farrar (1984)
St Vincent Basin, SA	L	5	400–700	Fardy, McOrist and Farrar (1984)
Leigh Creek, SA	H-B	80	300–1500	CSIRO (1966)
Collie, WA	S	22	15–331	Davy and Wilson (1984)
Northern area, NSW	BI	1000	<20–700	Clark and Swaine (1962a)
Southern area, NSW	BI	800	15–1500	Clark and Swaine (1962a)
Western area, NSW	BI	50	23–60	Swaine, Godbeer and Morgan (1984b)
NSW and Qnsld	BI	65	6–249	Knott and Warbrooke (1983)
NSW and Qnsld	BI	45	34–270	Fardy, McOrist and Farrar (1984)
Bowen Basin, Qnsld	BI	200	20–1000	Brown and Swaine (1964)
West Moreton area, Qnsld	BI	150	40–300	Brown and Swaine (1964)
Callide area, Qnsld	BI	100	40–250	Brown and Swaine (1964)
Fingal, Tasmania	BI	2	113, 238	Knott and Warbrooke (1983)
Baralaba, Qnsld	S-A	20	40–900	CSIRO (unpublished)
BRAZIL:				
Rio Grande do Sul	BI	13	44–206	Bellido and Arezzo (1987)
BULGARIA:				
4 areas	L	126	(95)	Eskenazi and Minceva (1989)
8 areas	BR	122	(132)	Eskenazi and Minceva (1989)
2 areas	BI	31	(43)	Eskenazi and Minceva (1989)
CANADA:				
Saskatchewan	L	8	201–552	Landheer, Dibbs and Labuda (1982)
Moose River Basin	L	17	<1–1000	Van der Flier-Keller and Fyfe (1987b)
Hat Creek No.1, BC	L, S	14	<60–2200	Goodarzi (1987b)
Hat Creek No.2, BC	L, S	32	35–316	Goodarzi and van der Flier-Keller (1988)
Alberta and BC	S	23	23–374	Landheer, Dibbs and Labuda (1982)
'Typical'	S, BI	55	8–735	Jervis, Ho and Tiefenbach (1982)
Nova Scotia	BI	186	76–87	Hawley (1955)
Fording Mine, BC	BI	22	1.4–306	Goodarzi (1988)
CHINA:				
'Selection'	–	15	13–330	Sun and Jervis (1986)
From 110 mines	–	–	27–894	Chen et al. (1986)
COLOMBIA:				
3 regions	BI	14	10–80	Rincon et al. (1978)
GERMANY EAST:				
Niederrhein	BR	62	17–187	Pietzner and Wolf (1964)

Table 5.30 (*contd*)

Sample details	Rank	Samples	Content	Reference
GERMANY WEST:				
From 1 power station	BR	7	28–81	Heinrichs (1982)
Mostly Ruhr	BI	27	15–265	Kautz, Kirsch and Laufhutte (1975)
NEW ZEALAND:				
Taranaki area	S	–	180–1280	Purchase (1985)
Waikato area	S	–	25–430	Purchase (1985)
Southland	S	–	160–300	Purchase (1985)
Westland	S-BI	–	0.7–80	Purchase (1985)
NIGERIA:				
'Major deposit'	L	1	31	Hannan et al. (1982)
'Major deposit'	S	6	1.2–49	Hannan et al. (1982)
'Major deposit'	BI	1	31	Hannan et al. (1982)
POLAND:				
Belchatow mine	BR	3	59–197	Tomza (1987)
7 mines	BI	31	17–450	Tomza (1987)
SOUTH AFRICA:				
	BI	11	<10–740	Watling and Watling (1976)
	BI	14	49–514	Willis (1983)
SWITZERLAND:				
	BR, L	9	94–740	Hügi et al. (1989)
	A	7	<26–1454	Hügi et al. (1989)
THAILAND:				
1 deposit	L	3	200–400	CSIRO (unpublished)
TURKEY:				
5 zones	L	5	44–125	Ayanoglu and Gunduz (1978)
UK:				
S. Yorkshire	S	8	23–116	Ward, Kerr and Otsuka (1986)
N. England, S. Wales	BI	26	<2–500	D.J. Swaine (unpublished)
23 collieries	BI	23	15–280	Hislop et al. (1978)
USA:				
Gulf Province	L	34	7–700	Swanson et al. (1976)
Texas	L	102	20–580	White, Edwards and Du Bose (1983)
Fort Union region	L	80	150–1500	Hatch and Swanson (1976)
Eastern USA	BI	617	3.4–1400	Zubovic et al. (1979)
Illinois	BI	114	10–130	Gluskoter et al. (1977)
Western USA	–	28	93–500	Gluskoter et al. (1977)
Alaska, N. Slope	L-BI	84	15–5000	Affolter and Stricker (1987)
Pennsylvania	A	53	5–700	Swanson et al. (1976)
YUGOSLAVIA:				
Kosova Basin	BR	2	100	Daci, Berisha and Gashi (1983)

Table 5.31 Results for tantalum (as ppm Ta)

Sample details	Rank	Samples	Content	Reference
AUSTRALIA:				
Latrobe Valley, Vic.	BR	28	<0.03–0.73	Fardy, McOrist and Farrar (1984)
St Vincent Basin, SA	L	5	0.18–2.1	Fardy, McOrist and Farrar (1984)
Western area, NSW	BI	11	0.49–0.84	R.H. Filby (1984, pers. commun.)
NSW and Qnsld	BI	7	0.09–0.33	Cahill and Mills (1983)
NSW and Qnsld	BI	45	0.05–0.78	Fardy, McOrist and Farrar (1984)
BRAZIL:				
Rio Grande do Sul	BI	13	0.41–1.1	Bellido and Arezzo (1987)
BULGARIA:				
	–	–	(0.02)	Kostadinov and Djingova (1980)
CANADA:				
Hat Creek No.1, BC	L, S	14	<0.07–0.7	Goodarzi (1987b)
Hat Creek No.2, BC	L, S	32	0.2–0.8	Goodarzi and van der Flier-Keller (1988)
'Typical'	S, BI	41	0.02–1.2	Jervis, Ho and Tiefenbach (1982)
Fording Mine, BC	BI	22	0.1–1.1	Goodarzi (1988)
CHINA:				
From 110 mines	–	–	0.07–4.5	Chen et al. (1986)
'Selection'	–	15	0.19–1.0	Sun and Jervis (1986)
NIGERIA:				
'Major deposit'	L	1	0.82	Hannan et al. (1982)
'Major deposit'	S	6	0.23–0.37	Hannan et al. (1982)
'Major deposit'	BI	1	0.81	Hannan et al. (1982)
PAKISTAN:				
'Local coal'	–	–	(0.29)	Chaudhary et al. (1984)
SOUTH AFRICA:				
Witbank coalfield	BI	146	0.31–1.0	Hart and Leahy (1983)
SWITZERLAND:				
	BR, L	9	<0.02–0.37	Hügi et al. (1989)
	A	9	0.04–1.2	Hügi et al. (1989)
TURKEY:				
5 zones	L	5	0.08–1.0	Ayanoglu and Gunduz (1978)
USA:				
Virginia	BI	83	<0.002–0.44	Henderson et al. (1985)
Illinois Basin	BI	114	0.07–0.3	Gluskoter et al. (1977)
Eastern USA	BI	23	0.12–1.1	Gluskoter et al. (1977)
Western USA	–	28	0.04–0.33	Gluskoter et al. (1977)

of relevant information, no assessment can be made of the effects of Te in the environment, but there is no reason to believe that coal burning can release harmful amounts of Te.

5.2.40 Thallium

There has been some interest in Tl in coal, probably because of the known toxic properties of Tl compounds. There used to be a rat poison containing Tl, which was administered once to remove a recalcitrant husband! The low concentrations in coal are not conducive to causing ill-effects during coal mining or usage. The first indication, albeit indirect, of Tl in coal was its presence in flue dust from a Yorkshire, UK, coal (Ramage, 1927). Up to 5 ppm Tl were found in coal ash by Goldschmidt (1937). The presence of Tl in an Australian bituminous coal was inferred because 200–800 ppm Tl were found in a superheater deposit in a stoker-fired boiler using a coal with <3 ppm Tl (Brown and Swaine, 1964). Studies of Tl in coal and in a host of minerals established that Tl is strongly enriched in pyrite from coals in Middle Asia (Voskresenskaya, 1968). It was felt that Tl was not combined organically because complexes of Tl with humic and fulvic acids are soluble and would not retain Tl during coalification (Voskresenskaya, Timofeyeva and Topkhana, 1962). These workers also found that pyrite with small inclusions of MoS_2 was very high in Tl. Finkelman (1981a) proposed that Tl is present in sulfides, probably mainly pyrite, surely a reasonable suggestion.

The determination of Tl has been carried out by AES, AAS, XRFS and spectrophotometrically. Perhaps an AAS technique is the method of choice. Shan, Yuan and Ni (1986) used a direct determination by EAAS, whereas Berndt et al. (1981) used preconcentration prior to the measurement by EAAS or ICPAES. There is a good review of AAS methods for Tl (Leloux, Lich and Claude, 1987).

Results for Tl are given in Table 5.32. Although there are probably insufficient results to estimate the range, a consideration of mean values and the number of results <0.2 ppm Tl leads to an estimate of about <0.2–1 ppm Tl for most coals. The mean of 1 ppm Tl for the nine countries using coal for power production in the EC (Sabbioni et al., 1983) seems high. Volcanoes emit Tl to the atmosphere, as has been shown by Patterson and Settle (1987). Although Zitko (1975) stated that 'thallium discharged in wastes from mines, ore-processing, and coal-burning plants, is contaminating the environment', it is difficult to justify that coal burning is having deleterious effects. After all, contamination *per se* need not mean serious consequences! More information on the fate of Tl during power production is needed to allay fears that could stem from Zitko's statement. There is a thorough review of all aspects of Tl by Smith and Carson (1977a).

5.2.41 Thorium

Thorium contributes to the radioactivity in coal and is therefore of interest biologically and environmentally. However, there are few references to Th in coal before the early 1970s, the exception being Burkser, Shapiro and

Table 5.32 Results for thallium (as ppm Tl)

Sample details	Rank	Samples	Content	Reference
AUSTRALIA:				
Latrobe Valley, Vic.	BR	16	<0.02	CSIRO (unpublished)
Leigh Creek, SA	H-B	2	0.08, 0.33	Knott and Warbrooke (1983)
Collie, WA	S	5	0.3–1	CSIRO (unpublished)
Western area, NSW	BI	130	<0.6–3	Swaine, Godbeer and Morgan (1984b)
NSW and Qnsld	BI	65	<0.02–0.60	Knott and Warbrooke (1983)
West Moreton area, Qnsld	BI	150	<0.2–2	CSIRO (unpublished)
Fingal, Tasmania	BI	2	0.07, 0.19	Knott and Warbrooke (1983)
GERMANY WEST:				
From 1 power station	BR	7	0.015–0.034	Heinrichs (1982)
Various	BI	10	0.01–0.60	Heinrichs (1975b)
	BI	2	0.51, 0.72	Brumsack, Heinrichs and Lange (1984)
NEW ZEALAND:				
Waikato area	S	5	<0.1–0.1	Lynskey, Gainsford and Hunt (1984)
Westland	S-BI	–	0.6–6.7	Purchase (1985)
SWITZERLAND:				
	BR, L	9	<0.2–15	Hügi et al. (1989)
	A	9	<0.2–2	Hügi et al. (1989)
USSR:				
Middle Asia	–	12	<0.1–2.3	Voskresenskaya, Timofeyeva and Topkhana (1962)
UK:				
20 collieries	BI	20	0.6–1.7	A.B. Nichols (1983, pers. commun.)
USA:				
Virginia	BI	83	0.51–2.8	Henderson et al. (1985)
Eastern Province	BI	600	(0.73)	Magee, Hall and Varga (1973)
Illinois Basin	BI	114	0.12–1.3	Gluskoter et al. (1977)
Interior Province	BI	123	(0.38)	Magee, Hall and Varga (1973)
Western States	–	104	(0.05)	Magee, Hall and Varga (1973)

Kondoguri (1931) who analysed some coals from the Donets Basin, USSR. The value of 100 ppm Th in coal ash (Scott, 1954) may be high because of interference from a Zr spectral line on the line used for Th, namely 283.7 nm. Palmer and Filby (1984) found that Th was associated with mineral matter, notably monazite and zircon in a US coal, while Finkelman (1981a) detected monazite, zircon and occasionally xenotime in several coals. Perhaps a small amount of Th could be associated with iron oxides and clays, as is the case with soils, but organically bound Th is unlikely. Hence, it is clear, not surprisingly, that Th is associated with mineral matter in most coals, perhaps mainly as monazite, with lesser amounts in zircon and xenotime.

The determination of Th in coal is usually done by INAA or XRFS and sometimes by SSMS. Results for Th are given in Table 5.33. Most coals

Table 5.33 Results for thorium (as ppm Th)

Sample details	Rank	Samples	Content	Reference
AUSTRALIA:				
Latrobe Valley, Vic.	BR	5	0.04–3	CSIRO (unpublished)
Latrobe Valley, Vic.	BR	28	<0.01–3.5	Fardy, McOrist and Farrar (1984)
St Vincent Basin	SA	5	2.1–17	
Leigh Creek, SA	H-B	3	1–7	CSIRO (unpublished)
Collie, WA	S	22	<0.4–6	Davy and Wilson (1984)
Northern area, NSW	BI	16	0.5–6	Swaine (1977a)
Southern area, NSW	BI	4	2–8	Swaine (1977a)
Western area, NSW	BI	55	4.5–12	Swaine, Godbeer and Morgan (1984b)
NSW and Qnsld	BI	65	0.9–13	Knott and Warbrooke (1983)
NSW and Qnsld	BI	45	0.57–7.9	Fardy, McOrist and Farrar (1984)
Bowen Basin, Qnsld	BI	11	0.5–8	Swaine (1977a)
West Moreton area, Qnsld	BI	8	1–7	Swaine (1977a)
Callide area, Qnsld	BI	5	<0.2–4	Swaine (1977a)
Fingal, Tasmania	BI	2	8, 19	Knott and Warbrooke (1983)
BRAZIL:				
Rio Grande do Sul	BI	12	3.0–8.9	Bellido and Arezzo (1987)
BELGIUM:				
From 12 mines	BI, A	48	0.5–10	Block and Dams (1975)
BULGARIA:				
'Some coals'	–	–	(2.4)	Kostadinov and Djingova (1980)
CANADA:				
Saskatchewan	L	7	1.0–6.6	Landheer, Dibbs and Labuda (1982)
Moose River Basin	L	17	0.14–6.5	Van der Flier-Keller and Fyfe (1987b)
Hat Creek No.1, BC	L, S	14	<0.1–4.0	Goodarzi (1987b)
Hat Creek No.2, BC	L, S	32	1.0–3.8	Goodarzi and van der Flier-Keller (1988)
Alberta and BC	S	12	1.8–9.0	Landheer, Dibbs and Labuda (1982)
'Typical'	S, BI	55	0.3–14	Jervis, Ho and Tiefenbach (1982)
Nova Scotia and New Brunswick	BI	6	1.9–6.1	Landheer, Dibbs and Labuda (1982)
Fording Mine, BC	BI	22	0.7–3.8	Goodarzi (1988)
CHINA:				
'Selection'	–	15	2.7–14	Sun and Jervis (1986)
From 110 mines	–	–	0.09–25	Chen et al. (1986)
FRANCE:				
For power production	–	1	8.5	Sabbioni et al. (1983)
GERMANY WEST:				
	BI	2	2.3, 4.4	Brumsack, Heinrichs and Lange (1984)
For power production	BI	5	1.6–3.6	Sabbioni et al. (1983)

Table 5.33 (contd)

Sample details	Rank	Samples	Content	Reference
JAPAN:				
	–	10	1.6–4.8	Nakaoka et al. (1982)
NEW ZEALAND:				
Waikato area	S	5	<1.0–1.5	Lynskey, Gainsford and Hunt (1984)
Westland	S-BI	–	0.23–1.2	Purchase (1985)
NIGERIA:				
'Major deposit'	L	1	5.9	Hannan et al. (1982)
'Major deposit'	S	6	1.8–2.2	Hannan et al. (1982)
'Major deposit'	BI	1	6.6	Hannan et al. (1982)
POLAND:				
Belchatow mine	BR	3	1.8–3.0	Tomza (1987)
7 mines	BI	31	0.16–12	Tomza (1987)
SOUTH AFRICA:				
Witbank coalfield	BI	146	2.7–16	Hart and Leahy (1983)
	BI	14	4.0–21	Willis (1983)
SWITZERLAND:				
	BR, L	9	0.21–5.7	Hügi et al. (1989)
	A	9	0.51–5.7	Hügi et al. (1989)
TURKEY:				
5 'kinds'	L	5	0.9–8.2	Akcetin, Ayca and Hoste (1973)
5 zones	L	5	0.29–8.5	Ayanoglu and Gunduz (1978)
USSR:				
Donets Basin	–	–	1–10	Burkser, Shapiro and Kondoguri (1931)
UK:				
23 collieries	BI	23	0.7–6.7	Hislop et al. (1978)
USA:				
Northern Great Plains	L	226	0.28–14	Zubovic, Hatch and Medlin (1979)
Gulf Province	L	34	<3–28	Swanson et al. (1976)
Texas	L	45	0.8–22	White, Edwards and Du Bose (1983)
Fort Union region	L	80	<2–9.4	Hatch and Swanson (1976)
Northern Great Plains	S	855	0.08–42	Zubovic, Hatch and Medlin (1979)
Rocky Mt Province	S	466	0.11–54	Zubovic, Hatch and Medlin (1979)
Virginia	BI	83	0.80–10.0	Henderson et al. (1985)
Interior Province	BI	687	0.04–79	Zubovic, Hatch and Medlin (1979)
Rocky Mt Province	BI	366	0.18–18	Zubovic, Hatch and Medlin (1979)
Alaska, N. Slope	L-BI	84	<0.1–37	Affolter and Stricker (1987)
Pennsylvania	A	53	2.8–14	Swanson et al. (1976)

would probably be in the range 0.5–10 ppm Th, the main exception being some South African coals, where a range of about 2–20 ppm Th is probable. The mean of 4 ppm Th for the coals burnt in nine countries of the EC (Sabbioni et al., 1983) is probably a reasonable mean for most coals. The main interest in Th in coal is its contribution to the total radioactivity (see Chapter 8).

5.2.42 Tin

There are few references to Sn in coal prior to 1960. The first reports of Sn in coal were by Sinnatt and Baragwanath (1938) who detected Sn in the ash of three brown coals from Victoria, Australia, and by Borovik and Ratynskii (1944), who found up to 300 ppm Sn in ashes from coals of the Kuznetsk Basin, USSR. Goldschmidt (1954) reported up to 400 ppm Sn in ash and some other determinations on ash were carried out. The first results for coal seem to be those of Hawley (1955). The mode of occurrence of Sn in most coals is probably in the mineral matter, although an organic association has been reported for some coals. Otte (1953) favoured organic and mineral matter as the forms of Sn in German coals. No organic bonding of Sn in US lignites was found by Given and Miller (1987). A number of Sn-bearing minerals were detected in Waynesburg, USA, coal, including five Sn oxides, two Sn sulfides, one Ni–Sn oxide and one Sn–Fe–Cu sulfide, while occasional particles of Sn–Cu–Pb oxides and Sn in some carbonate and silicate minerals were also seen (Finkelman, 1981a). This is another example of the range of mineral particles that can be found by careful examination using SEM–EDX techniques. Such an array is surely indicative of detrital origin. It is clear that Sn would be associated with mineral matter in most coals.

The determination of Sn has been done mainly by AES and recently by XRFS. Results for Sn in coal are given in Table 5.34. The most extensive survey was done on Australian coals (Clark and Swaine, 1962a). The general range for Sn in most coals is probably 1–10, with a mean of about 2 ppm Sn. Some of the higher values from Canada and Nigeria could possibly be related to local mineralization. It is not expected that coal usage would cause any problems from Sn.

5.2.43 Titanium

Many analyses of coal ash for Ti have been carried out as part of the normal ash analysis and there must be thousands of results reported in company files and the like. Among early reports are those by Wait (1896), Baskerville (1899) and Fieldner, Hall and Field (1918) who found up to 1.6% Ti in coal ash, up to 0.3% Ti in peat ash and up to 2.35% Ti in the ash of US coals. Although several coals have more than 1000 ppm Ti, it was decided on the basis of mean values to include Ti as a trace element, a decision that may not receive general approval! The mode of occurrence of Ti in coal has been discussed widely. Organic bonding has been suggested by McIntyre et al. (1985), Miller and Given (1987b) and Kojima and Furusawa (1986). Eskenazi (1972) had concluded that Ti 'is attached to peat and coals predominantly by an ion-exchange mechanism'. Several

Table 5.34 Results for tin (as ppm Sn)

Sample details	Rank	Samples	Content	Reference
ANTARCTICA:				
Thermally altered	BI	7	<2–5	Brown and Taylor (1961)
AUSTRALIA:				
Latrobe Valley, Vic.	BR	16	<0.8–0.45	Brown and Swaine (1964)
Latrobe Valley, Vic.	BR	20	<0.03–0.45	Bone and Schaap (1981a)
St Vincent Basin, SA	L	3	0.1–3	CSIRO (unpublished)
Esperance, WA	L	3	0.5–1	CSIRO (unpublished)
Leigh Creek, SA	H-B	80	<2–10	CSIRO (1966)
Collie, WA	S	22	<0.2–2	Davy and Wilson (1984)
Northern area, NSW	BI	1000	0.7–30	Clark and Swaine (1962a)
Southern area, NSW	BI	800	<1–5	Clark and Swaine (1962a)
Western area, NSW	BI	150	1–7	Swaine, Godbeer and Morgan (1984b)
NSW and Qnsld	BI	65	<1–32	Knott and Warbrooke (1983)
Bowen Basin, Qnsld	BI	200	0.6–20	Brown and Swaine (1964)
West Moreton area, Qnsld	BI	150	<0.7–8	Brown and Swaine (1964)
Callide area, Qnsld	BI	100	0.9–15	Brown and Swaine (1964)
Fingal, Tasmania	BI	2	3, 4	Knott and Warbrooke (1983)
Baralaba, Qnsld	S-A	20	<1.5–20	CSIRO (unpublished)
CANADA:				
Hat Creek No.1, BC	L, S	4	<80–107	Goodarzi (1987b)
'Typical'	S, BI	25	8–267	Jervis, Ho and Tiefenbach (1982)
Nova Scotia	BI	186	2–15	Hawley (1955)
COLOMBIA:				
3 regions	BI	14	<2–2	Rincon et al. (1978)
GERMANY WEST:				
	BI	2	3.6, 3.9	Brumsack, Heinrichs and Lange (1984)
NEW ZEALAND:				
Taranaki area	S	–	<0.1–1.9	Purchase (1985)
Waikato area	S	–	0.4–17	Purchase (1985)
Southland	S	–	0.6–5.4	Purchase (1985)
Westland	S-BI	–	0.8–7.5	Purchase (1985)
NIGERIA:				
'Major deposit'	L	1	98	Hannan et al. (1982)
'Major deposit'	S	6	35–239	Hannan et al. (1982)
'Major deposit'	BI	1	98	Hannan et al. (1982)
SOUTH AFRICA:				
	BI	14	<1–11	Willis (1983)
SWITZERLAND:				
	BR, L	9	0.09–3	Hügi et al. (1989)
	A	9	0.3–2	Hügi et al. (1989)
THAILAND:				
1 deposit	L	3	15–250	CSIRO (unpublished)
UK:				
S. Yorkshire	S	8	47–98	Ward, Kerr and Otsuka (1986)
N. England, S. Wales	BI	26	0.3–10	D.J. Swaine (unpublished)
20 areas	BI	232	1–75	Taylor (1973)

Table 5.34 (contd)

Sample details	Rank	Samples	Content	Reference
USA:				
Texas	L	43	0.4–68	White, Edwards and Du Bose (1983)
Virginia	BI	83	0.07–3.4	Henderson et al. (1985)
Illinois Basin	BI	114	0.2–51	Gluskoter et al. (1977)
Western USA	–	28	0.10–15	Gluskoter et al. (1977)
Alaska	L-BI	–	0.5–10	Merritt (1988)

Ti minerals were seen by Finkelman (1981a), notably rutile and anatase and an association with clays was found by Minkin, Chao and Thompson (1979) and Steinmetz, Mohan and Zingaro (1988), who also found some Ti associated with quartz. Some doubts about organically bound Ti were cast by Robbat, Finseth and Lett (1984), who proposed an explanation based on highly dispersed Ti minerals in coal. Perhaps this view was prompted by the observations of Wong et al. (1983) who found organo-Ti in addition to anatase. It would seem that the mode of occurrence of Ti in most coals is partly organic, probably dominantly so in low-rank coals, and partly inorganic, especially as rutile, anatase, other Ti-rich minerals and associated with clays.

The determination of Ti in coals can be done by XRFS, AES, INAA, AAS and spectrophotometry. There is a standard method (ASTM, 1983d) for determining Ti in coal ash. The coal is ashed at 750 °C, fused with lithium tetraborate ($Li_2B_4O_7$) followed by dissolution of the melt in dilute HCl. The measurement is made by FAAS using a nitrous oxide–acetylene flame and the line at 364.3 nm with background correction. A more recent standard (AS, 1985) uses digestion with HF + HCl in a PTFE-lined bomb or polypropylene bottle, followed by FAAS.

Results for Ti in coal are given in Table 5.35. Although several fairly high results are reported for some coals in most countries, the general range is probably 10–2000 ppm Ti. No untoward effects have been reported or could be envisaged for Ti in coal.

5.2.44 Tungsten

The earliest reports of W in coal are by Nunn, Lovell and Wright (1953) who found up to 90 ppm W in the ash of anthracites from Pennsylvania, USA, and by Headlee and Hunter (1955b) who found up to 150 ppm W in the ash of coals from West Virginia, USA. The first indication of W in an Australian coal came from some boiler deposits with up to 4000 ppm W, formed from a coal with about 9 ppm W (Brown and Swaine, 1964). There is one special paper on W in coal (Eskenazi, 1982b) and the investigations therein of Bulgarian coals indicated organic-bonding in brown and bituminous coals. It is postulated that W is brought into the coal swamp in freshwater, thermal water and particulate matter, thereby

Table 5.35 Results for titanium (as ppm Ti)

Sample details	Rank	Samples	Content	Reference
ANTARCTICA:				
Thermally altered	BI	7	380–780	CSIRO (unpublished)
ARGENTINA:				
Rio Turbio	–	6	1100–1500	A. Berset (1982, pers. commun.)
AUSTRALIA:				
Yallourn, Vic.	BR	16	6–40	Baragwanath (1962)
Latrobe Valley, Vic.	BR	16	10–320	Brown and Swaine (1964)
Latrobe Valley, Vic.	BR	20	10–1900	Bone and Schaap (1981a)
Latrobe Valley, Vic.	BR	28	<100–2000	Fardy, McOrist and Farrar (1984)
St Vincent Basin, SA	L	5	650–3700	Fardy, McOrist and Farrar (1984)
Esperance, WA	L	3	40–200	CSIRO (unpublished)
Leigh Creek, SA	H-B	80	350–4400	CSIRO (1966)
Collie, WA	S	22	60–1100	Davy and Wilson (1984)
Northern area, NSW	BI	1000	300–2500	Clark and Swaine (1962a)
Southern area, NSW	BI	800	400–1500	Clark and Swaine (1962a)
Western area, NSW	BI	150	250–>2000	Swaine, Godbeer and Morgan (1984b)
NSW and Qnsld	BI	65	400–3400	Knott and Warbrooke (1983)
NSW and Qnsld	BI	45	480–4200	Fardy, McOrist and Farrar (1984)
Bowen Basin, Qnsld	BI	200	200–2000	Brown and Swaine (1964)
West Moreton area, Qnsld	BI	150	740–4200	Brown and Swaine (1964)
Callide area, Qnsld	BI	100	600–2000	Brown and Swaine (1964)
Fingal, Tasmania	BI	2	2010, 2100	Knott and Warbrooke (1983)
CANADA:				
Saskatchewan	L	8	282–891	Landheer, Dibbs and Labuda (1982)
Moose River Basin	L	17	60–5000	Van der Flier-Keller and Fyfe (1987b)
Hat Creek No.1, BC	L, S	14	202–3830	Goodarzi (1987b)
Hat Creek No.2, BC	L. S	32	808–4000	Goodarzi and van der Flier-Keller (1988)
Alberta and BC	S	23	194–3080	Landheer, Dibbs and Labuda (1982)
'Typical'	S, BI	60	50–1950	Jervis, Ho and Tiefenbach (1982)
Nova Scotia and New Brunswick	BI	7	77–1005	Landheer, Dibbs and Labuda (1982)
Fording Mine, BC	BI	22	315–1900	Goodarzi (1988)
CHINA:				
'Selection'	–	15	531–2642	Sun and Jervis (1986)
COLOMBIA:				
3 regions	BI	14	300–3600	Rincon et al. (1978)
CZECHOSLOVAKIA:				
	–	–	960–2360	Valeska and Havlova (1959)
GERMANY EAST:				
Niederrhein	BR	62	6–163	Pietzner and Wolf (1964)

Table 5.35 (*contd*)

Sample details	Rank	Samples	Content	Reference
GERMANY WEST:	BI	2	440, 620	Brumsack, Heinrichs and Lange (1984)
NEW ZEALAND:				
Mokau area	S	–	80–1860	Gainsford (1985)
Waikato area	S	5	48–1240	Lynskey, Gainsford and Hunt (1984)
NIGERIA:				
'Major deposit'	L	1	962	Hannan et al. (1982)
'Major deposit'	S	6	809–1187	Hannan et al. (1982)
'Major deposit'	BI	1	1499	Hannan et al. (1982)
Enugu, 3 areas	S	3	621–635	Ndiokwere, Guinn and Burtner (1983)
POLAND:				
Belchatow mine	BR	3	298–703	Tomza (1987)
7 mines	BI	31	54–2870	Tomza (1987)
SOUTH AFRICA:				
Transvaal, Natal	BI, A	45	700–2500	Kunstmann and Gass (1957)
	BI	11	157–350	Watling and Watling (1976)
	BI	14	65–3800	Willis (1981)
SWITZERLAND:				
	BR, L	9	60–2000	Hügi et al. (1989)
	A	9	120–5450	Hügi et al. (1989)
TURKEY:				
5 'kinds'	L	5	220–1100	Akcetin, Ayca and Hoste (1973)
USSR:				
Central districts	–	–	1000–3000	Golovko (1960)
UK:				
N. England, S. Wales	BI	26	20–1500	D.J. Swaine (unpublished)
23 collieries	BI	23	100–1500	Hislop et al. (1978)
S. Wales (12 seams)	BI	300	–24 120(2500)	Chatterjee and Pooley (1977)
Midlands	BI	84	120–1700	Thorne et al. (1983)
USA:				
Gulf Province	L	34	200–7500	Swanson et al. (1976)
Fort Union region	L	80	100–780	Hatch and Swanson (1976)
S. and E. Arkansas	L	53	310–7500	Hilderbrand, Clardy and Holbrook (1981)
Eastern USA	BI	474	95–1320	Swanson et al. (1976)
Virginia	BI	83	230–3800	Henderson et al. (1985)
Illinois Basin	BI	114	200–1500	Gluskoter et al. (1977)
Western USA	S	183	<15–1200	Swanson et al. (1976)
Alaska, N. Slope	L-BI	84	40–5200	Affolter and Stricker (1987)
Pennsylvania	A	53	140–5300	Swanson et al. (1976)

Table 5.36 Results for tungsten (as ppm W)

Sample details	Rank	Samples	Content	Reference
AUSTRALIA:				
Latrobe Valley, Vic.	BR	28	<0.1–1.2	Fardy, McOrist and Farrar (1984)
St Vincent Basin, SA	L	5	1.1–8.0	Fardy, McOrist and Farrar (1984)
Leigh Creek, SA	H-B	2	3, 5	Knott and Warbrooke (1983)
Northern area, NSW	BI	1000	<0.6–20	Clark and Swaine (1962a)
Southern area, NSW	BI	800	<1–2	Clark and Swaine (1962a)
Western area, NSW	BI	40	0.7–6	Swaine, Godbeer and Morgan (1984b)
NSW and Qnsld	BI	65	0.5–8	Knott and Warbrooke (1983)
NSW and Qnsld	BI	45	0.8–13	Fardy, McOrist and Farrar (1984)
Bowen Basin, Qnsld	BI	200	<4–7	Brown and Swaine (1964)
West Moreton area, Qnsld	BI	150	<4–6	Brown and Swaine (1964)
Callide area, Qnsld	BI	100	1–2	Brown and Swaine (1964)
Fingal, Tasmania	BI	2	<2	Knott and Warbrooke (1983)
BELGIUM:				
From 12 mines	BI, A	48	0.2–1.5	Block and Dams (1975)
BRAZIL:				
Rio Grande do Sul	BI	13	0.33–1.6	Bellido and Arezzo (1987)
BULGARIA:				
14 deposits	L, BI	359	0.01–784	Eskenazi (1982b)
CANADA:				
Moose River Basin	L	17	1.0–3.0	Van der Flier-Keller and Fyfe (1987b)
Hat Creek No.2, BC	L, S	22	0.4–1.1	Goodarzi and van der Flier-Keller (1988)
CHINA:				
'Selection'	–	15	<0.6–2.4	Sun and Jervis (1986)
From 110 mines	–	–	0.22–31	Chen et al. (1986)
GERMANY WEST:				
Mostly Ruhr	BI	27	<1.4–10	Kautz, Kirsch and Laufhutte (1975)
POLAND:				
Belchatow mine	BR	3	0.22–4.1	Tomza (1987)
7 mines	BI	31	0.04–2.4	Tomza (1987)
SOUTH AFRICA:				
	BI	11	78–739	Watling and Watling (1976)
	BI	14	<2.0–8.1	Willis (1983)
SWITZERLAND:				
	BR, L	9	<1.0–9.8	Hügi et al. (1989)
	A	9	1–31	Hügi et al. (1989)
UK:				
S. Yorkshire	S	8	0.33–1.1	Ward, Kerr and Otsuka (1986)
USA:				
Virginia	BI	83	0.009–0.40	Henderson et al. (1985)
Illinois Basin	BI	114	0.04–4.2	Gluskoter et al. (1977)
Western USA	–	28	0.13–3.3	Gluskoter et al. (1977)

170   Contents of trace elements in coals

enabling the organic matter to concentrate W. Any W in the particulate matter and any grains of W minerals (scheelite, wolframite) will dissolve as organic complexes, which then become the source of W for coaly matter. It seems that W occurs in Ghugus coals, India, associated with carbonate minerals (Singh, Singh and Chandra, 1983).

The determination of W in coal is usually done by INAA, but Eskenazi (1982b) used an AES technique with a low limit of detection (1 ppm W in ash, although some values of 0.2 and 0.5 ppm W in ash are given in the statement of results).

Results for W in coals are given in Table 5.36. There are some notably high values for some Bulgarian coals, which may be near areas of W mineralization, although no specific W minerals were found in these coals. Some high-W coals were also reported by Watling and Watling (1976). Most coals are probably in the range 0.5–5 ppm W. No reports of problems from W during coal mining or usage have been found, and there is no reason to expect ill-effects from W from coal burning or other operations.

5.2.45 Uranium

There is a continuing interest in U in coal, because it is a source of radioactivity and because it may be an economic source of U. It is just 200 years since the discovery of U by M.H. Klaproth, although it was not isolated until 1841 by E.M. Peligot (Weeks, 1945). The first detection in coal was by Berthoud (1875) who found up to 2% U in coal from near Denver, USA. Subsequent field studies have proven several areas with high-U coals, especially in the United States, mainly in the Dakotas, Wyoming, Montana, Colorado and New Mexico (Vine, 1956). The volume of published work on U in coal ranks with that on Ge and cannot be dealt with in detail here. Useful information is given by Davidson and Ponsford (1954), Abernethy and Gibson (1963) and Bouska (1981); there is an annotated bibliography by Kehn (1957) and geochemical aspects are discussed by Breger (1958). It seems that U is carried into the coal swamp in solution as carbonate complexes (Breger, Deul and Meyrowitz, 1955), which then release uranyl ions to form uranyl–organic complexes. Manskaya and Drozdova (1968) suggest that 'humic acids, depending on their degree of polymerization and the pH of the solution, may either transport or deposit uranium'. In many coals, especially low-U coals, U is predominantly organically bound (Breger, Deul and Rubinstein, 1955). Studies of a U-rich bog led to the same conclusion, with the proposal that the U is bound by complexation and sorption by functional groups (Idiz, Carlisle and Kaplan, 1986). Adsorption by humic acid-like matter is favoured by Szalay (1964) who later had evidence for the role of cation exchange by insoluble humic acid (Szalay, 1969), a mechanism also favoured by Voskresenskaya (1960) and by Jeczalik (1970). An association between U and clays was found in two Canadian coals (van der Flier and Fyfe, 1985) and Kakimi et al. (1969) proposed that U was probably 'carried mainly by syngenetic coprecipitation with humus and clay fractions', where 'carried' seems to mean 'occurred'. Perhaps a small amount of U could be associated with clay by adsorption. There is evidence for U-bearing minerals in some U-rich coals (Denson, 1959; Breger, 1974), examples being uraninite, coffinite, autunite, torbernite and carnotite. In coals from Northumberland, UK, Asuen (1987) found U associated with carbonate minerals. No enrichment of U in sulfide or sulfate minerals was found in coals from Nova Scotia, Canada (Zodrow and Zentilli, 1979). Finkelman (1981a) studied some low-U coals and found U associated with zircons, REE phosphates and uraninite, and sometimes apatite, rutile and calcite. In some cases, very finely divided U-containing minerals may be intimately associated with the organic coaly matter, making identification difficult. Clearly, the modes of occurrence of U in coal are diverse, but organic bonding seems general, together with associations with mineral matter. Further information is given in Chapter 3.

The determination of U has been carried out by INAA, XRF and by several chemical methods, as mentioned in Chapter 4. Nowadays the two instrumental methods are used widely. Results for U in coal are given in Table 5.37, where the stress is on coals other than U-rich coals. The range for most coals is probably 0.5–10, with a mean of about 2 ppm U, which is also the mean for the nine coals used in the nine EC countries for power production (Sabbioni et al., 1983). The relevance of U in coal to emissions from coal burning is discussed in Chapter 8.

5.2.46 Vanadium

Interest in V in coal hinges on possible corrosion effects during utilization and health effects at high levels. Early reports of over 20% V in some South American coal ashes (Clarke, 1916) have been put into perspective by Hewett (1909) and Hess (1932), who suggested that the coals were really asphaltites. The earliest determination of V reported for a coal was Mingaye (1903) who found 100–700 ppm V, probably in the ash of a coal from near Branxton, NSW, Australia, and later found 15 ppm V in a Japanese coal (Mingaye, 1907). Jorissen (1905) detected V in soot from the combustion of a Belgian coal and Simpson (1912) found <100–1200 ppm V in the ash of coal from Collie, western Australia. An investigation of 500 coals from the Soviet Union showed a wide range of values from 0.01 to 4.9% V in ash (Zilbermintz, 1935). Almassy and Szalay (1956) and Uzunov and Karadzhova (1968) favoured organic bonding for V in Hungarian and Bulgarian coals, while Manskaya and Drozdova (1968) suggested that V may be concentrated in coal in 'the form of complexes with phenolic compounds', but they did not infer the presence of such complexes in any coals. The careful work of Bonnett and Czechowski (1981) did not lead to the identification of V-porphyrins in coal, a finding confirmed by Maylotte et al. (1981), who found that V^{3+} is in octahedral oxygen coordination in a high-V coal (1000–1800 ppm V). This could indicate the presence of roscoelite. Although Bethell (1963) suggested carnotite as a 'most favoured mode of extrinsic combination', this does not seem likely for many coals. From a study of Polish coals Marczak and Lewinska-Ochwat (1987) concluded that 98% of the total V was present in mineral matter. Palmer and Filby (1984) found V to be concentrated in clays, but other associations were not ruled out, at least for the US coal

Table 5.37 Results for uranium (as ppm U)

Sample details	Rank	Samples	Content	Reference
AUSTRALIA:				
Latrobe Valley, Vic.	BR	5	0.01–1.8	CSIRO (unpublished)
Latrobe Valley, Vic.	BR	20	<0.04–1.4	Bone and Schaap (1981a)
St Vincent Basin, SA	L	5	0.43–3.8	Fardy, McOrist and Farrar (1984)
Leigh Creek, SA	H-B	3	0.6–4.3	CSIRO (unpublished)
Collie, WA	S	22	<1	Davy and Wilson (1984)
Northern area, NSW	BI	18	0.5–4.1	Swaine (1977a)
Southern area, NSW	BI	5	1.8–5.0	Swaine (1977a)
Western area, NSW	BI	55	1.3–4.5	Swaine, Godbeer and Morgan (1984b)
NSW and Qnsld	BI	65	0.3–3.0	Knott and Warbrooke (1983)
NSW and Qnsld	BI	45	0.28–2.5	Fardy, McOrist and Farrar (1984)
Bowen Basin, Qnsld	BI	11	1.0–3.6	Swaine (1977a)
West Moreton area Qnsld	BI	8	0.4–3.2	Swaine (1977a)
Callide area, Qnsld	BI	5	0.4–3.0	Swaine (1977a)
Fingal, Tasmania	BI	2	3, 4	Knott and Warbrooke (1983)
BRAZIL:				
Rio Grande do Sul	BI	12	2.7–19	Bellido and Arezzo (1985)
BULGARIA:				
'Some coals'	–	–	(4.0)	Kostadinov and Djingova (1980)
CANADA:				
Saskatchewan	L	8	0.8–2.9	Landheer, Dibbs and Labuda (1982)
Moose River Basin	L	17	0.14–7.2	Van der Flier-Keller and Fyfe (1987b)
Hat Creek No.1, BC	L, S	14	0.2–2.4	Goodarzi (1987b)
Hat Creek No.2, BC	L, S	32	0.7–2.2	Goodarzi and van der Flier-Keller (1988)
Alberta and BC	S	23	0.49–2.0	Landheer, Dibbs and Labuda (1982)
'Typical'	S, BI	60	0.1–4.0	Jervis, Ho and Tiefenbach (1982)
Nova Scotia and New Brunswick	BI	7	<0.1–1.8	Landheer, Dibbs and Labuda (1982)
Fording Mine, BC	BI	22	0.4–2.4	Goodarzi (1988)
CHILE:				
Arauco Basin	BI	399	1.6–4.8	Collao et al. (1987)
CHINA:				
'Selection'	–	15	0.97–97	Sun and Jervis (1986)
From 110 mines	–	–	0.16–21	Chen et al. (1986)
FRANCE:				
For power production	–	1	3.9	Sabbioni et al. (1983)
GERMANY WEST:				
Mostly Ruhr	BI	27	<1–13	Kautz, Kirsch and Laufhutte (1975)
For power production	BI	5	0.9–2.0	Sabbioni et al. (1983)
34 mines	BI	50	0.25–2.2	Riepe (1986)

Table 5.37 (contd)

Sample details	Rank	Samples	Content	Reference
INDIA:				
Different mines	–	14	1.1–3.6	Chakarvarti and Nagpaul (1980)
JAPAN:				
From Ube	–	–	0.8–25	Inagaki and Tokuga (1959)
	–	10	0.48–1.3	Nakaoka et al. (1982)
NEW ZEALAND:				
Taranaki area	S	–	0.38–0.46	Purchase (1985)
Waikato area	S	–	0.015–0.39	Purchase (1985)
Westland	S-BI	–	0.07–0.34	Purchase (1985)
NIGERIA:				
'Major deposit'	L	1	1.3	Hannan et al. (1982)
'Major deposit'	S	6	0.45–0.62	Hannan et al. (1982)
'Major deposit'	BI	1	0.93	Hannan et al. (1982)
POLAND:				
For power production	BI	2	(1.5)	Sabbioni et al. (1983)
ROMANIA:				
Several regions	–	–	17–29	Atanasiu and Soroiu (1960)
SOUTH AFRICA:				
Witbank coalfield	BI	146	1.2–3.7	Hart and Leahy (1983)
	BI	14	3.0–7.3	Willis (1983)
SPAIN:				
5 locations	L	69	<4–75	Martin and Garcia-Rosell (1971)
SWITZERLAND:				
	BR, L	4	1.6–9.2	Hügi et al. (1989)
	A	9	<0.26–8.0	Hügi et al. (1989)
TURKEY:				
5 'kinds'	L	5	1.4–64	Akcetin, Ayca and Hoste (1973)
5 zones	L	5	0.21–3.3	Ayanoglu and Gunduz (1978)
UK:				
15 main fields	BI	15	1.1–3.0	Sabbioni et al. (1983)
USA:				
Northern Great Plains	L	226	0.21–13	Zubovic, Hatch and Medlin (1979)
Gulf Province	L	34	0.5–17	Swanson et al. (1976)
Texas	L	102	0.4–6.2	White, Edwards and Du Bose (1983)
Fort Union region	L	80	<0.1–5.1	Hatch and Swanson (1976)
Northern Great Plains	S	855	0.09–16	Zubovic, Hatch and Medlin (1979)
Rocky Mt Province	S	466	0.06–76	Zubovic, Hatch and Medlin (1979)
Appalachian Region	BI	1478	0.10–19	Zubovic, Hatch and Medlin (1979)
Interior Province	BI	687	0.20–59	Zubovic, Hatch and Medlin (1979)
Rocky Mt Province	BI	366	0.13–42	Zubovic, Hatch and Medlin (1979)
Alaska, N. Slope	L-BI	84	0.3–10	Affolter and Stricker (1987)
Pennsylvania	A	53	0.3–25	Swanson et al. (1976)

they examined. Rao *et al.* (1980) proposed an inorganic association for V in some Indian coals. Finkelman (1981a) found V in clays, especially illite. Although there are conflicting reports, it is felt that there may be a small proportion of organically bound V in some coals, together with inorganic V in clays and other specific minerals.

The determination of V in coal has been achieved by XRFS, AES, AAS and also by INAA and spectrophotometry. There are two standard methods (ASTM, 1983a; AS, 1986) which use FAAS, the conditions being given under Be (Section 5.2.4). The line at 318.0 nm is used with background correction. Results for V in coal are given in Table 5.38. Most coals would be expected to be in the range 2–100 ppm V. The mean value for Australian, South African and US coals is probably about 25 ppm V, whereas that for West German and UK coals is around 60 ppm V. This latter estimate is confirmed by the mean of 65 ppm V for the nine coals burnt in power stations in nine countries of the EC (Sabbioni *et al.*, 1983).

Although V in some coals, burnt under certain conditions may be conducive to causing corrosion, there are no other problems, nutritional or environmental, to be expected from coal mining and usage. The severe health problems that may arise from compounds high in V are not found with coal because of the generally low concentrations.

5.2.47 Yttrium

The earliest determinations of Y in coal are those by Goldschmidt (1937) who found up to 800 ppm Y, Butler (1953) with <30–530 ppm Y and Arnautov and Shipilov (1960) with 20–200 ppm Y, all these values being in ash. The earliest extensive survey of Y in coals was carried out in Australia on a wide range of coals (Clark and Swaine, 1962a). Ershov (1961) favoured the occurrence of Y in Soviet coals as organometallic compounds, and Given and Miller (1987) found Y to be partly complexed with organic matter in US lignites. Partly organic, but mostly inorganic are suggested by Ratynskii and Glushnev (1967a). An association with mineral matter seems most likely, based on studies of various coals by Finkelman (1981a), who identified several grains of xenotime in at least 25% of samples. This seems a good basis for postulating fine-grained xenotime as a prime mode of occurrence of Y in many coals.

The determination of Y may be effected by AES and XRFS. Results for Y in coal are given in Table 5.39. Apart from a few exceptions, such as the high values for two local areas in Texas and Bulgaria, the range for most coals is probably 2–50, with a mean of about 15 ppm Y. There is no apparent reason to expect any untoward effects from Y in coal during mining and usage. It is therefore surprising to see Y listed among 'the elements of chief environmental concern in coal' (Merritt, 1988), although it is listed by PECH (1980) as an element of 'no immediate concern'.

5.2.48 Zinc

Its essentiality in nutrition makes Zn an important element biologically and environmentally, but, as with many elements, excess can be detrimental. There are reports of Zn being detected in UK coal seams by

Table 5.38 Results for vanadium (as ppm V)

Sample details	Rank	Samples	Content	Reference
ANTARCTICA:				
Thermally altered	BI	7	8–320	Brown and Taylor (1961)
ARGENTINA:				
Rio Turbio	–	6	30–45	A. Berset (1982, pers. commun.)
AUSTRALIA:				
Latrobe Valley, Vic.	BR	16	<0.2–1	Brown and Swaine (1964)
Latrobe Valley, Vic.	BR	20	<0.3–24	Bone and Schaap (1981a)
Latrobe Valley, Vic.	BR	15	0.1–3.4	Bolger (1989)
St Vincent Basin, SA	L	5	4.3–49	Fardy, McOrist and Farrar (1984)
Esperance, WA	L	3	<3–7	CSIRO (unpublished)
Leigh Creek, SA	H-B	80	2.5–60	CSIRO (1966)
Collie, WA	S	22	1–17	Davy and Wilson (1984)
Northern area, NSW	BI	1000	3–80	Clark and Swaine (1962a)
Southern area, NSW	BI	800	15–100	Clark and Swaine (1962a)
Western area, NSW	BI	150	10–150	Swaine, Godbeer and Morgan (1984b)
NSW and Qnsld	BI	65	7–93	Knott and Warbrooke (1983)
NSW and Qnsld	BI	45	13–90	Fardy, McOrist and Farrar (1984)
Bowen Basin, Qnsld	BI	200	2–60	Brown and Swaine (1964)
West Moreton area, Qnsld	BI	150	8–150	Brown and Swaine (1964)
Callide area, Qnsld	BI	100	7–80	Brown and Swaine (1964)
Fingal, Tasmania	BI	2	17, 34	Knott and Warbrooke (1983)
Baralaba, Qnsld	S-A	20	8–40	CSIRO (unpublished)
BELGIUM:				
From 12 mines	BI, A	48	10–100	Block and Dams (1975)
CANADA:				
Saskatchewan	L	8	4.7–22	Landheer, Dibbs and Labuda (1982)
Hat Creek No.1, BC	L, S	14	23–267	Goodarzi (1987b)
Hat Creek No.2, BC	L, S	32	47–300	Goodarzi and van der Flier-Keller (1988)
Alberta and BC	S	23	0.94–108	Landheer, Dibbs and Labuda (1982)
'Typical'	S, BI	59	2.4–87	Jervis, Ho and Tiefenbach (1982)
Nova Scotia	BI	186	7–28	Hawley (1955)
Fording Mine, BC	BI	22	6.3–69	Goodarzi (1988)
CHINA:				
'Selection'	–	15	19–255	Sun and Jervis (1986)
CHILE:				
Arauco Basin	BI	399	10–44	Collao et al. (1987)
COLOMBIA:				
3 regions	BI	14	3–300	Rincon et al. (1978)
DENMARK:				
	BR	–	2.7–7	Bogvad and Nielsen (1945)
INDIA:				
	L	3	<1–16	Ghosh et al. (1987)
	BI	50	0.4–147	Ghosh et al. (1987)

Table 5.38 (*contd*)

Sample details	Rank	Samples	Content	Reference
JAPAN:				
Various parts	–	27	0.3–110	Iwasaki and Ukimoto (1942)
GERMANY WEST:				
From 1 power station	BR	7	5.1–15	Heinrichs (1982)
Mostly Ruhr	BI	27	31–179	Kautz, Kirsch and Laufhutte (1975)
NEW ZEALAND:				
Taranaki area	S	41	1–36	Black (1981)
Waikato area	S	–	1.0–19	Purchase (1985)
Southland	S	–	1.6–63	Purchase (1985)
Westland	S-BI	–	0.1–34	Purchase (1985)
NIGERIA:				
'Major deposit'	L	1	46	Hannan et al. (1982)
'Major deposit'	S	6	8.5–14	Hannan et al. (1982)
'Major deposit'	BI	1	47	Hannan et al. (1982)
Enugu, 3 areas	S	3	9.4–11	Ndiokwere, Guinn and Burtner (1983)
POLAND:				
Belchatow mine	BR	3	13–29	Tomza (1987)
	BI	–	5–180	Widawska-Kusmierska (1981)
7 mines	BI	31	1.5–109	Tomza (1987)
SOUTH AFRICA:				
Power-station coals	BI	21	23–132	Kunstmann and Harris (1966)
	BI	11	48–118	Watling, Watling and Wardale (1976)
	BI	14	17–43	Willis (1983)
SWITZERLAND:				
	BR, L	9	3.6–330	Hügi et al. (1989)
	A	9	4.7–155	Hügi et al. (1989)
THAILAND:				
1 deposit	L	3	5–8	CSIRO (unpublished)
TURKEY:				
5 'kinds'	L	5	14–114	Akcetin, Ayca and Hoste (1973)
USSR:				
Moscow Basin	–	3	2	Laktionova et al. (1987)
South Yakutia	–	–	22–98	Egorov, Laktionova and Borts (1981)
Pechora	–	4	13–35	Blomqvist (1983)
UK:				
S. Yorkshire	S	8	0.31–96	Ward, Kerr and Otsuka (1986)
N. England, S. Wales	BI	26	3–80	D.J. Swaine (unpublished)
23 collieries	BI	23	8–150	Hislop et al. (1978)
USA:				
Northern Great Plains	L	226	0.90–110	Zubovic, Hatch and Medlin (1979)
Gulf Province	L	34	7–100	Swanson et al. (1976)
Texas	L	84	0.1–140	White, Edwards and Du Bose (1983)
Fort Union region	L	80	1–30	Hatch and Swanson (1976)

Table 5.38 (contd)

Sample details	Rank	Samples	Content	Reference
Northern Great Plains	S	855	0.91–370	Zubovic, Hatch and Medlin (1979)
Rocky Mt Province	S	466	0.14–330	Zubovic, Hatch and Medlin (1979)
Appalachian Region	BI	1478	1.5–150	Zubovic, Hatch and Medlin (1979)
Interior Province	BI	687	1.1–350	Zubovic, Hatch and Medlin (1979)
Rocky Mt Province	BI	366	1.9–120	Zubovic, Hatch and Medlin (1979)
Alaska, N. Slope	L-BI	84	2–300	Affolter and Stricker (1987)
Pennsylvania	A	53	2–70	Swanson et al. (1976)

Table 5.39 Results for yttrium (as ppm Y)

Sample details	Rank	Samples	Content	Reference
ANTARCTICA:				
Thermally altered	BI	7	4–32	Brown and Taylor (1961)
AUSTRALIA:				
Latrobe Valley, Vic.	BR	16	<0.2–2	Brown and Swaine (1964)
St Vincent Basin, SA	L	3	<3–10	Brown and Swaine (1964)
Esperance, WA	L	3	<3	CSIRO (unpublished)
Leigh Creek, SA	H-B	80	3–40	CSIRO (1966)
Collie, WA	S	5	<0.4–7	CSIRO (unpublished)
Northern area, NSW	BI	1000	1–80	Clark and Swaine (1962a)
Southern area, NSW	BI	800	3–50	Clark and Swaine (1962a)
Western area, NSW	BI	150	5–30	Swaine, Godbeer and Morgan (1984b)
NSW and Qnsld	BI	65	3–48	Knott and Warbrooke (1983)
Bowen Basin, Qnsld	BI	200	1–60	Brown and Swaine (1964)
West Moreton area, Qnsld	BI	150	4–40	Brown and Swaine (1964)
Callide area, Qnsld	BI	100	1–15	Brown and Swaine (1964)
Fingal, Tasmania	BI	2	16, 21	Knott and Warbrooke (1983)
Baralaba, Qnsld	S-A	20	1–8	CSIRO (unpublished)
BULGARIA:				
4 deposits	BR, L	11	2–126	Eskenazi (1987b)
COLOMBIA:				
3 regions	BI	14	2–40	Rincon et al. (1978)
GERMANY WEST:				
From 1 power station	BR	7	2.0–4.1	Heinrichs (1982)
NEW ZEALAND:				
Taranaki area	S	–	2.0–5.3	Purchase (1985)
Waikato area	S	–	1.3–7.0	Purchase (1985)
Southland	S	–	0.5–2.7	Purchase (1985)
Westland	S-BI	–	0.1–5.0	Purchase (1985)
SOUTH AFRICA:				
	BI	14	14–30	Willis (1983)

Table 5.39 (contd)

Sample details	Rank	Samples	Content	Reference
SWITZERLAND:	BR, L	9	1–20	Hügi et al. (1989)
	A	9	7–30	Hügi et al. (1989)
USSR:				
South Yakutia	–	–	6.3–33	Egorov, Laktionova and Borts (1981)
UK:				
N. England, S. Wales	BI	26	0.6–15	D.J. Swaine (unpublished)
USA/				
Gulf Province	L	34	2–50	Swanson et al. (1976)
Texas	L	84	3–725	White, Edwards and Du Bose (1983)
Fort Union region	L	80	<1–15	Hatch and Swanson (1976)
Eastern USA	BI	617	0.78–69	Zubovic et al. (1979b)
Alaska, N. Slope	L-BI	84	0.5–30	Affolter and Stricker (1987)
Pennsylvania	A	53	1–30	Swanson et al. (1976)

Louis (1901–02), Crook (1913) and Moss, Hirst and Needham (1929–1930), but the first determinations preceded these by about 20 years, as Jensch (1887) had found 700–9000 ppm Zn in six German coal ashes. Goldschmidt (1937) reported up to 1% Zn in German coal ashes. The mode of occurrence of Zn in coals has been suggested frequently. There are advocates for organic bonding, surely very reasonable for low-rank coals (Chapter 3), but there have been many detections of sphalerite dating from almost a century ago (Binns and Harrow, 1897; Crook, 1913; Dove, 1921) and this is a prime source of Zn in many coals, being prevalent in cleats.

The determination of Zn in coals has been done mainly by AES, AAS, XRFS and INAA, although the latter method has some limitations; ICPAES is also a suitable method. There are two standard methods (ASTM, 1983a; AS, 1986), the coal sample being treated as outlined for Be (Section 5.2.4); an air–acetylene flame is used with the line at 213.9 and background correction.

Results for Zn are given in Table 5.40. Apart from some relatively high values, most coals should be in the range 5–300 ppm, but mean values for areas where significant numbers of samples have been analysed vary widely. It seems that a mean value of about 25 represents Australian and some US coals, but European coals have higher mean values, as represented by the coals used for power production in nine EC countries, where the mean values range from 57 to 172, with an overall mean value of 111 ppm Zn (Sabbioni et al., 1983). It is not envisaged that deleterious effects would arise from Zn during coal mining and usage.

5.2.49 Zirconium

Among early results for Zr are up to 5000 ppm Zr in some German coal ashes (Goldschmidt, 1937), 20–1000 ppm Zr in Spitsbergen coal ashes

Table 5.40 Results for zinc (as ppm Zn)

Sample details	Rank	Samples	Content	Reference
ARGENTINA:				
Rio Turbio	–	6	80–100	A. Berset (1982, pers. commun.)
AUSTRALIA:				
Latrobe Valley, Vic.	BR	16	1–30	Brown and Swaine (1964)
Latrobe Valley, Vic.	BR	20	0.6–87	Bone and Schaap (1981a)
Latrobe Valley, Vic.	BR	15	5–129	Bolger (1989)
Latrobe Valley, Vic.	BR	28	0.5–13	Fardy, McOrist and Farrar (1984)
St Vincent Basin, SA	L	5	15–40	Fardy, McOrist and Farrar (1984)
Leigh Creek, SA	H-B	80	15–200	CSIRO (1966)
Collie, WA	S	22	1–72	Davy and Wilson (1984)
Northern area, NSW	BI	1000	9–300	Clark and Swaine (1962a)
Southern area, NSW	BI	800	15–100	Clark and Swaine (1962a)
Western area, NSW	BI	150	10–150	Swaine, Godbeer and Morgan (1984b)
NSW and Qnsld	BI	65	5–55	Knott and Warbrooke (1983)
NSW and Qnsld	BI	45	15–45	Fardy, McOrist and Farrar (1984)
Bowen Basin, Qnsld	BI	200	8–500	Brown and Swaine (1964)
West Moreton area, Qnsld	BI	150	4–1000	Brown and Swaine (1964)
Callide area, Qnsld	BI	100	7–1000	Brown and Swaine (1964)
Fingal, Tasmania	BI	2	23, 26	Knott and Warbrooke (1983)
Baralaba, Qnsld	S-A	20	100–150	CSIRO (unpublished)
BELGIUM:				
From 12 mines	BI, A	48	10–320	Block and Dams (1975)
BRAZIL:				
Rio Grande do Sul	BI	13	34–430	Bellido and Arezzo (1987)
CANADA:				
Saskatchewan	L	9	6.7–138	Landheer, Dibbs and Labuda (1982)
Moose River Basin	L	17	5.2–1311	Van der Flier-Keller and Fyfe (1987b)
Hat Creek No.1, BC	L, S	14	7.6–48	Goodarzi (1987b)
Hat Creek No.2, BC	L, S	32	12–83	Goodarzi and van der Flier-Keller (1988)
Alberta and BC	S	23	4.9–111	Landheer, Dibbs and Labuda (1982)
Nova Scotia	BI	186	13–64	Hawley (1955)
Fording Mine, BC	BI	22	8–49	Goodarzi (1988)
CHINA:				
From 110 mines	–	–	0.56–192	Chen et al. (1986)
COLOMBIA:				
3 regions	BI	14	<20–130	Rincon et al. (1978)
GERMANY WEST:				
From 1 power station	BR	7	3.2–7.1	Heinrichs (1982)
Mostly Ruhr	BI	27	17–210	Kautz, Kirsch and Laufhutte (1975)
Various	BI	10	14–1742	Heinrichs (1975b)

Table 5.40 (*contd*)

Sample details	Rank	Samples	Content	Reference
NEW ZEALAND:				
Taranaki area	S	41	<1–90	Black (1981)
Waikato area	S	–	1–56	Purchase (1985)
Southland	S	–	1–80	Purchase (1985)
Westland	S, BI	–	0.4–80	Purchase (1985)
POLAND:				
Belchatow mine	BR	3	28–198	Tomza (1987)
	BI	–	5–300	Widawska-Kusmierska (1981)
7 mines	BI	31	12–4047	Tomza (1987)
ROMANIA:				
'Several types'	–	–	1.6–48	Savul and Ababi (1958)
SOUTH AFRICA:				
	BI	11	6–19	Watling and Watling (1976)
	BI	14	3.2–16	Willis (1983)
SWITZERLAND:				
	BR, L	9	10–80	Hügi *et al.* (1989)
USSR:	A	9	2.5–65	Hügi *et al.* (1989)
Pechora	–	4	14–24	Blomqvist (1983)
UK:				
S. Yorkshire	S	8	1.4–17	Ward, Kerr and Otsuka (1986)
N. England, S. Wales	BI	26	10–400	D.J. Swaine (unpublished)
20 areas	BI	232	3–1700	Taylor (1973)
23 collieries	BI	23	8–150	Hislop *et al.* (1978)
S. Wales (12 seams)	BI	300	−261(26)	Chatterjee and Pooley (1977)
USA:				
Northern Great Plains	L	226	1.0–88	Zubovic, Hatch and Medlin (1979)
Gulf Province	L	34	5.4–201	Swanson *et al.* (1976)
Texas	L	144	1.0–486	White, Edwards and Du Bose (1983)
Fort Union region	L	80	1.0–28	Hatch and Swanson (1976)
Northern Great Plains	S	855	0.88–220	Zubovic, Hatch and Medlin (1979)
Rocky Mt Province	S	466	0.92–910	Zubovic, Hatch and Medlin (1979)
Appalachian Region	BI	1478	1.3–1100	Zubovic, Hatch and Medlin (1979)
Interior Province	BI	687	1.2–5100	Zubovic, Hatch and Medlin (1979)
Rocky Mt Province	BI	366	0.85–120	Zubovic, Hatch and Medlin (1979)
Alaska, N. Slope	L-BI	84	2–67	Affolter and Stricker (1987)
Pennsylvania	A	53	0.1–65.0	Swanson *et al.* (1976)
YUGOSLAVIA:				
Kosova Basin	BR	2	35, 85	Daci, Berisha and Gashi (1983)

Detailed information for individual trace elements 181

(Butler, 1953) and 23–106 ppm Zr in the ash of three coals from West Germany and the United Kingdom (Degenhardt, 1957). There is an extensive review by Smith and Carson (1978) of all aspects of Zr. The earliest extensive survey was for Australian coals (Clark and Swaine, 1962a). On the basis of density separations, Ward (1980) suggested a predominantly organic affinity of Zr in most Australian coals he studied. However, this conclusion would be nullified if very finely divided zircons were embedded in the organic coaly matter. For Bulgarian coals, Eskenazi (1987a) concluded from her detailed study that Zr was mainly associated with mineral matter, except in some low-ash coals. Hoehne (1957) identified zircon in coals and Kemezys and Taylor (1964) found it in some Australian coals. Zircon was also seen in several coals by Finkelman (1981a) who regarded it as the prime site of Zr there. Butler (1953) inferred that detrital zircon was in Spitsbergen coals. Golovko (1960) suggested that Zr was mainly associated with the heavy fraction of some Soviet coals but that some adsorption on clay particles could have occurred. For US lignites, Given and Miller (1987) decided that Zr was 'pretty consistently associated with minerals'. It seems that zircon is regarded as the main mode of occurrence of Zr in most coals.

The determination of Zr is usually carried out by AES or XRFS. Results for Zr in coal are given in Table 5.41. The range for most Australian coals is probably about 5–300, with a mean value of about 100 ppm Zr, whereas other coals would be mostly in the range 5–200, with a mean value of about 30 ppm Zr. The general insolubility of Zr compounds makes it a rather innocuous element, with no known deleterious effects at the concentrations found in coal.

In the tables, specific references to results given on an ash basis have only been made occasionally, but there are many publications listing such data. Some of these are in the following references: for Australia (Clark and Swaine, 1962a), for Austria (Brandenstein, Janda and Schroll, 1960), for Belgium (Block and Dams, 1976), for Canada (Kronberg et al., 1981; Fyfe, Kronberg and Brown, 1982), for China (Sun and Jervis, 1986), for Czechoslovakia (Bouska and Havlena, 1959; Mechacek, 1972, 1976), for Germany (Otte, 1953; Leutwein and Rösler, 1956; Radmacher and Schmitz, 1965; Kessler, Malan and Valeska, 1965; Krejci-Graf, 1983), for Hungary (Benko and Szadeczky-Kardoss, 1957; Odor, 1969; Varga, Bella and Benocs, 1972), for India (Mukherjee and Ghosh, 1977; Malik and Ahmad, 1979; Pareek and Bardhan, 1985), for New Zealand (DSIR, 1949; Sim and Lewin, 1975; Soong and Berrow, 1979), for Poland (Kuhl and Ziolkowski, 1954; Idzikowski and Trzebiatowski, 1960), for Portugal (de Brito, 1955), for Spain (Lopez de Azcona and Camunas Puig, 1947), for the United States (Stadnichenko, 1953; Headlee and Hunter, 1953, 1955b; Deul and Annell, 1956; Zubovic, Stadnichenko and Sheffey, 1964, 1966; Abernethy, Peterson and Gibson, 1969; Swanson et al., 1976; Zubovic et al., 1979, 1980), for Yugoslavia (Hamrla, 1959).

The survey of published results (Tables 5.1–5.41) is just that, and results for any area or coal type cannot necessarily be taken as representative of

Table 5.41 Results for zirconium (as ppm Zr)

Sample details	Rank	Samples	Content	Reference
ANTARCTICA:				
Thermally altered	BI	7	54–400	Brown and Taylor (1961)
ARGENTINA:				
Rio Turbio	–	6	10–15	A. Berset (1982, pers. commun.)
AUSTRALIA:				
Latrobe Valley, Vic.	BR	16	0.3–4	Brown and Swaine (1964)
St Vincent Basin, SA	L	3	<8–30	CSIRO (unpublished)
Leigh Creek, SA	H-B	80	25–300	CSIRO (1966)
Collie, WA	S	22	3–67	Davy and Wilson (1984)
Northern area, NSW	BI	1000	5–600	Clark and Swaine (1962a)
Southern area, NSW	BI	800	10–400	Clark and Swaine (1962a)
Western area, NSW	BI	150	15–300	Swaine, Godbeer and Morgan (1984b)
NSW and Qnsld	BI	65	26–219	Knott and Warbrooke (1983)
Bowen Basin, Qnsld	BI	200	6–250	Brown and Swaine (1964)
West Moreton area, Qnsld	BI	150	20–300	Brown and Swaine (1964)
Callide area, Qnsld	BI	100	10–200	Brown and Swaine (1964)
Fingal, Tasmania	BI	2	152, 238	Knott and Warbrooke (1983)
Baralaba, Qnsld	S-A	20	10–80	CSIRO (unpublished)
BRAZIL:				
Rio Grande do Sul	BI	13	67–221	Bellido and Arezzo (1987)
CANADA:				
Saskatchewan	L	9	34–334	Landheer, Dibbs and Labuda (1982)
Moose River Basin	L	17	<1–530	Van der Flier-Keller and Fyfe (1987b)
Alberta and BC	S	–	25–476	Landheer, Dibbs and Labuda (1982)
Nova Scotia and New Brunswick	BI	7	2.6–104	Landheer, Dibbs and Labuda (1982)
COLOMBIA:				
3 regions	BI	14	10–40	Rincon et al. (1978)
GERMANY EAST:				
3 areas	BR	100	(<0.4–3.5)	Malberg (1961)
GERMANY WEST:				
From 1 power station	BR	7	5.4–14	Heinrichs (1982)
NEW ZEALAND:				
Taranaki area	S	–	4–80	Purchase (1985)
Waikato area	S	–	0.6–50	Purchase (1985)
Southland	S	–	1.1–22	Purchase (1985)
Westland	S-BI	–	0.6–60	Purchase (1985)
SOUTH AFRICA:				
	BI	14	47–356	Willis (1983)
SWITZERLAND:				
	BR, L	9	<5–50	Hügi et al. (1989)
	A	9	9–250	Hügi et al. (1989)
THAILAND:				
1 deposit	L	3	5–8	CSIRO (unpublished)

Table 5.41 (*contd*)

Sample details	Rank	Samples	Content	Reference
UK:				
N. England, S Wales	BI	26	0.7–80	D.J. Swaine (unpublished)
23 collieries	BI	23	4–30	Hislop *et al.* (1978)
USA:				
Gulf Province	BI	34	7–200	Swanson *et al.* (1976)
Texas	BI	98	6–316	White, Edwards and Du Bose (1983)
Fort Union region	BI	80	2–70	Hatch and Swanson (1976)
Eastern USA	BI	617	1.5–310	Zubovic *et al.* (1979)
Western USA	–	28	12–170	Gluskoter *et al.* (1977)
Alaska, N. Slope	L-BI	84	2–500	Affolter and Stricker (1987)
Pennsylvania	A	53	7–50	Swanson *et al.* (1976)

the area or type. However, experience with Australian coals indicates that a proper sampling of a limited number of coals from a particular seam may well establish the ranges of values for trace elements in that seam. Occasional high values occur, usually because of an exceptional mineral assemblage and may be taken as possibly indicative of mineralization in the source area. High concentrations of some elements may forewarn of possible problems during usage, for example coals high in Cd, Hg and U may not be suitable for power production. When detailed information is required, the relevant references should be consulted.

Chapter 6

Comparisons of coal with shale and soil

Although it is difficult to make general comparisons, especially using mean values, it is worth while comparing the ranges of values for most coals (from each section in Chapter 5), mean values for some coals, ranges and mean values for soils, and mean values for shales. This gives some idea of the trace-element status of coal relative to two other major sedimentary materials, and is helpful in keeping certain sensitive elements in perspective. Of course, if comparisons are to be made specifically for a particular coal, then determinations must be made of trace elements in that coal, local soil and nearby shale.

6.1 Comparisons of coal with soil

Data given in Table 6.1 have been gathered from several sources. The ranges for coal are those that were assessed from the data in Chapter 5. Where possible, four mean values are given, namely for Australia (CSIRO publications), the European Communities (Sabbioni et al., 1983) or the United Kingdom (Hislop et al., 1978), the United States (Zubovic, Hatch and Medlin, 1979; Zubovic et al., 1980) and for world coals (Yudovich et al., 1972). The soil ranges are mostly from Swaine (1955) with some from Bowen (1979), while the mean values are from Bowen (1979) and Shacklette and Boerngen (1984), as well as Cd values from Bewers, Barry and MacGregor (1987) and the Tl value for shale from Heinrichs, Schulz-Dobrick and Wedepohl (1980). The only element with a higher range and mean value for coal is B. Several elements have lower ranges (Cd, Cr, Ga, Li, Mn, Ni, Sc, Th, V, Zr), the other elements having about the same ranges. On the basis of mean values, As, Be, Ge, Pb, Mo and Se are slightly higher or higher in coal, except that Australian coals are lower in As and Pb than soil and Sb is slightly lower than soil. Uranium is marginally lower in coal. For most trace elements the ranges and mean values for coal are not markedly different from those for soil.

6.2 Comparisons of coal with shale

The data in Table 6.1 show that shale is higher than coal in 16 elements, including As, F, Th and Zn, and slightly higher in Sb, Be, B and U, while

Table 6.1 **Ranges and mean values for coal, soil and shale** (as ppm)

Element	Coal		Soil		Shale
	Range	Means	Range	Mean	Mean
Antimony	0.05–10	0.5, 2.3, 1.0, –	0.2–10	0.7	1.5
Arsenic	0.5–80	1, 5, 11, 15, 29	1–50	7	13
Barium	20–1000	70–300, 120	100–3000	500	550
Beryllium	0.1–15	1.5–2.0, 3	<5–40	0.9	3
Boron	5–400	30–60, 80	2–100	30	130
Cadmium	0.1–3	0.08, 0.5, 0.5, –	0.02–10	0.6	0.3
Chlorine	50–2000	150	8–1800	100	160
Chromium	0.5–60	6, 41, 20, 8	5–1000	55	90
Cobalt	0.5–30	4–8, 4	1–40	9	19
Copper	0.5–50	15, 23, 17, 7	2–100	25	39
Fluorine	20–500	150	20–700	300	800
Gallium	1–20	4, 6, 6, 2.3	5–70	17	23
Germanium	0.5–50	6, 7	0.1–50	1	2
Lead	2–80	10, 54, 30, 59	2–100	19	23
Lithium	1–80	~20	5–200	24	76
Manganese	5–300	130, 70, 50, 70	200–3000	550	850
Mercury	0.02–1	0.10, 0.30, 0.15, –	0.01–0.5	0.1	0.18
Molybdenum	0.1–10	1.5, 5.2, 3, 4	0.2–5	1	2.6
Nickel	0.5–50	15, 41, 20, 13	5–500	20	68
Phosphorus	10–3000	–	35–5300	500	700
Scandium	1–10	4	<10–25	9	13
Selenium	0.2–1.4	0.8, 1.7, 2.7, –	0.1–2	0.4	0.5
Strontium	15–500	~100, 240	50–1000	240	300
Thallium	<0.2–1	–	0.1–0.8	0.2	0.4
Thorium	0.5–10	2.7, 4, 4, –	1–35	9	12
Tin	1–10	2, 1.3	1–20	2	6
Titanium	10–2000	900, 700, 800, 400	1000–10 000	3000	4600
Uranium	0.5–10	2	0.7–9	2.7	3.7
Vanadium	2–100	20, 60, 30, 19	20–500	80	130
Zinc	5–300	25, 57–172, 40, 20	10–300	70	120
Zirconium	5–200	100, 16, 25, 28	60–2000	250	160

Cd and Hg show no real differences. Of the remaining elements, Pb, Mo and Se are higher in coal than in shale, the exception being the low Pb in Australian coal.

It is seen from the above comparisons that coal tends to be higher in Ge, Pb, Mo and Se than in soil and shale. Otherwise coal is less than or much the same as soil and shale in most trace-element contents. It is stressed that these statements are generalizations and that particular coals may reverse these trends. Perhaps the main usefulness of these comparisons is to show that coal is not notably different in trace-element contents from the other two sedimentary materials. There is an interesting compilation of data on trace elements in shales and other rocks, based on 45 000 analyses (Ronov, Bredanova and Migdisov, 1988). The results show changes in mean values for shales depending on the age of the shale. In the above comparisons the ages of the coal, soil and shale were not taken into account.

The results for coal may be seen in another way by comparing them with the contents of trace elements in fertilizers. Bowen (1966) has summarized the vast array of data compiled by Swaine (1962c) for nitrogen (N), potassium (K) and phosphorus (P) fertilizers. For example, K fertilizers have 15 ppm As and 1 ppm Cd, while N fertilizers have 100 ppm As and 5 ppm Se. For P fertilizers there are several trace elements with higher contents than most coals, namely 10–1000 ppm As, 200 ppm Cr, 20 000 ppm F, 4 ppm Mo, 100 ppm Pb, 30 ppm Se, 50 ppm U and 150 ppm Zn.

Chapter 7

Variations within seams

Variations in trace-element contents are expected in coals from different areas and coal basins (geographical variations). There are also variations in seams forming part of a stratigraphic sequence, as well as vertical and lateral variations within a particular seam.

7.1 Geographical variations

These can be ascertained from compilations of large numbers of results from several areas, for example in the United States (Gluskoter et al., 1977; Swanson et al., 1976; Zubovic et al., 1979, 1980), in South Africa (Willis and Hart, 1986) and in Australia (Clark and Swaine, 1962a). Variations in the contents of some trace elements in coals from three main areas in the United States are shown in Table 7.1. The Appalachians are

Table 7.1 **Variations in trace-element contents in coals from 3 regions in USA** (Swanson et al., 1976; values as ppm, with mean values in parentheses)

	Appalachians	Northern Great Plains	Rocky Mountains
Ba	7–700(100)	15–2000(500)	3–700(200)
B	1–100(30)	30–200(70)	10–150(70)
Cr	<0.5–70(20)	0.7–30(5)	0.5–70(5)
Pb	1–70(15)	1.4–42(5)	0.9–19(5.5)
Sr	7–700(100)	15–700(150)	15–700(100)
V	2–150(20)	1.5–50(10)	2–100(15)
Zr	2–300(50)	3–50(15)	3–70(20)

highest in Cr, Pb and Zr, but lowest in B, while Ba is highest in the Northern Great Plains. The three regions have about the same mean content of Sr and V. Of course, these are generalizations and particular coals may vary, as shown by the ranges of values. In NSW, Australia, coals from the southern area are higher in Be, Ge, Mn, Ni, Pb and Zr than those from the northern area, but B is higher in the northern area (Clark and Swaine, 1962a).

Sometimes areas notably higher in a particular trace element are delineated by considering results in this way. For example, Harvey and Ruch (1986) found that the mean contents of Cl in coals from Rocky Mountains, Northern Great Plains and Appalachian regions were <200, <100 and <200 ppm Cl, whereas the corresponding value for coals from the Illinois Basin was 1700 ppm Cl. Van der Sloot (1981) did a cluster analysis on data for Illinois coals (Gluskoter et al., 1977), and found definite similarities between samples taken at different locations and times within a particular area. He also found some similarities of trace-element contents, for example from different countries, based on cluster analysis for 17 elements.

Variations in trace-element contents from the edges to the centre of a basin need to be studied in detail. However, there is some evidence for a decrease towards the centre, probably because of removal of ions from incoming waters, thereby giving the central areas access to lower concentrations of trace elements.

7.2 Stratigraphic variations

Since there is a direct relationship between the stratigraphy of coal seams and the environment of deposition during coalification, changes in trace-element contents are to be expected in coals from different stratigraphic stages. Goodarzi and van der Flier-Keller (1988) have plotted results for 21 trace elements and ten REE for a 500 m stratigraphic sequence in the Hat Creek coal deposit No. 2, British Columbia, Canada. There is a change in rank in the coals from the four stratigraphic zones from lignite to subbituminous. Some distinct effects are evident, for example As, B and Sb tend to be higher in the top two zones A and B, while there is little change for Cs, Mn and V. The constancy for Mn is perhaps surprising, but seems to be fairly well matched with the patterns for Fe and Ca. The changes for the REE are much the same, except for Lu which tends to be fairly constant. There are probably minor variations that are obscured because the diagrams use very small scales, necessary to accommodate the data for so many elements in 32 samples.

Although there are results for several stratigraphic sequences from the Sydney Basin and the Ipswich coal measures, Australia, they have not been appraised. No doubt, similar situations occur elsewhere also. It is salutary to mention that the examination of large numbers of results from different aspects may show unexpected effects.

7.3 Vertical variations

Although Gregorowicz (1957) did not find any relationship between the Ge content of some Polish coals and depth, there are several reports of high values for Ge in the top and bottom sections of a seam (e.g. Headlee and Hunter, 1951). It is possible that the enrichment of some elements at the top of seams may have occurred during the post-burial stage (Zubovic, Stadnichenko and Sheffey, 1964), presumably when the organic matter had high bonding and adsorption capacities. Although some trace elements were concentrated at the top or bottom of some seams from Nova Scotia, Canada, there were irregular lateral variations (Hawley and Rimsaite, 1954). Kunstmann and Hamersma (1955) studied subsections from four seams in Natal, South Africa (1–2 m thick). In two seams there was marked accumulation of Ge in the top of the seams, adjacent to the sandstone roof, whereas the other two seams showed the highest concentrations at the bottom of the seam, adjacent to the sandstone or carbonaceous shale floor. In Illinois, USA, some coals show enrichments of Sb, Ge, Mo, Se, U and V at the top of seams (Gluskoter et al., 1977), possibly because of overlying black shales (Hatch, 1983). Results for trace elements in 44 vertical sections from the Sydney Basin, NSW, Australia, showed variable distributions of trace elements, although sometimes there was a tendency for higher concentrations of some elements near the roof or floor, for example with Ge in some seams (Clark and Swaine, 1962a). A detailed study of three vertical sections, each about 30 m thick, from Blair Athol, Queensland, Australia (CSIRO, 1960), involved the determination of about 20 trace elements in 87 coal samples and 22 dirt-band samples (i.e. samples with >35% ash yields). Highest values in roof and/ or floor coal subsections were generally found for Co, Ni and Ge, occasionally for B, Sn, Pb and V, and for Be, but not markedly so. Highest values for Ga and Zr were towards the bottom of the seam, while Cr values fluctuated very little. In most cases, values for dirt bands were higher than those for adjacent coals. A detailed study of trace elements in samples of subsections from five vertical sections from the Herrin (No. 6) coal, Illinois, USA, showed variable distributions but top and bottom subsections were the preferred sites for several elements (Gluskoter et al., 1977). Harris, Barrett and Kopp (1981) examined variations in concentrations of several trace elements in two seams, 7 m and 10 m thick, from East Tennessee, USA. The only marked concentration effect was that of Ni in each case. Arsenic tended to be higher towards the top of each seam, while Sr had opposite trends in the two seams. Marked fluctuations were evident for Mn and Rb. The trends in one seam were cited as evidence for possible marine influence; it was a pity that B was not determined.

Variations in the contents of trace elements in a 18 m section in a mine in British Columbia, Canada, showed some high values near the top of the seam, but mostly scattered maxima, often in dirt-bands and probably 'primarily related to lithology' (Goodarzi and van der Flier-Keller, 1989). Inagaki and Tokuga (1959) found that U tended to be concentrated near the top and bottom of some Japanese seams. The marked fluctuations that are a feature of the above studies are not surprising, as they would reflect changing conditions of deposition and of mineral and trace-element inputs.

7.4 Lateral variations

No doubt, examination of data for large numbers of coals from restricted areas would enable some information to be gleaned about lateral variations in trace-element contents. However, there seems to have been only one study specifically carried out for this purpose (Swaine, Godbeer and

Morgan, 1984b). Samples of coal were taken over a period of four years from nine locations in the Lithgow seam, NSW, Australia. In this area, stable basement conditions have prevailed since Permian times and hence any effects of folding and faulting should be minimal. Up to 50 trace elements were determined in a maximum of 133 samples, taken over about 100 km of seam. The variability can be assessed well by calculating the variance ratio, namely the ratio of the maximum to the minimum value, a concept which seems to have been introduced initially by Headlee and Hunter (1953), who used it with their results for 16 seams in West Virginia, USA, and by Zubovic (1960) for results from several locations in two seams from Indiana and Kentucky, USA. Values of variance ratios for 33 samples from one location compared with those for 100 samples from eight other locations in the Lithgow seam are given in Table 7.2. A few obvious

Table 7.2 Variance ratios for trace elements in one coal seam (variance ratio = maximum divided by minimum value)

	1 location	8 locations
As	2.8	5.8
B	2	2
Be	1.7	2.2
Cd	1.8	2.1
Cu	3	3
F	1.2	1.4
Ge	2.5	3.8
Hg	2.3	3.1
Mn	4.4	16
Pb	1.6	2.3
Se	1.6	1.7
Th	1.8	2.4
U	1.6	1.8
Zn	2.5	3.0
Zr	1.7	2.5

outliers were omitted in calculating the variance ratios. For the one location, most trace elements had variance ratios from 1.2 (for F) to 2.5 (for Zn). Values for As, Cu and Mn were higher (2.8, 3, 4.4). For the eight locations most elements had values from 1.4 (for F) to 3.1 (for Hg), but there were higher values for Ge, As and Mn (3.8, 5.8, 16). The greater spread of values for Mn was related to variations in the amounts of carbonate minerals. If two locations are omitted from the calculations, then the variance ratios are lowered to agree closely with those for the one location. The uniformity of these results is in keeping with the narrow ranges of 0.5–0.7% for total S and 16–23% for ash yield, and it would seem likely that stable conditions prevailed during most of the coalification period. This investigation has been discussed elsewhere (Swaine, 1984c; Fardy and Swaine, 1985).

In a study of trace elements in coals from a Czechoslovakian mine, Bouska and Havlena (1959) did not find any clear relationships between trace element contents in coals from the same seam at distances of 50–100 m. It should not be assumed that the results for the Lithgow seam are in any way typical; indeed they may well indicate the least variability that can be expected in any one seam. However, samples taken from short distances may well give low variance ratios. As van der Sloot *et al.* (1982b) commented 'coal handling may be regarded as a large scale homogenization process'. They sampled feed coal at a Netherlands power station over a three-day period and obtained variance ratios of 1.2–1.8 for several trace elements which were determined with good precision. These variance ratios are reasonably similar to those found for coals from the one location in the Lithgow seam referred to above.

Chapter 8
Radioactivity and coal

There is a continuing interest in radioactivity from coal, especially that connected with combustion. An attempt will be made here to review several aspects of radioactivity related to coal. This is a task fraught with difficulty because of the inherent pitfalls in the real meaning of radioactivity and because there has been much contention about the relevance of coal in this matter. Despite former efforts to malign coal compared with nuclear energy as a source of radiation, sanity now prevails. Indeed, the situation has been stated sensibly by Zimmermeyer (1978) as 'comparisons between the radiation releases from nuclear and coal-fired power stations in normal service have little contribution to make to the debate on nuclear energy since, in each case, the hazard is negligible'.

The main sources of radioactivity in coal are U and Th. For Hungarian coals, Szalay and Almassy (1956) found that U was the prime source, as it was also for Greek lignites (Papastefanou and Charalambous, 1979). Most coals have contents of U and Th that are in the same ranges as those for most granites (KHM, 1983) and Davidson and Ponsford (1954) stated that 'coal formations are usually among the least radioactive of the common sedimentary strata'. Ericsson (1983) has concluded that 'coal in general contains less natural activity than ordinary soil and rock'. Hence, it seems fair to assess coal as an insignificant source of radiation compared with total natural background and medical uses of radiation. It is well known that superphosphate is more radioactive than soils and limestones (Turner, Radley and Mayneord, 1958; Swaine, 1962c) and has been used as a source of U. According to Cohen (1979) 'the effects of radioactivity releases by one year of present annual operations are 10 times larger for phosphate mining than for coal burning'.

The measurement of general radioactivity and of specific radionuclides should be carried out by specialists. An evaluation of the techniques by Abel (1986) was the basis of recommended procedures which were 'a combination of gamma and X-ray spectrometry, along with radiochemical separations and alpha energy analysis to yield a complete radionuclide characterization'. It is not necessary to measure every radionuclide, as this would entail measuring 35 radionuclides in the three decay series from ^{238}U, ^{235}U and ^{232}Th. There are still disagreements about the number of radionuclides that should be considered, one approach being to measure 15 and another to measure seven, these latter being the ones regarded as biologically important. Untried and Boeck (1982) measured the natural radioactivity of some Austrian coals, eight radionuclides being determined. Fardy, McOrist and Farrar (1984) used high-resolution gamma spectrometry to measure ^{226}Ra, ^{210}Pb and ^{40}K and low-level radiochemistry to determine ^{230}Th, ^{228}Ra, ^{210}Pb and ^{210}Po. The levels of ^{238}U and ^{232}Th were determined by INAA. They examined seven bituminous coals from New South Wales and Queensland, while Cooper and Leith (1983) examined 24 NSW bituminous coals for six radionuclides.

The natural radioactivity of 180 Polish run-of-mine coal samples was measured for three radionuclides by Skowronek (1986). The behaviour of natural radionuclides in 15 coal samples from western United States was studied in detail by Coles, Ragaini and Ondov (1978) who observed 13 gamma rays from ^{238}U(8), ^{235}U(1) and ^{232}Th(4). Salmon, Toureau and Lally (1984) measured the activity from ^{238}U, ^{230}Th, ^{226}Ra and ^{232}Th from 20 UK coals.

Various studies have been carried out in connection with the combustion of coal in power stations. As expected, flyash tends to have higher levels of radionuclides than the parent coal and some of this flyash reaches the atmosphere in the stack gases, while most of it is contained in the flyash removed by electrostatic precipitators or other particle-attenuation devices. Nevertheless, the radionuclide concentrations in plume air from Japanese power stations at the points of maximum concentrations were less than 0.5% of those of natural origin (Nakaoka et al., 1982). Measurements of the levels of ^{210}Pb and ^{210}Po in soil and grass samples from near the Didcot, UK, power station showed no 'statistically significant increases' (Smith-Briggs, 1984), while Robson et al. (1981) found that the contributions of ^{210}Pb from UK power stations were 'at the most, a small fraction of the naturally occurring level' and did not pose additional risks for human health, animals or vegetation, even near the power stations. A study of a power station in a rural area in Bavaria showed that the amount of radioactivity emitted by the coal burning was 'too low to significantly change the natural variation of these radionuclides in the soil', the nuclides being ^{210}Pb, ^{210}Po, ^{226}Ra, ^{238}U and ^{232}Th (Rosner et al., 1984). This finding was supported by analyses of vegetation from the area around the power station. The import of these results is summarized well by Blackburn (1985) as 'although the total radioactivity in a large mass of mildly radioactive coal or lignite may be considerable it does not present either an external or an ingestion radiation hazard'.

However, there are reports that coal-burning releases more radioactivity than nuclear energy sources. This view was put forward by Eisenbud and Petrow (1964) and partly supported by a study of four power stations in India (Mishra, Lalit and Ramachandran, 1980). In another study of an Indian power station burning coal with 14–100 ppm U (Bauman and Horvat, 1981) some chromosome aberrations were found in some workers at the power station. As well as the relatively high U-content of the coal, it was stated that 'the radioactive dust penetrates everywhere; removal of dust is not possible'. This is surely an unusual situation. On the basis of measurements of ^{226}Ra, ^{228}Ra, ^{228}Th, ^{210}Pb, ^{235}U and ^{238}U in flue gas emissions and particulates from a 2000 MW Polish power station, burning lignite, Glowiak and Pacyna (1980) stated that 'radioactive contamination caused by coal-fired power plants is about 30 times higher than that caused by atomic power stations'. However, Pacyna (1980) found that the doses of radionuclides to different organs in the body were about the same for coal-fired and nuclear power stations, the former being higher for bone and the latter for thyroid. Nevertheless, he still maintained that radionuclides from coal firing 'create a serious environmental hazard'. Although the radiation from coal firing was regarded as an order of magnitude higher than from an equivalent nuclear plant, Mishra, Lalit and Ramachandran (1984) pointed out that these doses were in the range of natural background. Okamoto (1979) believed that flue gas from power stations posed a greater radioactive hazard than nuclear reactors. The radioactivity from a Greek power station burning lignite indicated that the atmospheric inputs were greater than those from nuclear power plants (Papastefanou and Charalambous, 1980). Although an assessment of the relative inputs of radioactivity from coal and nuclear power stations was not the aim of McBride et al. (1978), a plea was made for 'evaluation of the radiological impact on the environment' when assessing coal-fired and nuclear-power plants (McBride et al., 1977). Camplin (1980) supported a limited environmental investigation for areas near power stations in order to establish the levels of radionuclides in animal foods and the like.

In view of the above statements about coal radioactivity, it is necessary to put the matter into perspective by citing contrary views. Wagner et al. (1980) found that radioactivity in coal was generally about the same as that in most rocks and soils. Radiation exposure from coal-fired and nuclear power plants corresponds to <1% of the average natural background radiation (Jacobi, Schmier and Schwibach, 1982) and represents 'a very small augmentation of the natural background' (Corbett, 1983), a finding confirmed by Salmon, Toureau and Lally (1984). Another statement regarding coal is 'radioactivity was insignificant' (Kühn, 1960) and the careful assessment by Beck and Miller (1980) which led to the conclusion that emissions from coal firing 'will not result in significant perturbations on natural environmental radiation levels or significant increases in doses to any individual'. Coals from western United States do 'not present any significant radiological hazard' (Abbott, Styron and Casella, 1983). Martin et al. (1971) felt that there was no need to stress the radiological aspects of fossil fuel plants because their public health significance is negligible. After a careful appraisal of published information KHM (1983) concluded that 'radioactivity fed into the surroundings by emission into the atmosphere from combustion of coal is very small in comparison with the radioactivity occurring naturally'.

The overall situation has been assessed by Robson (1984) who studied coal and nuclear systems in the United Kingdom and concluded that the 'radiological impact of the routine generation of electricity is less than, usually by a substantial margin, variations in natural levels of exposure as well as international safety limits. Whatever the mode of generation, estimated collective dose from the complete fuel cycle is only in the order of 0.2% of that due to natural radiation.'

A study of ^{226}Ra in bones of Polish residents and in skeletons (Jaworowski and Bilkiewicz, 1982) showed that there was less ^{226}Ra in the contemporary bones than in those from the eleventh to the nineteenth centuries.

Perhaps leaching from flyash disposal areas could contaminate nearby water (Blackburn and Gueran, 1979), but the limited study by Wagner and Greiner (1981) indicated that leachates from coals and related solids contained 'very little radioactivity of environmental concern'. Indeed, the leachability of flyash from western US coals was so low that Styron *et al.* (1981) found that the flow from the ash ponds diluted the concentrations of radionuclides in the aquifer below the ponds.

It is clear that there is a divergence of opinion about the relevance of radioactivity connected with coal usage. It seems that most coals would be in the category such that the radioactivity released during burning in 'modern plants meeting EPA's particulate emission standards is not a matter of concern' (Beck and Miller, 1980). As Corbett (1983) stated 'comparisons between radiation doses from coal burning and nuclear power are of limited value'. There was too much stress put on such comparisons, perhaps in an effort to discredit coal. The complexities of comparing the two sources of radioactivity were an underlying difficulty. However, the matter has been clarified and there seems to be general agreement that radioactivity from coal burning and nuclear-power production is only a small fraction of background radiation and is below permissible limits. It would be unwise to burn coal with more than about 30 ppm U, without checking the emissions and solid waste products. In any case, the radioactivity of flyash used in building materials should be checked. It is sobering to recall the discussion by Brown (1987) on the topic 'does a little radiation do you good?' At present, there is no definite answer to this question, but the importance of several elements at trace levels makes it unwise to categorically dismiss the possibility of a positive answer to Brown's question.

An unusual practical application of the differences in radioactivity between low-ash brown coals from Victoria, Australia, and the associated interseam sediments and overburden has been described by Aylmer, Holmes and Brockway (1987). They developed a natural gamma-ray gauge which is fitted to the front of a coal-dredger, so that the operator is warned of sediments and high-ash materials, thereby diverting them to a waste dump. This ensures that the coal removed for use has a low-ash yield. The natural gamma-ray activities of the brown coal and of the associated sediments are very low, but the difference is enough to discriminate between them (Brockway, Aylmer and Holmes, 1984).

Chapter 9

Relevance of trace elements in coal

Reported or possible beneficial or untoward effects of some trace elements in coal have been referred to in Chapter 1 and under separate elements in Chapter 5. There are several other aspects of trace elements in coal which will now be dealt with in detail.

9.1 Effects of beneficiation

'Beneficiation' is the general term for processes that aim at removing, or lessening, unwanted constituents from run-of-mine coal and for improving coal quality. Such processes are often termed 'coal cleaning' or 'coal washing'. There are various chemical and physical methods of beneficiation and there are some published results of their effects on trace-element contents. It is clear that particle size and association with the organic-coaly matter are important factors controlling the effects of beneficiation processes on trace-element contents. Relatively large mineral particles, for example pyrite and sphalerite, or clusters of minerals can be readily removed, but finely divided particles, especially those embedded in coaly matter, are very difficult or impossible to remove by the usual processes used for large-scale beneficiation, which are based on density separations in water. The in-depth study of mineral matter in coal by Finkelman (1981a) showed the presence of many trace elements associated with minerals, mainly clays, and as fine-grained discrete minerals. Separations based on specific gravity differences are bedevilled by unexpected results because many fine-grained minerals are encapsulated in organic-coaly matter (Finkelman and Gluskoter, 1983). Prior to carrying out large-scale beneficiation of a particular coal it is helpful to know something about the nature, particle size and distribution of mineral matter and *ipso facto* of many trace elements. Such a study of two low-rank coals from Wyoming and Montana, USA, helped to explain some results of float–sink separations (Straszheim *et al.*, 1988). Fine grinding to release mineral particles is limited because very fine particles may well float with the organic-rich fraction. Various new techniques, chemical and physical, have been investigated, and the effects of such processes on the redistribution of trace elements would probably differ from the situation with the conventional wet-gravity methods (NAS, 1979).

An investigation of trace elements in samples of an Illinois, USA, coal which had been treated with Na_2CO_3 solutions at different temperatures and pressures and with molten NaOH–KOH mixtures, showed marked reductions in the levels of several elements (Norton, Araghi and Markuszewski, 1985). Early experiments by Geer, David and Yancey (1943) indicated that the reduction in P content by coal cleaning depended on the mode of occurrence of P. Separation of carbon-rich material by treating finely ground coal by an oil agglomeration technique resulted in the substantial removal of several trace elements (Capes et al., 1974). Trace elements that are associated with pyrite are removed with the pyrite in most coal-cleaning procedures. For example, up to 40% of the Hg in some US coals may be removed with the heaviest fraction containing pyrite (Schultz, Hattman and Booher, 1973) and three Illinois coals lost half their total Hg with the heavy fractions (Ruch, Gluskoter and Kennedy, 1971). The results of a study of nine US coals, which were separated into different specific gravity fractions, were shown on washability curves, that is, plots of ppm trace element versus % recovery (Gluskoter et al., 1977). Another study of four Illinois coals was carried out by Harvey et al. (1983). The results of a further study of ten US coals were shown diagrammatically as cumulative trace-element content versus specific gravity of separation (Cavallaro et al., 1978). These plots show clearly the extent of removal of trace elements in different specific gravity fractions. The distribution of trace elements into cleaned coal, coal wastes and process waters produced by a $25 t h^{-1}$ demonstration coal preparation plant was examined by Conzemius, Chriswell and Junk (1988).

The effects of different beneficiation procedures on the removal of trace elements from two NSW, Australia, coals were investigated by Knott, Thompson and Ruch (1985). The procedures were conventional float–sink, selective oil agglomeration, froth flotation and a special fine-coal centrifugal separation. No advantage was found by fine grinding of coal prior to cleaning.

Although beneficiation is carried out primarily to reduce the contents of mineral matter, especially pyrite, there is the added advantage of decreasing the concentrations of several trace elements. Not only are relatively high concentrations decreased, but very low concentrations may also be decreased, as is shown in Table 9.1, which gives results for three coals from the Sydney Basin, Australia, which had been washed commercially (1984, K.W. Riley, W.C. Godbeer, pers. commun.). The reductions in concentrations of Cd and Hg are interesting. The marked lowering of F contents has been confirmed by similar results for washed coals from ten other seams in the Sydney Basin (McGlynn and Rice, 1982; 1986, W.C. Godbeer, pers. commun.).

Trace-element contents are usually decreased by beneficiation, the extent being variable for different elements and coals. In general, those elements associated with the mineral matter are more readily removed than those that are mainly organically bound. In the commonly used coal-washing methods, many trace elements are concentrated in the heavy fraction (sink), as Meserole et al. (1979) and others have found for a wide range of coals. Particle size is important and experiments are required to find the optimum size for a particular coal. It is clear that very fine particles

Table 9.1 The effects of washing on the trace-element contents of three coals from the Sydney Basin, Australia (results as ppm)

Sample	As	Cd	F	Hg	Sb
Saxonvale:					
run-of-mine	2.1	0.09	255	0.026	0.4
washed	0.5	0.06	64	0.024	0.3
reject	4.5	0.21	502	–	0.4
Wongawilli:					
run-of-mine	4.0	0.29	243	0.062	1.1
washed	1.1	0.02	98	0.041	0.7
Lithgow:					
run-of-mine	1.1	0.37	121	0.028	1.3
washed	0.5	0.15	90	0.017	0.7

embedded in the coaly matter will remain there and it is usually impracticable to remove these by very fine grinding. Although it is often advantageous to remove some trace elements, especially those of environmental significance, from some coals, there is a concomitant increase in the rejects from beneficiation. In the disposal of these washery rejects there should be an awareness of their trace-element contents, especially the possibility of releases of soluble ions. For example, pyrite is present in washery rejects and under some conditions the pyrite may be oxidized giving solutions of low pH which could leach some trace elements from the disposal area.

9.2 Seam correlation and discrimination

The correlation of seams within a basin or similar area has been carried out by using petrological characteristics (Taylor and Warne, 1960; Smyth, 1967; Falcon, 1978). Trace elements were used by Samuel (1933) to identify an anthracite seam in Wales and by Pendias (1964) to correlate some seams in the Walbrzych Basin, Poland. Smith (1941b) suggested that the high P in a South African seam could be used for correlation.

Coal seams in the Sydney Basin, Australia, which are 100–150 km apart, with very deep seams in the intervening area, have been examined by Swaine (1983) using what he termed as trend curves. These curves are prepared by plotting the ranges of trace-element contents and mean values for seams arranged vertically, that is, in stratigraphic order (Figure 9.1). Seams from the southern area are on the left and those from the northern area are on the right. Assuming that the trends are similar in each area, then the curves for Cu and V indicate correlations between Balgownie–Great Northern and Wongawilli–Fassifern. This is put forward as one approach, but it cannot be regarded as absolute.

The use of trace elements as an aid in seam discrimination was investigated by Doolan et al. (1984b) who used 103 samples from four well-separated seams from the northern area of the Sydney Basin, Australia.

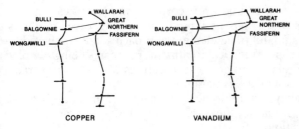

Figure 9.1 Trend curves for copper and vanadium in coal seams from the Sydney Basin, Australia (―― range of values, ● mean value; reproduced from Swaine, 1983, with permission of Theophrastus Publications)

Analytical data for 30–40 trace elements and major elements were treated statistically using mainly discriminant and cluster analysis methods. Their main conclusions were that no single element could be used for seam discrimination, that no evident relationships existed between plies within a seam and that interelement correlations were of varying relevance for each seam. However, discrimination was found by using statistically derived functions involving eight constituents, namely, Al, Ba, Cu, Ga, Mg, K, Ti and Zr expressed on a dry organic-matter-free basis (Doolan et al., 1985a). Data for B, F, Ge, Ni, P, Sr and V gave added support. This in-depth investigation depended on reliable analytical data for carefully selected samples and the development of relevant discriminant functions.

There is a continuing interest in an inorganic approach to correlating Canadian coal seams, since Newmarch (1950, 1953) used Ba, Mg, Na, Si and V to identify seams in British Columbia. Nichols and D'Auria (1981) determined 12 elements in coals from the Hat Creek and Crowsnest areas in British Columbia. Discriminant analysis of their data made it possible to differentiate between samples from seams in the same location and between seams from different locations, the respective success rates being 80% and 93%. Seams in the Byron Creek area, British Columbia, were identified by their contents of B, Na, Th and U (Goodarzi, 1987c), while Goodarzi and Cameron (1987) suggested that 'differences in element assemblages and relative concentrations among seams could be useful in seam correlation, provided lateral variation is not great'. Following a study of 179 coal samples from 12 boreholes in the Lublin Basin, Poland, Cebulak and Rozkowska (1983) decided that the correlation of coal seams there required data for at least 13 elements, but preferably 22 elements, for 'conclusive results'. Among the elements used were As, Ba, Co, Ge, Li, Nb, Ni, Pb, Rb, Sr, Ti, Y and Zr. Samples from two seams in an area of 120 km^2 in South Africa were 'distinguished clearly' by their concentrations of Br and Co (Willis and Hart, 1986). The differentiation of coals in the Keelung–Taipei region, Taiwan, was helped by considering the contents of B, Ba, Cl, Mn, P and Ti (Youh, 1978).

Bouska (1981) reviewed published work in Eastern Bloc countries apropos the distribution of trace elements in coal seams relevant to identification and correlation. Several examples are given of useful indications from trace-element contents. The overall situation has been stated well, namely 'that geochemistry can be justifiably regarded as a promising means for the solution of correlation, identification and stratigraphic problems, but it has its limitations as have other methods' (Bouska, 1981). A meaningful trace-element approach should be based on sufficient samples for proper statistical treatment, the seams being sampled carefully, especially in relation to geological features. As with most trace-element matters, proper analysis is paramount.

It is worth while making efforts to achieve seam identification because of relevance to correlation, to the assessment of deposits and to the planning of mine operations.

9.3 Degree of marine influence

Landergren (1945) suggested that B concentrations in marine and nonmarine sediments were meaningfully different, and Nicholls (1963) reviewed several relevant investigations. Several trace elements in a range of sediments were considered by Landergren and Manheim (1963) who concluded that 'the only geochemical indicator which has been found to be consistently sensitive to varying degrees of salinity is boron'. Experimental studies have shown that clays remove B from water (Fleet, 1965; Lerman, 1966), the most efficient one being illite. The contents of B in illite from Illinois coals showed the distribution of palaeosalinity in the Herrin (No. 6) coal member, Illinois, USA (Bohor and Gluskoter, 1973). An investigation of B in dirt-bands, including roof and floor subsections, from coals in the Sydney Basin, Australia, showed that mean values for B ranged from 43 ppm B in freshwater-influenced areas to 164 ppm B in marine-influenced areas (Swaine, 1962b). Detailed examination of these samples in terms of their contents of clay, quartz and carbonate minerals indicated that the effects were only found in clay-rich bands, the dominant clay being kaolinite (Swaine, 1969a).

It seemed worth while to ascertain the usefulness, or otherwise, of the B contents of coals as indicators of the degree of salinity prevailing when the coals were being formed. Results for coals from the Sydney Basin, Australia, indicated increasing mean B contents from 18 ppm (Illawarra) to 54 ppm (Newcastle) to 91 ppm (Tomago) to 164 ppm (Greta). These results were interpreted as indicating that the Greta coal measures were influenced by marine conditions, the Tomago by brackish or intermittent marine conditions and the Illawarra by freshwater conditions (Swaine, 1962b). The Newcastle coal measures seemed to have been influenced by conditions somewhere between brackish and freshwater. Some support for one of the above postulates was given by a geological study that proposed that 'throughout the Tomago Coal Measures marine, or at least a brackish influence remained' (Diessel, 1980). This approach was used for coals from the Bowen Basin, Australia, where freshwater conditions predominated except for one area where brackish to marine conditions were indicated by B contents in coal (Swaine, 1971), as shown in Figure 9.2. It is considered that these latter coals were affected by a marine incursion during the early stages of coal formation. It has been found that the spread of B results even in a marine-influenced area, such as the Greta coal measures, may include some very low results (Swaine, 1967). Perhaps

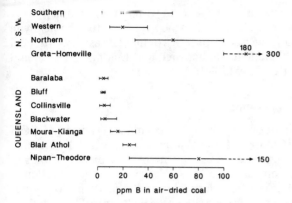

Figure 9.2 Ranges and mean values for boron in coals from seams in NSW and Queensland (reproduced from Swaine, 1983, with permission of Theophrastus Publications)

there could have been some areas which were less affected by the marine incursion, for example raised swamps, as postulated by McCabe (1984). The relevance of other trace elements in coal to changes in salinity conditions was examined for the Sydney Basin. Although Ga and Co, and to a lesser extent Ni and Pb, tended to be higher in freshwater-influenced than in marine-influenced coals, at most these elements were only broadly in keeping with the indications given by B contents (Swaine, 1967). In a study of Kittanning coals, USA, Bailey (1981) found Zn higher in marine-influenced coals, while Cr, Cu and Ti were higher in fluvial-influenced coals.

The use of B contents in coal is not proposed as an indicator of palaeosalinity, but, rather, as an indicator of the degree of marine influence on coals, the effects probably occurring during the early stages of coal formation when the organic matter had an enhanced capacity for removing B from solution. It is not clear whether this was by adsorption or by chemical fixation. There is indirect evidence that B may be organically combined in some coals, for example it seems that B in the Greta (marine-influenced) coals is organically combined, whereas this does not seem to be the case for the Illawarra (freshwater-influenced) coals in the Sydney Basin (Chapter 3). The discrimination of marine and nonmarine sediments is not simple and no one parameter can be expected to be definitive. The use of B contents in coal has limitations that are probably covered by the statement of Nicholls (1963), namely 'the present status of boron as a salinity indicator may be fairly described as one in which further work is required to establish whether it can be considered reliable in the absence of palaeontological evidence'.

9.4 Environmental aspects

Perhaps the main current emphasis on trace elements in coal is on their effects in the environment. The vexed question of radioactivity is dealt with in Chapter 8. Other aspects are connected with mining, rehabilitation after mining, and combustion. Health effects are referred to specifically in Section 9.5. There is an overview of environmental aspects of coal utilization, including trace elements (Chadwick and Lindman, 1982), and a review by Gibson and Kennedy (1980).

9.4.1 Mining and related operations

General information on underground and surface mining is given in PECH (1980) and environmental controls for coal mining are covered by Hannan (1980). Underground mining does not usually affect trace elements in coal, although there could be some local effects. However, some abandoned mines may be a source of acid waters derived from the oxidation of pyrite. The oxidation of pyrite depends on several factors including pH, oxidation-reduction potential, microorganisms and the nature of the pyrite (Swaine, 1978a), framboidal pyrite being very reactive (Swaine, 1986a) because of its large surface area. Such acid waters may leach some trace elements from rock and coal in the mine.

There is more likelihood of oxidizing conditions during surface mining where freshly exposed surfaces are more readily aerated and more prone to microbiological effects that catalyse the oxidation of ferrous to ferric iron, a key step in the oxidation of pyrite. The resultant lowering of pH may enable the aqueous solutions to leach some trace elements from the overburden. However, the presence of carbonate minerals will increase the pH, thereby favouring the removal of trace elements by adsorption on clay and iron oxide particles. It has been found in the United States that surface mining produces much less acid mine drainage than underground mining (PECH, 1980). Modern methods of mining take into consideration the need to control water movements and to avoid upsetting aquifers and local waterways. In particular, solid wastes need to be disposed of properly.

Beneficiation may be carried out in the vicinity of mines, thereby adding another solid waste material and large volumes of water containing soluble ions and finely ground coal and rock particles. Problems from trace elements in leachates should be minimal if there is a proper selection of emplacement sites, correct techniques of handling wastes and treatment of leachates, if required. It may be economical to treat wastes in order to produce useful fill material and to generate some power by combustion of the coal in the wastes. Environmental aspects of coal wastes are discussed in NAS (1979).

The main source of concern with wastes from coal mining is pyrite, and any methods of controlling this will reduce the mobilization of trace elements.

9.4.2 Rehabilitation after mining

An important part of coal mining, especially surface mining, is the rehabilitation of disturbed land. General aspects of rehabilitation after coal mining are discussed by Rummery and Howes (1978), Hannan (1980), PECH (1980) and Chadwick and Lindman (1982). The last reference gives a classification of land disturbance from underground mining, surface mining and waste deposits, prior to outlining methods of reclamation used in several countries. There has been little consideration of the relevance of trace elements to rehabilitation. However, when the land is ultimately used for agricultural purposes, trace-element availability should be taken into account. Many trace elements in reclaimed land will be useful for the growth of a wide range of vegetation. Some caution is needed with B and Mn which may limit the growth of some plants under certain conditions of pH and drainage. Although too much available Mo and Se does not affect plant growth, an excess of these elements may affect grazing animals. There are few reports of untoward effects from trace elements in areas disturbed by coal mining. Two reports refer to possible harm from Mn (Rees and Sidrak, 1956) and from B (Holliday et al., 1958), but such effects would decrease as weathering proceeds (Jones and Lewis, 1960). It is wise to carry out pot and field trials of plants on particular areas prior to establishing vegetation. In general, trace elements in overburden and the like are probably more likely to be useful to plants rather than detrimental, and in some cases there may even be deficiencies of trace elements. However, this may not be the case with washery rejects which have variable trace-element contents that need to be evaluated (Knott, Thompson and Lee, 1985). It seems that Agricola in the sixteenth century was aware of the need for reclamation after metal mining, when he stated 'where woods and glades are cut down, they may be sown with grain . . . so that they repair the losses which the inhabitants suffer' (Agricola, 1556). Nowadays, reclamation is an integral part of mining.

9.4.3 Combustion for power production

The main products of the combustion of pulverized coal are carbon dioxide, nitrogen oxides (NO_x) and sulfur oxides (SO_x) together with water vapour. The decomposition of pyrite and carbonate minerals also yields some SO_x and CO_2. The mineral matter is mostly decomposed to give residual solid materials that are found in bottom ash (from up to about 20% of the original mineral matter) and flyash (from up to about 80% of the original mineral matter). The bottom ash falls out of the combustion area in the furnace, whereas the flyash is carried along with the flue gases. In a modern power station burning pulverized coal, more than 99% of the flyash should be removed from the flue gases by some means of particle attenuation, usually electrostatic precipitation or the use of fabric filters. This means that most trace elements are also removed, the separation efficiency being 95–99+% (KHM, 1983). Most of the bottom ash and flyash are removed to ash disposal ponds, but some flyash is used, for example as a partial replacement for cement in concrete (Swaine, 1981b). A small proportion of flyash, mostly fine particles (say, <10 μm diameter) reaches the atmosphere with the stack gases. The distribution of trace elements during coal combustion is shown in Figure 9.3 (Swaine, 1978d). Varying proportions of the halogens, Hg and Se may reach the atmosphere in the vapour state, as well as associated with fine flyash particles. Flyash is composed mostly of glassy aluminosilicates, together with some crystalline phases, namely mullite, iron oxides and quartz. Perhaps the

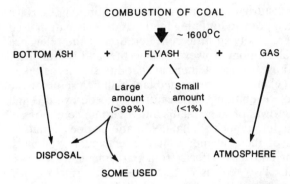

Figure 9.3 The distribution of trace elements during the combustion of coal in a modern power station

glassy aluminosilicates are formed by reactions between the decomposition products of calcite and siderite, namely CaO and FeO, and of clays (Swaine, 1978e). Since complete combustion is difficult to achieve, flyash contains some carbon, usually of the order of a few per cent. This carbon is probably similar to a char or coke (Swaine, 1969b).

The distribution of trace elements and their concentrations in bottom ash, flyash and the gas phase depends on the concentrations in coal, on the chemical properties of the element, especially volatility, on the boiler type and load, on the operating conditions and on the operation of the particle-attenuation equipment. The concentrations of several trace elements increase as the particle size decreases (Davison *et al.*, 1974; Gladney *et al.*, 1976; Swaine, 1977a, 1978e), the mechanism of enrichment probably being volatilization followed by condensation or adsorption on previously formed particles as the flue gas temperature decreases. Perhaps compound formation may occur in the surface layers on flyash particles, as suggested for Pb and Se (Swaine, 1977a, 1978c). An alternative mechanism was put forward by Hulett and Weinberger (1980), namely that 'there is also the possibility that surface enrichment can occur by diffusion of trace elements from the inside of the particle'. Smith (1980) has reviewed the trace-element chemistry of coal during combustion, including mechanisms for the formation of flyash and for surface enrichment of trace elements. There are several reports of investigations of trace elements in coal, bottom ash, flyash and stack gases, the aim being to calculate mass balances; for example, the detailed study of trace-element distributions at three West German power stations (Gerhard *et al.*, 1985). These calculations depend on representative samples, accurate analysis and proper measurements of process flow rates, surely very difficult to achieve. In particular, the sampling of stack gases poses problems. Smith (1987) has assessed the situation well in her statement 'the use of assumed mass balances to infer the stack emissions is unreliable'. A novel approach was used by Brumsack, Heinrichs and Lange (1984) who calculated emissions from West German power stations by means of a model based on element partition coefficients between bottom ash, flyash and stackflyash. There are two good reviews of trace elements from coal combustion (Lim, 1979; Smith, 1987), together with two bibliographies (Knox, 1980; Harter, 1982) and information on the behaviour of inorganic constituents in coal combustion (Raask, 1985a). Slack (1981) has reviewed various aspects of particulate matter from combustion processes, including methods of removal and of the control of gaseous emissions (SO_x and NO_x).

As expected, bottom ash contains the same trace elements as coal, but elements that can be regarded as volatile are depleted relative to flyash (Heinrichs, Brumsack and Lange, 1984; Tomza and Kaleta, 1986). However, some very volatile elements have been found in bottom ash, for example Bi, Br, I and Tl (Brumsack, Heinrichs and Lange, 1984) and Hg (Kaakinen et al., 1975; Swaine, Godbeer and Morgan, 1984a). Several elements, especially volatile elements are enriched in flyash (Block and Dams, 1976; Swaine, 1977b). There is a bimodal particle-size distribution in flyash (Haynes et al., 1982) with peaks at about $0.1\,\mu m$ and at about $3\,\mu m$ for some US flyashes (McElroy et al., 1982). Data for trace elements in flyash are given by Furr et al. (1977), Swaine (1977b) and Meij, Janssen and van der Kooij (1986). Swaine, Godbeer and Morgan (1984a) and Swaine (1985b) investigated variations in the contents of trace elements in flyash from four electrostatic precipitators at a power station, over a period of 3 years. Changes in the ratios of outlet to inlet values with time showed a general constancy, but there were two peaks for some volatile elements (Cu, Ge, Pb), perhaps caused by changes in operating conditions affecting the temperatures of electrostatic precipitation (Swaine, 1985b). The contents of several trace elements, especially volatile ones, were highest in flyash from the final precipitator (outlet), as expected because of the relatively high proportion of fine particles. This means that the flyash emitted with the stack gases has higher contents of volatile trace elements than the bulk flyash removed by the first precipitator (inlet). Meij et al. (1984) believe that the composition of flyash in outlet samples 'is indicative of the composition of the emitted flyash'.

The collection of representative samples of particulates from stack gases is difficult. Varying proportions of As, Br, Hg, I and Se are in the vapour state in stack gases (Germani and Zoller, 1988). Values for eight trace elements in flue gases emitted from UK power stations which have been calculated by Halstead (1981) show wide variations, for example from $3-23\,\mu g\,m^{-3}$ for Sb and Se to $4-230\,\mu g\,m^{-3}$ for As, 70% of the flyash being $<10\,\mu m$. Based on these estimates Halstead (1981) concluded that 'it is thus reasonable to assume that particulate emissions from the stacks of modern coal-fired power stations contribute only a minor part of the ambient concentrations of the trace elements considered in the atmosphere at ground level'. Data for daily stack emission rates for 13 trace elements from three US power stations showed higher values for most elements from the burning of eastern (probably bituminous) coals compared with western (probably subbituminous) coals; the daily variations at one power station were as low as a factor of about 2 for several elements (Que Hee et al., 1982). In the study by Kautz et al. (1984) some trace elements were determined in flue gas samples, and Meij, Janssen and van der Kooij (1986) determined the annual emissions ($t\,y^{-1}$) of B, Br, Cl, F, Hg and Se in the vapour phase from Dutch power stations.

Environmental and climatic impacts of coal burning are dealt with in depth by Singh and Deepak (1980).

The distribution of trace elements during other processes of deriving energy from coal has not been sufficiently well investigated. There is some information on gasification processes (NAS, 1979) and fluidized-bed combustion (Smith, 1987). Evidence from small-scale fluidized-bed combustors indicates variations in enrichments on small particles, perhaps because of changes in residence times in the hot zone and rates of cooling (Littlejohn, 1984). Some of the conditions of the operation of fluidized-bed combustors favour better retention of several volatile elements than in conventional combustors (Smith, 1987). There is some evidence that the use of flue gas desulfurization systems can improve the retention of trace elements, even those passing particle-attenuation systems (Smith, 1987), although it has been stated that such systems 'may well, in fact, increase particulate emissions' (Gibson and Kennedy, 1980).

9.4.4 Deposition of trace elements from the atmosphere

Trace elements in the atmosphere are mostly present in association with particulates, although some may be at least partly in the elemental state, for example Hg (Lindberg, 1986). Natural sources of atmospheric trace elements include the weathering of rocks and soils, volcanism, erosion of metal-rich surface deposits, reactions at water surfaces, thermal springs, forest fires and vegetation. Among the anthropogenic sources are the combustion of coal, oil and wood, mining and industrial operations, waste incineration, agricultural operations, vehicle emissions, including tyre- and engine-wear and cremation. There is an interesting comment on Hg in Sweden, namely 'at the beginning of the 1990s the national emission from coal firing will be of the same order as emissions from crematoria today' (KHM, 1983). Prime sources vary for particular trace elements at different locations, and most of the sources will contribute to background levels found in remote places. Meszaros (1981) estimated that natural sources account for 4–5 times more than anthropogenic sources and Jaworowski, Bysiek and Kownacka (1981), after studying the flows of metals into the atmosphere, concluded that 'the anthropogenic contribution is a small fraction of the flows, which are dominated by natural processes'. Local sources will influence the levels of atmospheric trace elements in nearby areas. The measurement of trace elements in deposition from the atmosphere at remote locations (background) requires careful sampling, handling of samples and analysis; and information should be given about 'distance from sources and the prevailing meteorological air masses' (Barrie et al., 1987). There is a good review of 14 trace elements associated with airborne particulates (Schroeder et al., 1987); these elements are described as 'toxic', although ten of them also have positive biological functions. The long-range transport of trace elements has been studied by Pacyna and Ottar (1985) and Pacyna et al. (1985) who found variations between summer and winter. Among recent studies of emissions of trace elements to the atmosphere are those by Millancourt et al. (1986) on Cl and F and by Lightowlers and Cape (1988) on HCl.

There are two approaches to considerations of trace elements in the atmosphere, termed 'source' and 'receptor'. Knowledge of the composition of sources is important and is a requirement for the development of receptor models which seek to apportion the contributions of particular trace elements to various sources. Undue emphasis on the composition of sources *per se* is not a realistic approach to ascertaining composition of aerosols at a distance from the source. Recent studies related to receptor modelling have been carried out by Pacyna (1986) using statistical trajectory sector analysis, by Sanchez Gomez and Ramos Martin (1987) using cluster analysis, by Lowenthal and Rahn (1988) and by Lowenthal, Wunschel and Rahn (1988) using regional elemental tracers, and by Hopke (1988) using target transformation factor analysis. Gordon (1988) has reviewed the status of receptor models and concluded that 'after 20 years of study, receptor models have moved into regulatory use'. The measurement of trace elements in samples of material deposited from the atmosphere has been carried out by various methods, for example deposit gauges, sticky surfaces, impactors and moss bags and envelopes. The concentrations of trace elements in samples of air is carried out by low-vol and high-vol sampling. Information on some of these methods is given by Schroeder *et al.* (1987), together with comments on the advantages and disadvantages of several filter substrates. Tramontano, Scudlark and Church (1987) give details of an automated method for collecting samples of atmospheric precipitation. The use of live moss and later of moss held in bags as collectors of trace elements in atmospheric deposition seems to originate from studies by Rühling and Tyler (1970) and Goodman and Roberts (1971) respectively. Detailed information on the moss technique is given by Little and Martin (1974) and Goodman *et al.* (1979), while Martin and Coughtrey (1982) have reviewed all aspects of biological monitoring of trace elements from the atmosphere. Much has been written about the fate of trace elements in the atmosphere and how they are deposited and there is a relevant concise summary of the situation up to 1984 (Swaine, Godbeer and Morgan, 1984a). Trace elements are removed from the atmosphere by dry deposition and wet deposition which is a two-stage process involving scavenging by rain (in-cloud) and precipitation (sub-cloud). It is not easy to measure wet and dry deposition separately or to assign the proportions of trace elements reaching the earth by each of these processes. Most methods measure the sum of wet and dry deposition. Graedel (1982) summed up the situation: 'numerical estimates for wet and dry deposition are too uncertain to permit reliable trace component budgets to be constructed, even if source identification was more definitive than is the case'.

The use of moss to collect trace elements in deposition from the rural environs of an Australian 1240 MW power station, burning pulverized bituminous coal and fitted with electrostatic precipitators, was investigated in detail by Swaine, Godbeer and Morgan (1984a). The moss was held in fine-mesh envelopes, each containing 2 g purified *Sphagnum* moss. Because the amounts of most trace elements were very low, even in samples collected over three month periods, the moss had to be cleaned thoroughly and contamination had to be kept down and ascertained by frequent analyses of blanks. Up to 50 trace elements were determined by

Table 9.2 Deposition as mg m^{-2} (ranges for 12 results over 3 y period; mean values are in parentheses)

Distance from power station (km)	1.8	6.6	15.3	27.4
Boron	1.7–13 (5.8)	0.25–2.4 (1.4)	0.4–1.6 (0.92)	0.15–0.55 (0.33)
Cadmium	0.066–0.117 (0.022)	0.002–0.037 (0.014)	0.002–0.030 (0.011)	0.002–0.022 (0.009)
Copper	0.50–3.4 (1.6)	0.12–1.1 (0.57)	0.01–0.67 (0.34)	<0.01–0.47 (0.18)
Manganese	2.6–11 (7.7)	0.73–9.3 (5.9)	0.07–13 (3.7)	<0.01–10 (2.4)
Zinc	2.1–8.4 (4.5)	0.59–4.2 (2.2)	0.61–3.5 (1.6)	<0.01–2.2 (0.63)

Table 9.3 Annual deposition around the Australian power station compared with annual inputs from rock weathering, litter decay and fertilizers (as mg m^{-2})

Element	Location (km)	Annual deposition	Rock weathering	Litter decay	Fertilizers
Arsenic	1.8	0.80	0.04	0.20	2.4
	6.6	0.12			
Cadmium	1.8	0.09	0.003	0.15	0.02
	6.6	0.06			
Copper	1.8	6.3	1.3	8	0.16
	6.6	2.3			
Manganese	1.8	31	26	400	5
	6.6	23			
Molybdenum	1.8	0.91	0.04	0.15	0.07
	6.6	0.19			
Zinc	1.8	18	2	40	3
	6.6	8.8			

AES, FAAS, EAAS and INAA on samples of coal, flyash, bottom ash and soil, as well as the moss, over a period of four years. Results of this in-depth study are given by Godbeer, Morgan and Swaine (1981, 1984), Swaine, Godbeer and Morgan (1983, 1989), Filby and Swaine (1983), Nguyen et al. (1983) and Swaine (1984a). Some of the results for B, Cd, Cu, Mn and Zn in the deposition at four locations are shown in Table 9.2. As expected, the amounts of trace elements deposited decrease with distance from the power station. These results are lower than published results for rural areas in other countries and much lower than those for urban and industrial areas (Swaine, Godbeer and Morgan, 1984a, 1989) and indicate that the amounts of trace elements being added to soils are insignificant compared with the natural contents in soils. Bunzl, Rosner and Schmidt (1983) found that the emissions of Co, Ni and Pb from a power station 'are obviously too small to change the local distribution pattern of these elements in the soils'. These results can be put into perspective by comparing the annual deposition of trace elements at two locations (1.8 and 6.6 km from the power station) with annual inputs from rock weathering, litter decay and fertilizers (Table 9.3), where the rock weathering estimates are from Bowen (1979), assuming an annual weathering of 27 g of crustal rock per m^2 of soil, the litter decay values are from Bowen (1979) and the fertilizer values are from Bowen (1966), the latter being based on data in Swaine (1962c). Inputs of Cu, Mn and Zn from litter decay are higher than those from the annual deposition near the power station. The As values are most interesting with the highest inputs being from fertilizers, and the Cd values are very low overall. For Mo, the annual deposition at the 1.8 km location is more than the other inputs but this is only an addition of about 0.3% to local soils.

The examination of the *Sphagnum* moss using SEM showed that the structure is made up of small holes, about 10 μm across, and folds (Swaine, Godbeer and Morgan, 1983, 1984a), and the moss after exposure shows the retention of two kinds of particles, angular ones (soil and rock) and spherical ones (flyash). This means that the main sources of trace elements are most probably soil and rock particles and flyash. It is known that moss also retains some elements by cation exchange on cell walls (Martin and Coughtrey, 1982). It has been suggested by Straughan *et al.* (1981) that the difference between the ratio of ^{87}Sr to ^{86}Sr in flyash and soil may be useful as a tracer for flyash in the environment. It is surely desirable to be able to estimate the relative amounts of these two kinds of particles and, knowing the composition of each, to estimate amounts of each trace element. There are two ways of doing this, one by using the favourable ratio of Ge in outlet flyash to that in soil (75 : 1.5 ppm), and the other by estimating the amounts of mullite in samples of deposition, knowing that mullite occurs in flyash but not in soil. In Figure 9.4, changes in the levels of Cu and Zn at a location 1.8 km from the Australian power station are shown for a three year period. Values for total deposition are shown by the unbroken line and those for flyash (calculated from Ge-contents) are

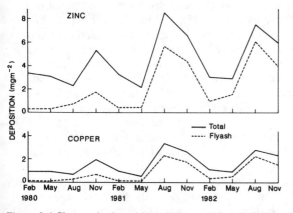

Figure 9.4 Changes in deposition of copper and zinc at a location 1.8 km from an Australian power station (reproduced from Swaine, Godbeer and Morgan, 1989, with permission of New Zealand Trace Elements Group)

shown by the broken line. It is seen that the proportion of flyash in the total deposition varies greatly, sometimes being less than 10%. The marked seasonal variations in total deposition are caused by changes in wind (speed and direction), atmospheric stability and the nature of the nearby terrain. The results of the overall investigation showed that there were different patterns of distribution of trace elements at each location. It is clear that meaningful results cannot be expected from short-time sampling at a few locations.

It is suggested that the moss-envelope technique is useful for monitoring the atmospheric deposition around power stations. Temple *et al.* (1981) have summarized the usefulness of moss for this purpose and Martin and Coughtrey (1982) made the pertinent statement 'that if any discrepancies arise between measurements obtained by deposit gauges and by moss bags, the cause of such discrepancies should not necessarily be attributed to moss bags rather than to deposit gauges'.

9.5 Health aspects

Because coal contains some trace elements that may be deleterious to health under certain conditions, there is a continuing interest in such elements in coal. It is imperative that considerations of the biological effects of trace elements should take into account the fact that most common trace elements may be essential or hazardous. Essential elements may show deficiency or toxicity depending on concentration and availability, the form of the element being important, that is, whether it is present as an ion or as a simple or complex inorganic or organic compound. In the sixteenth century Paracelsus appraised the situation well: 'All substances are poisons; there is none which is not a poison. The right dose differentiates a poison and a remedy.' For essential elements there is an optimal range of concentration for healthy growth of plants and animals, and this range may depend on pH, oxidation-reduction conditions and other factors. Thus the assessment of the status of a trace element in coal in regard to health is not a simple matter and should not be the subject of generalizations. The exceptions are Cd, Hg and Pb which have no known essential roles in plant or animal nutrition. Hence, these elements should be kept in mind when assessing health aspects of coal mining and usage. However, their generally very low concentrations in coal mean that untoward effects are unlikely. Similarly with As, B, F, Mn, Mo and Se, which are more important for their essentiality, although under some conditions there may be undesirable effects. For example, leachates from some flyash disposal areas may have more of some elements than is permitted by local regulations. Most of these elements are designated by PECH (1980) as 'elements of greatest concern' and by Swaine (1982) as of 'prime environmental interest' (Table 1.1).

Health aspects of trace elements in coal have been reviewed by Lim (1979) and NAS (1979), and PECH (1980) deals with possible health effects of trace elements during mining, beneficiation, combustion and synthetic fuel production. Caution should be used in interpreting results of epidemiological studies because of the difficulty of isolating the variables. For example, how could a study of the health of people living in the environs of a power station be related to trace elements in stack emissions, unless there was a high concentration of an element? There were some estimates of concentrations of trace elements in air near a modern power station which indicated that the concentrations 'were several orders of magnitude below standards recommended for inhalation exposure' (Lim, 1979). The question of radioactivity from coal combustion is discussed in Chapter 8, the conclusion being that there is no radioactivity health hazard from the burning of most coals in modern power stations. However, it is salutary to recall the statement by Piperno (1975) that 'no distinct causal relationships have been shown to exist between a pathological entity and coal-related trace-element pollution; this of course does not mean there are none'. However, after reviewing environmental and health aspects of increased coal utilization Gehrs *et al.* (1981) concluded that 'trace element emissions including radioactive elements, from coal combustion in power plants are generally considered an unimportant source of exposure to trace elements'.

In regard to vegetation, KHM (1983) concluded from 'cultivation experiments that the fly ash emitted from installations with good dust separation will not demonstrably raise levels in our useful plants. Even in the case of woodland the deposition of metals from coal combustion will lie far below the levels where any negative effects can occur.' This comment is based on an in-depth assessment of coal which would be burnt in Sweden. Kohno, Takanashi and Fujiwara (1982) concluded that 'it appears that atmospheric releases of trace elements are not likely to have significant and hazardous effects on soil and vegetation' near power stations.

Some flyash is used as a partial replacement for cement in concrete. There may be a slight increase in the emission of radon compared with other building materials, but this is generally within accepted regulations. If flyash is used as a soil amendment, some trace elements should be taken into account. Undue concentrations of B may affect some plants adversely (Holliday *et al.*, 1958), but continued weathering may reduce such concentrations to tolerable levels (Jones and Lewis, 1960). The uptake of B by plants on flyash-amended soils depends on several factors, including type of soil, rate of application of flyash and plant species (Elseewi *et al.*, 1981). In some cases, B-deficiency in plants may be corrected by flyash additions. Before making final decisions about flyash additions to soils, the relevant factors should be examined using greenhouse and laboratory experiments. Under some conditions Mn may affect some plants (Rees and Sidrak, 1956), and Mo and Se uptakes from flyash amended soils should be considered in relation to the health of animals eating plants growing on such areas. The properties of Australian flyashes relevant to their agronomic use have been studied by Aitken, Campbell and Bell (1984).

In general, it seems that trace elements cannot be regarded as a hazard during coal mining and usage, assuming that due care is taken to ensure proper operational conditions. Indeed, with some elements that are essential for nutrition, there may be slight advantages from minor additions to, say, soils. It has been stated that 'there is a danger that the common stress on toxicity may shift the emphasis away from essentiality' (Swaine, 1989), surely a most undesirable state of affairs. The above discussion has dealt with coals that are used for power production and the like. There are examples of local occurrences of coal with high contents of one or other trace element, for example the high-Se coals in Enshi province, China (Yang et al., 1983). In order to avoid the continued use of such coals perhaps 'serious consideration should be given to regulation of upper limits in the trace element composition of fuels that are burned' (Wallace and Berry, 1979). It is interesting that Se-deficiency problems exist on a large scale in some areas of China, where Keshan disease and Kashin-Beck disease may affect millions of people.

9.6 Coal as a possible source of metals and nonmetals

The occurrence of occasional high contents of some trace elements in coal has prompted many suggestions for extracting these elements from coal ash, flyash or other coal-based materials. Small-scale recovery of Au and Ag was carried out on some USA coals containing a few ppm Au (Stone, 1912) and a peat deposit in Wales was used as a Cu ore (Andrews and Fuge, 1986). A mixed vein of barite and witherite in a colliery at Durham, UK, was a major source of these minerals for many years (Briggs, 1934). Various suggestions for coal as a source of useful elements have been reviewed by Swaine (1962a). Headlee and Hunter (1955b) suggested that coal ashes with an enrichment ratio (coal ash to earth's crust) of greater than 10 might be a source of Ag, As, B, Bi, Ga, Ge, Hg, La, Pb, Sb, Zn and Zr, while Leutwein and Rösler (1956) indicated that B, Be, Ge, and V could be extracted from coal ash. Since Goldschmidt (1930) suggested coal as a source of Ge there has been a continuing interest in the extraction of Ge from certain by-products of coal, but this does not seem to have stimulated large-scale production of Ge.

Because there has been so much work done on Ga and Ge as sources of these elements in coal, relevant publications will be reviewed. Treatments used to extract Ga and Ge from flue dusts are discussed by Reynolds (1950). Morgan (1935) treated coal ash with concentrated HCl and was able to separate $GaCl_3$ from $GeCl_4$ by distillation, the $GaCl_3$ remaining in the still. Flue dusts from producer gas units yield recoverable amounts of Ga and Ge (Powell, Lever and Walpole, 1951) and Ga and Ge can be recovered from some flue dusts from gas manufacture (Powell, 1954). A small-scale plant was made to extract Ga from some Japanese flue dusts and gas liquors (Inagaki, 1956). Another method was based on heating coal with iron humates to produce gaseous Ga compounds, which were treated to yield a solution containing 5–20 g Ga per litre from which Ga was separated electrolytically (Ryczek et al., 1970). Zhou and Ren (1981) regarded a certain Chinese coal as 'rich enough in gallium to make gallium extraction profitable'. Flyash from South African coal is proposed as a source of Ga by Marshall, Robert and Burden (1987), and Shpirt et al. (1984) feel that the recovery of Ga and Ge from flyash should be practicable. Research leading to the production of Ge in the United Kingdom from flue dusts has favoured reduction processes (Chirnside,

1950; Chirnside and Cluley, 1952; de Merre, 1956). Japanese processes were based on gas liquors, flue dusts and similar materials (Inagaki, 1955; Inagaki, Oikawa and Moriya, 1955; Ono, Inada and Konno, 1955). In one process, 30 g GeO_2 were produced from 100 t coal (Inagaki, 1955). Research on Soviet coal and related materials enabled the Soviet Union to 'completely eliminate the import of germanium for production of semiconductors' (Krichko, 1984). It is clear that some Ge has been produced commercially in the United Kingdom, Japan and the Soviet Union. Investigations were carried out on the potential value of Ge in flue dusts from US coals (Brown and Carman, 1954; Corey et al., 1959). Seidl and Stamberg (1960) showed that small amounts of Ge in solutions of low pH may be isolated by using a chelate resin made from fluorone derivatives and formaldehyde. A chemical extraction technique was used by Chaudhuri, Gupta and Khathing (1979) to remove Ge from some Indian coals.

Suggestions of coal as a source of other trace elements have been made by Jedwab (1960) for Be, by Kuhl (1957) for Co, by Correns (1956) for I, by Dvornikov (1967e) for Hg from coking processes, by Reid (1981) for Se and by O'Gorman and Walker (1971) for Cu and Sn. Following discoveries of coals enriched in U, for example, in western United States, it was realized that these coals could be regarded as low-grade U ores (Davidson and Ponsford, 1954). Local deposits may contain several per cent U in coal (Denson, 1959). Taupitz (1984) refers to U-rich coals from Macedonia, Greece, which could yield some U as a by-product of coal combustion. However, care should be taken that the combustion would not be detrimental to the local environment. The occurrence of sphalerite in some coals from the Illinois and Forest City Basins, USA, 'may represent an exploitable by-product of coal mining and combustion', namely Zn (Whelan, Cobb and Rye, 1988).

It is clear that flue dusts from processes in which coal is subjected to reducing conditions may have enhanced contents of some trace elements, notably Ga and Ge. Enrichments in several trace elements may occur during grate firing, for example up to 0.8% As, 5% B, 0.2% Bi, 0.7% Cu, 0.4% Ga, 1% Ge, 0.3% Mo, 2% Pb, 0.25% Sb, 0.08% Tl and 1% Zn (Brown and Swaine, 1964). There is some enhancement of several trace elements in certain flyashes, especially in samples from the outlet electrostatic precipitator. For example, Swaine, Godbeer and Morgan (1984a) found enhanced contents of As, Ga, Ge, Hg, Pb, Sb, Sn and Zn in samples from the outlet precipitator at a power station burning an Australian bituminous coal, but these contents were much lower than those that can be obtained in some flue dusts. The glassy nature of about 70% of most flyashes makes the extraction of trace elements difficult. 'The use of coal or coal by-product as starting material for the production of a particular element depends mainly on economic factors, which may change quickly.' (Swaine, 1962a.) In a suitable economic milieu, methods of upgrading relatively low-grade material using specific chemical reagents, for example phenylfluorone for Ge, may be applicable.

Trace elements should be considered during coal mining, rehabilitation after mining and during coal usage, especially power production. Adverse health effects from trace elements during these operations are not expected from coals that are normally mined and used. The possibility of using coal or coal-related products as sources of some elements should be kept in mind, especially in relation to changing economic conditions. The current interest in environmental aspects of coal demands that trace elements must be taken into account, if only to allay fears that may be fuelled by lack of proper information.

Chapter 10

Concluding remarks

It is clear that much work has been carried out on various aspects of trace elements in coal, perhaps the main current interest being in environmental aspects. In the preceding chapters, relevant published information has been reviewed concerning the origin, mode of occurrence, methods of analysis, contents, seam variations, radioactivity and various practical applications of trace elements in coal. Although the sampling and analysis of coal for trace-element determinations are not easy, suitable methods are available. The modern trend is towards multielement methods, but there is still a place for AAS methods. In the hands of competent analytical chemists these methods can yield reliable results but it must not be assumed that such results can be obtained by tyros. The data in Chapter 5 should be a useful guide for assessing results, especially those from newly mined areas.

It is pertinent to point out several topics which are worthy of further investigations. Although much is known about the associations of trace elements with mineral matter, more work is required to elucidate the nature of finely divided minerals in the submicrometre range, often embedded in macerals, and of the associations of trace elements with clays. In most coals, trace elements are primarily associated with the mineral matter, except for some low-rank coals. The 'determination of the structure of coals is one of the most difficult and most important problems facing chemists' (Green et al., 1982), and a better understanding of coal constitution may improve our knowledge of how some trace elements are associated with the organic matter. Better knowledge of the modes of occurrence of trace elements in coal is needed to improve our understanding of the fate of the trace elements during beneficiation, mining, rehabilitation and combustion. The variability of trace-element contents in seams and between seams may well depend primarily on the distribution of mineral matter, especially fine-grained particles, and is therefore relevant to seam correlation. In any case, the successful use of trace elements for seam correlation probably depends on large numbers of determinations for several elements, so that there are enough data to use the necessary statistical methods.

The environmental aspects of trace elements from the combustion of coal will probably continue to be the most important consideration. In this connection, more measurements are needed of the amounts of trace elements being deposited at different locations and times in the environs of power stations, as discussed in Section 9.4.4. As Lim (1979) pointed out, 'atmospheric pollution from coal should be placed in the proper perspective by comparing coal combustion emissions with emissions from other sources'. Hence, it is worth while estimating those proportions of trace elements from coal combustion that are part of deposition from the atmosphere, and the use of Ge for this purpose should be given further consideration. The monitoring of trace-element depositions must be done on a space–time basis, using several locations over a period of at least one year. The improvements in the efficiency of particle-attenuation devices, for example electrostatic precipitators and fabric filters, mean that flyash contains relatively higher contents of many trace elements and hence care must be taken to prevent the leaching of significant amounts of trace elements from ash disposal areas. More research is needed to determine what changes occur to trace elements in the atmosphere and the nature of the deposition process. It is well to recall the statement by Lim (1979) that 'in the context of atmospheric pollution, coal combustion is neither the only nor the most important source of trace metals'.

Rehabilitation after coal mining may require information about the availability of some trace elements to plants and ultimate effects on grazing animals (Section 9.4.2).

New methods of coal combustion, notably fluidized-bed, may alter the patterns of distribution of trace elements, thereby affecting the emissions to atmosphere. In these cases, the deposition around the source should be measured. Likewise, the effects on trace-element emissions of wet scrubbers for the removal of sulfur oxides should be examined. There is a need for research on the fate of F during combustion and its eventual status in the atmosphere.

There is a continuing interest in coal and its by-products as a possible source of valuable metals or nonmetals, especially Ga and Ge, and a viable situation 'depends mainly on economic factors, which may change quickly' (Swaine, 1962a). Hence, it is pertinent to keep these matters in mind when assessing trace-element data.

Health aspects of trace elements in coal should be seen in the proper perspective, avoiding the pitfall of stressing toxicity aspects of trace elements *per se*. The important factors of concentration and availability to an organism should be considered when an element is being discussed from a health point of view in a particular situation. In most cases, the essentiality of trace elements may well be more important than their supposed toxicity. Of course, the heating of a coal with an abnormally high concentration of a particular element, say F or Se, in a closed place is inadvisable, because of risk to people nearby, or the contamination of food. The question of radioactivity has been discussed in Chapter 8 and the conclusion of an in-depth study in Sweden summarizes the situation well, namely 'the radioactivity fed into the surroundings by emission into the atmosphere from combustion of coal is very small in comparison with the radioactivity occurring naturally' (KHM, 1983). The assessment of the effect of a particular trace element in a particular situation is always difficult because of possible interactive effects from other trace elements. Too often, blame is traced to a source which is only one of several sources.

Except for a few isolated cases, trace elements in coal should not be regarded as harmful to health or the environment. However, new coals should be examined to ensure that those with abnormally high levels of certain elements are not used in a way that might cause untoward effects.

In the immediate future, the use of coal for electricity production means that there will be further interest in trace elements in coal and flyash, mainly from an environmental point of view. Enough work has been done to provide the analytical methods and the approaches, for example for measuring deposition from the atmosphere, which can be used to obtain the reliable data that will be required, indeed demanded, for the proper assessment of trace elements during the mining and usage of coal.

Chapter 11
References

You will find it a very good practice always to verify your references, sir!
Martin Routh (1755–1854)

The abbreviations of the titles of journals are those used by Chemical Abstracts. When the original paper was not seen, the actual source of information is given, being separated from the original reference by a semicolon. For the sake of continuity Mc will be taken as Mac.

Abbey, S. (1981) Reliability in the analysis of rocks and minerals. *Anal. Chem.*, **53**, 529A–534A

Abbey, S. and Rousseau, R.M. (1985) Pragmatism vs rigour: a debate on the resolution of disparate analytical data on four Canadian iron-formation reference samples. *Geostand. Newsl.*, **9**, 1–16

Abbolito, E. (1960) Investigation of Ge in lignites of Umbria by X-ray methods. *Period. mineral.*, **29**, 9–19; *CA*, **54**, 19340 (1960)

Abbott, D.T., Styron, C.E. and Casella, V.R. (1983) Radionuclides in Western coal. *Rep. MLM-3026*, 28 pp.; *CA*, **100**, 141871 (1984)

Abel, K.H. (1986) Techniques for chemical analysis of radionuclides in fly ash: an evaluation. *Rep. EPRI-EA-4728*, 54 pp

Abel, K.H. and Rancitelli, L.A. (1975) Major, minor and trace element composition of coal and fly ash, as determined by instrumental neutron activation analysis. In Babu (1975), pp. 118–138

Abernethy, R.F. and Gibson, F.H. (1963) Rare elements in coal. *US Bur. Mines Inf. Circ.* No. 8163, 69 pp

Abernethy, R.F. and Gibson, F.H. (1968) Colorimetric method for arsenic in coal. *US Bur. Mines Rep. Invest.* No. 7184, 10 pp

Abernethy, R.F., Peterson, M.J. and Gibson, F.H. (1969) Spectrochemical analyses of coal ash for trace elements. *US Bur. Mines Rep. Invest.* No. 7281, 30 pp

ACM (1985) Latrobe Valley brown coal, largest deposit in the world. *Aust. Coal Miner*, **7**(8), 14–21

ACM (1987) Purest black coal in Australia mined from Collie fields of WA. *Aust. Coal Miner*, **9**(1), 15–16

ACS (1984) Chemistry of mineral matter and ash in coal. *Am. Chem. Soc. Div., Fuel Chem. Prepr.*, **29**(4), 1–346

ACYB (1987) *Australian Coal Year Book 1987*. Melbourne: Publishing and Marketing Australia, 235 pp

Adams, F. (1982) Recent advances in analytical spark source mass spectrometry. *Phil. Trans. R. Soc. London Ser. A*, **305**, 509–519

Adcock, F. and Muir, P.L. (1953) *BHP Res. Rep.*, No. 517; Pilkington (1957)

Admakin, L.A. (1974) Metal content of genetic types of coal in some fields of Transbaikal. *Dokl. Akad. Sci. Earth Sci. Sect.*, **217**, 195–197

Adolphi, P. and Störr, M. (1985) Glow discharge excited low temperature ashing – a new technique for separating mineral matter of coals. *Fuel*, **64**, 151–155

Affolter, R.H. and Hatch, J.R. (1984) Geochemical characterization of Rocky Mountain, Northern Great Plains, and Interior Province coals. *Bull. Am. Assoc. Pet. Geol.*, **68**, 447

Affolter, R.H. and Stricker, G.D. (1987) Geochemistry of coal from the Cretaceous Corwin and Chandler Formations, National Petroleum Reserve in Alaska (NPRA). In *Alaskan North Slope Geology* (eds. I. Tailleur and P. Wellmer), **1**, 217–224. Bakersfield: Society of Economic Palaeontologists

Affolter, R.H. and Stricker, G.D. (1989) Effects of paleolatitude on coal quality – a model for organic sulfur distribution in United States coal. *Abstr. 74th Ann. Convention Am. Assoc. Pet. Geol., San Antonio*

Agricola, G. (1556) *De Re Metallica*, Translated by H.C. Hoover and L.H. Hoover (1950). New York: Dover, 638 pp

Agterdenbos, J., van Elteren, J.T., Bax, D. and Terheege, J.P. (1986) The determination of selenium with hydride generation AAS-IV. Application to coal analysis. *Spectrochim. Acta, part B*, **41B**, 303–316

Ahrens, L.H. and Taylor, S.R. (1961) *Spectrochemical Analysis*. London: Pergamon, 454 pp

Aiken, G.R., McKnight, D.M., Wershaw, R.L. and MacCarthy, P. (eds.) (1985) *Humic Substances in Soil, Sediment and Water, Geochemistry, Isolation, and Characterization*. New York: Wiley Interscience, 692 pp

Airey, D. (1982) Contributions from coal and industrial materials to mercury in air, rainwater and snow. *Sci. Total Environ.*, **25**, 19–40

Aitken, R.L., Campbell, D.J. and Bell, L.C. (1984) Properties of Australian fly ashes relevant to their agronomic utilization. *Aust. J. Soil Res.*, **22**, 443–453

Akcetin, S., Ayca, E. and Hoste, E. (1973) Neutron activation analysis of Turkish lignites. *Radiochem. Radioanal. Lett.*, **15**, 13–28

Alekseev, L.S. (1960) Distribution of impurities in coals of the Urgal deposit. *Geol. Geofiz. Akad. Nauk SSSR, Sibir. Otdel*, No. 10, 69–77; *CA*, **55**, 11803 (1961)

Alexander, C.C., Thorpe, A.N. and Senftle, F.E. (1979) Basic magnetic properties of bituminous coal. *Fuel*, **58**, 857–863

Alkemade, C.T.J. and Milatz, J.M.W. (1955) Double-beam method of spectral selection with flames. *J. Opt. Soc. Am.*, **45**, 583–584

Allen, R.M. and van der Sande, J.B. (1984) Analysis of sub-micron mineral matter in coal via scanning transmission electron microscopy. *Fuel*, **63**, 24–29

Almassy, G. and Szalay, S. (1956) Analytical studies on vanadium and molybdenum content of Hungarian coals. *Mag. Tud. Akad. Kem. Tud. Oszt. Kozl.*, **8**, 39–45; *Fuel Abstr.*, **23**, 2972 (1958)

Altschuler, Z.S., Schnepfe, M.M., Silber, C.C. and Simon, F.O. (1983) Sulfur diagenesis in Everglades peat and the origin of pyrite in coal. *Science*, **221**, 221–227

Alvarado, J., Leon, L.E., Lopez, F. and Lima, C. (1988) Comparison of conventional and microwave wet acid digestion procedures for the determination of iron, nickel and vanadium in coal by electrothermal atomisation of atomic absorption spectrometry. *J. Anal. At. Spectrom.*, **3**, 135–138

AMC (1960) Report of Analytical Methods Committee. *Analyst (London)*, **85**, 643–656

AMC (1987) Recommendations for the definition, estimation and use of the detection limit. *Analyst (London)*, **112**, 199–204

Ananyan, V.L. (1962) Effect of soil air radioactivity on the plants. *Dokl. Akad. Nauk Arm SSR*, **34**, 113–116; *CA*, **57**, 7575 (1962)

Anderson, R.A. (1981) Nutritional role of chromium. *Sci. Total Environ.*, **17**, 13–29

Anderson, W. and Turner, R.C. (1956) Radon content of the atmosphere. *Nature (London)*, **178**, 203–204

Andrejko, M.J., Cohen, A.D. and Raymond, R. (1983) Origin of mineral matter in peat. In Raymond and Andrejko (1983), pp. 3–24

Andren, A.W., Klein, D.H. and Talmi, Y. (1975) Selenium in coal-fired steam plant emissions. *Environ. Sci. Technol.*, **9**, 856–858

Andrews, M.J. and Fuge, R. (1986) Cupriferous bogs of the Coed y Brenin area, North Wales and their significance in mineral exploration. *Appl. Geochem.*, **1**, 519–525

Antonovics, J., Bradshaw, A.D. and Turner, R.G. (1971) Heavy metal tolerance in plants. *Adv. Ecol. Res.*, **7**, 1–85

Arnautov, N.V. and Shipilov, L.D. (1960) Spectrographic determination of yttrium in coal ash. *Geol. Geofiz.* (1), 115–117; *CA*, **55**, 6822 (1961)

Aruscavage, P. (1977) Determination of arsenic, antimony, and selenium in coal by atomic absorption spectrometry with a graphite tube atomizer. *J. Res. US Geol. Surv.*, No. 5 405–408
AS (1970) Chemical analysis of materials by atomic adsorption spectroscopy. *Stand. Assoc. Aust.*, AS CK18–1970, 14 pp
AS (1971) Fly ash for use in concrete. *Stand. Assoc. Aust.*, AS 119 and 1130–1971, 16 pp
AS (1980a) Methods for the analysis and testing of coal and coke, Part 10 – arsenic in coal and coke. *Stand. Assoc. Aust.*, AS 1038, Part 10–1980, 14 pp
AS (1980b) Methods for the analysis and testing of coal and coke, Part 8 – chlorine in coal and coke. *Stand. Assoc. Aust.*, AS 1038, Part 8–1980, 12 pp
AS (1982) Wavelength dispersive X-ray fluorescence spectrometers – methods of test for determination of precision. *Stand. Assoc. Aust.*, AS 2563–1982, 13 pp
AS (1983a) Methods for the analysis and testing of coal and coke, Part 22 – Direct determination of mineral matter of hydration of minerals in hard coal. *Stand. Assoc. Aust.* AS 1038.22–1983, 21 pp
AS (1983b) Guide for the taking of samples from hard coal seams *in situ*. *Stand. Assoc. Aust.* AS 2617–1983, 12 pp
AS (1984a) Sampling of solid mineral fuels. Part 4 – Hard coal – sampling from stationary situations. *Stand. Assoc. Aust.*, AS 2646.4–1984, 18 pp
AS (1984b) Sampling of solid mineral fuels. Part 2 – Hard coal – sampling from moving streams. *Stand. Assoc. Aust.*, AS 2646.2–1984, 18 pp
AS (1984c) Sampling of solid mineral fuels. Part 6 – Hard coal – preparation of samples *Stand. Assoc. Aust.*, AS 2646.6–1984, 21 pp
AS (1985) Methods for the analysis and testing of coal and coke, Part 14.2 – Analysis of higher rank coal ash and coke ash (acid digestion – flame atomic absorption spectrometric method). *Stand. Assoc. Aust.*, AS 1038.14.2–1985, 13 pp
AS (1986) Methods for the analysis and testing of coal and coke. Part 10.1 – Determination of trace elements – determination of eleven trace elements in coal, coke and fly-ash – flame atomic absorption spectrometric method. *Stand. Assoc. Aust.*, AS 1038.10.1–1986, 13 pp
AS (1988) Methods for the analysis and testing of coal and coke. Part 10.3 – Determination of trace elements – coal, coke and fly-ash – determination of boron content – spectrophotometric method. *Stand. Assoc. Aust.*, AS 1038.10.3–1988, 7 pp
AS (1989) Methods for the analysis and testing of coal and coke. Part 10.4 – Determination of trace elements – coal, coke and fly-ash – determination of fluorine content – pyrohydrolysis method. *Stand. Assoc. Aust.*, AS 1038.10.4–1989, 14 pp
ASTM (1978a) Standard method of preparing coal samples for analysis. *Am. Soc. Test Mater.*, D2013-72 (reapproved 1978), 15 pp
ASTM (1978b) Standard test method for chlorine in coal. *Am. Soc. Test Mater.*, D2361-66 (reapproved 1978), 4 pp
ASTM (1979) Standard test method for total fluorine in coal by the oxygen bomb combustion/ion selective electrode method. *Am. Soc. Test Mater.*, D3761-79, 4 pp
ASTM (1982) Standard methods for collection of a gross sample of coal. *Am. Soc. Test Mater.*, D2234-82, 17 pp
ASTM (1983a) Standard test method for trace elements in coal and coke ash by atomic absorption. *Am. Soc. Test Mater.*, D3683-78 (reapproved 1983), 4 pp
ASTM (1983b) Standard test method for total chlorine in coal by the oxygen bomb combustion/ion selective electrode method. *Am. Soc. Test Mater.*, D4208-83, 4 pp
ASTM (1983c) Standard test method for total mercury in coal by the oxygen bomb combustion/atomic absorption electrode method. *Am. Soc. Test Mater.*, D3684- 78 (reapproved 1983), 4 pp
ASTM (1983d) Standard test method for major and minor elements in coal and coke ash by atomic absorption. *Am. Soc. Test Mater.*, D3682-78 (reapproved 1983), 8 pp
ASTM (1984) Standard methods of analysis of coal and coke ash. *Am. Soc. Test Mater.*, D2795–1984, 8 pp
ASTM (1986) Standard practice for use of the terms precision and bias in ASTM test methods. *Am. Soc. Test Mater.*, E177-86, 16 pp
Asuen, G.O. (1987) Assessment of major and minor elements in the Northumberland coalfield, England. *Int. J. Coal Geol.*, **9**, 171–186
Atanasiu, G. and Soroiu, M. (1960) The radioactivity of some coal deposits in Romania. *Acad. Repub. Pop. Rom. Inst. Fiz. At., Stud. Cercet. Fiz.*, **11**, 745–751; *CA*, **55**, 16958 (1961)

Atkinson, R.M., Dickinson, D. and Harris, F.J.T. (1950) Arsenical contamination of chicory during drying. *J. Sci. Food Agric.*, **1**, 264–266
Aubrey, K.V. (1952) Germanium in British coals. *Fuel*, **31**, 429–437
Aubrey, K.V. (1954) Frequency distribution of the concentrations of elements in rocks. *Nature (London)*, **174**, 141–142
Aubrey, K.V. (1958) Germanium in coal and some of its residual products. *Rev. Ind. Miner.*, 15 July, 51–61
Aubrey, K.V. and Payne, K.W. (1954) Volatilisation of germanium during the ashing of coal. *Fuel*, **33**, 20–25
Augustithis, S.S. (ed.) (1983) *The Significance of Trace Elements in Solving Petrogenetic Problems and Controversies*. Athens: Theophrastus Publ., 917 pp
Austin, L.S. and Millward, G.E. (1988) Simulated effects of tropospheric emissions on the global antimony cycle. *Atmos. Environ.*, **22**, 1395–1403
Averitt, P., Breger, I.A., Swanson, V.E., Zubovic, P. and Gluskoter, H.J. (1972) Minor elements in coal – a selected bibliography, July 1972. *US Geol. Surv. Prof. Pap.*, No. 800-D, D169–D171
Averitt, P., Hatch, J.R., Swanson, V.E., Breger, I.A., Coleman, S.L., Medlin, J.H., Zubovic, P. and Gluskoter, H.J. (1976) Minor and trace elements in coal – a selected bibliography of reports in English, January 1976. *US Geol. Surv. Open-File Rep.*, No. 76-481, 16 pp
Ayanoglu, S.F. and Gunduz, G. (1978) Neutron activation analysis of Turkish coals. I Elemental contents. *J. Radioanal. Chem.*, **43**, 155–157
Aylmer, J.A., Holmes, R.J. and Brockway, D.J. (1987) A gauge for the discrimination of overburden and sediments from low ash coal. *Annu. Conf. Australas. Inst. Min. Metall.*, pp. 285–292
Azambuja, D.S., Formoso, M.L.L. and Bristoti, A. (1981) Study of the association of some minor elements with the organic and inorganic fraction of coal from the Leas mine. *Min. Metall.*, **44**, 4–12
Baas Becking, L.G.M., Kaplan, I.R. and Moore, D. (1960) Limits of the natural environment in terms of pH and oxidation-reduction potentials. *J. Geol.*, **68**, 243–284
Babu, S.P. (ed.) (1975) Trace elements in fuel. *Adv. Chem. Ser.*, **141**, Washington, DC: Am. Chem. Soc., 216 pp
Bacon, C.A. (1986) Coal in Tasmania. *Proc. 20th Symp. Advances in the Study of the Sydney Basin*, University of Newcastle, Newcastle, NSW, pp. 4–6
Bae, E.S. and Kim, Y.W. (1980) Mean concentration of lead in coal and fuel oil. *Koryo Uikitae Chapchi*, **11**, 21–26; *CA*, **96**, 71580 (1982)
Bailey, A. (1975) Analysis of coal ash for trace metals by solvent extraction and atomic absorption spectrophotometry. *Proc. W. Va. Acad. Sci.*, **47**, 46–52
Bailey, A. (1981) Chemical and mineralogical differences between Kittanning coals from marine-influenced versus fluvial sequences. *J. Sediment. Petrol.*, **51**, 383–395
Bailey, A. and Kosters, E.C. (1983) Silicate minerals in organic-rich Holocene deposits in southern Louisiana. In Raymond and Andrejko (1983), pp. 39–51
Baker, W.E. (1973) The role of humic acids from Tasmanian podzolic soils in mineral degradation. *Geochim. Cosmochim. Acta*, **37**, 269–281
Baker, W.E. (1978) The role of humic acid in the transport of gold. *Geochim. Cosmochim. Acta*, **42**, 645–649
Ball, C.G. (1935) Mineral matter of No. 6 Bed coal at West Frankfort, Franklin County, Illinois. *Ill. State Geol. Surv. Rep. Invest.*, No. 33, 106 pp
Banerjee, N.N., Rao, H.S. and Lahiri, A. (1974) Germanium in Indian coals. *Indian J. Technol.*, **12**, 353–358
Baragwanath, G.E. (1962) Some aspects of the formation and nature of brown coal, and of the behaviour of brown coal ash in water tube boilers, with special reference to Victorian deposits. *Australas. Inst. Min. Metall. Proc.*, No. 202, 131–249
Barnard, W.R. and Nordstrom, D.K. (1982) Fluoride in precipitation. II. Implications for the geochemical cycling of fluorine. *Atmos. Environ.*, **16**, 105–111
Barnes, J.R., Clague, A.D.H., Clayden, N.J., Dobson, C.M. and Jones, R.B. (1986) The application of ^{29}Si and ^{27}Al solid state n.m.r. spectroscopy to characterising minerals in coals. *Fuel*, **65**, 437–441
Barnes, M.A., Barnes, W.C. and Bustin, R.M. (1984) Diagenesis 8. Chemistry and evolution of organic matter. *Geosci. Can.*, **11**, 103–114
Barrie, L.A., Lindberg, S.E., Chan, W.H., Ross, H.B., Arimoto, R. and Church, T.M.

(1987) On the concentration of trace metals in precipitation. *Atmos. Environ.*, **21**, 1133–1135

Baskerville, C. (1899) The occurrence of vanadium, chromium and titanium in peats. *J. Am. Chem. Soc.*, **21**, 706–707

Baucells, M., Lacort, G., Roura, M. and Rauret, G. (1984) Determination of boron in silicate geological material by an emission spectrographic method. *Appl. Spectros.*, **38**, 572–574

Bauman, A. and Horvat, D. (1981) The impact of natural radioactivity from a coal-fired power plant. *Sci. Total Environ.*, **17**, 75–81

Bayet, A. and Slosse, A. (1919) Arsenical intoxication in the coal and coal-derivative industries. *CR Acad. Sci.*, **168**, 704–706; *CA*, **13**, 2930 (1919)

Bayliss, R.J. and Whaite, H.M. (1966) A study of the radium alpha-activity of coal, ash and particulate emission at a Sydney power station. *Air Water Pollut.*, **10**, 813–819

BCRA (1988) Literature review – analysis of coal and coke. *Br. Coke Res. Assoc. Quart.*, No. 21, 36–66

BCRA (1989) Literature review – analysis of coal and coke. *Br. Coke Res. Assoc. Quart.*, No. 22, 43–70

Beck, H.L. and Miller, K.M. (1980) Some radiological aspects of coal combustion. *IEEE Trans. Nucl. Sci.*, **NS-27**, 689–694

Becker, D.A. (1977) Achieving accuracy in environmental measurements using activation analysis. *Nat. Bur. Stand. (US) Spec. Publ.*, No. 464, 43–46

Beeson, D.C. (1984) The relative significance of tectonics, sea level fluctuations, and paleoclimate to Cretaceous coal distribution in North America. Thesis, University of Colorado, 202 pp

Behringer, K. (1925) Contributions to the iodine question and investigations on the geochemistry of iodine. Dissertation, Technical College Stuttgart, 60 pp; CIEB (1956), 87

Beising, R. and Kirsch, H. (1974) The contents of the trace element fluorine in fossil fuels during combustion. *VGB Kraftwerkstechnik*, **54**, 268–286

Bekyarova, E.E. and Rouschev, D.D. (1971) Forms of binding of germanium in solid fuels. *Fuel*, **50**, 272–279

Belcher, C.B. and Brooks, K.A. (1963) The determination of strontium in coal ash by atomic absorption spectrophotometry. *Anal. Chim. Acta*, **29**, 202–205

Bellido, L.F. and Arezzo, B. de C. (1985) Uranium and thorium determination in Brazilian coals by epithermal neutron activation analysis. *J. Radioanal. Nucl. Chem.*, **92**, 151–158

Bellido, L.F. and Arezzo, B. de C. (1986) Non-destructive analysis of inorganic impurities in Brazilian coals by epithermal neutron activation. *J. Radioanal. Nucl. Chem.*, **100**, 21–29

Bellido, L.F. and Arezzo, B. de C. (1987) Non-destructive analysis of trace elements in coal by resonance neutron activation analysis. *Quim. Nova*, **10**(1), 42–44

Belopolskaya, T.L. and Serikov, I.V. (1969) Chemical and X-ray spectral determination of selenium in coals, lignites, and rocks rich in organic matter. *Khim. Tverd. Topl.* (6), 73–79; *CA*, **72**, 74461 (1970)

Belopolskii, M.P. and Popov, N.P. (1964) Determination of scandium in aluminosilicates, coal ash and minerals. *Zavod. Lab.*, **30**, 1441–1442; *CA*, **62**, 8382 (1965)

Bembrick, C.S. (1983) Stratigraphy and sedimentation of the Late Permian Illawarra Coal Measures in the Western Coalfield, Sydney Basin, New South Wales. *J. Proc. R. Soc. NSW*, **116**, 105–117

Bencko, V. and Symon, K. (1977) Health aspects of burning coal with a high arsenic content. *Environ. Res.*, **13**, 378–385

Bencko, V., Vasilieva, E.V. and Symon, K. (1980) Immunological aspects of exposure to emissions from burning coal of high beryllium content. *Environ. Res.*, **22**, 439–449

Bencko, V., Symon, K., Stalnik, L., Batora, J., Vanco, E. and Svandova, E. (1980) Rate of malignant tumor mortality among coal-burning-power-plant workers occupationally exposed to arsenic. *J. Hyg. Epidemiol. Microbiol. Immunol.*, **24**, 278–284

Benko, I. and Szadeczky-Kardoss, G. (1957) Spectrographic determination of trace elements in coal ash. *Magy. Kem. Foly.*, **63**, 78–84

Bennett, A.J.R. (1964) Origin and formation of coal seams. *Queensl. Gov. Min. J.*, **65**, 258–269

Bennett, B.G. (1981) Exposure commitment assessments of environmental pollutants. *MARC Rep.*, No. 23, 59 pp

Bennett, B.G. (1982) Exposure of man to environmental nickel – an exposure commitment assessment. *Sci. Total Environ.*, **22**, 203–212

Benson, S.A. and Holm, P.L. (1983) Comparison of inorganics in three low-rank coals. *Am. Chem. Soc. Div., Fuel Chem. Prepr.*, **28** (2), 234–239

Benson, S.A. and Holm, P.L. (1985) Comparison of inorganic constituents in three low-rank coals. *Ind. Eng. Chem. Prod. Res. Dev.*, **24**, 145–149

Benson, S.A., Falcone, S.K. and Karner, F.R. (1984) Elemental distribution and association with inorganic and organic components in two North Dakota lignites. *Am. Chem. Soc. Div., Fuel Chem. Prepr.*, **29** (4), 36–47

Berchtold, G.B. (1983) Fundamentals for sampling and analysing coal. *World Coal*, **9**(1), 38–41

Berger, H., Meyberg, F. and Dannecker, W. (1986) Effects of some typical environmentally relevant matrixes on trace element analysis by graphite-tube furnace AAS (atomic absorption spectrometry). *Fortschr. Atomspektrom. Spurenanal.*, **2**, 607–617; *CA*, **106**, 77911 (1987)

Berkovitch, I. (1956) Is coal a useful source of metals? *Chem. Process. Eng.*, **37**, 305–307

Berkowitz, N. (1979) *An Introduction to Coal Technology*. New York: Academic Press, 345 pp

Berkowitz, N. (1985) *The Chemistry of Coal*. Amsterdam: Elsevier, 513 pp

Bernard, J.H. and Padera, K. (1954) Bravoite from the Kladno–Rakonitz coal basin. *Geologie*, **3**, 155–169

Bernas, B. (1968) Method for decomposition and comprehensive analysis of silicates by atomic absorption spectrometry. *Anal. Chem.*, **40**, 1682–1686

Berndt, H., Messerschmidt, J., Alt, F. and Sommer, D. (1981) Determination of thallium in minerals and coal by AAS (injection method, platinum loop method, graphite cuvette) and ICP-AES. *Fresenius' Z. Anal. Chem.*, **306**, 385–393

Berns, E.G. and van der Zwaan, P.W. (1972) The prohydrolytic determination of fluoride. *Anal. Chim. Acta*, **59**, 293–297

Bernstein, L.R. (1985) Germanium geochemistry and mineralogy. *Geochim. Cosmochim. Acta*, **49**, 2409–2422

Berthoud, E.L. (1875) On the occurrence of uranium, silver, iron, etc., in the Territory Formation of Colorado Territory. *Proc. Nat. Acad. Sci., Philadelphia*, **27**, 363–365

Bertine, K.K. and Goldberg, E.D. (1971) Fossil fuel combustion and the major sedimentary cycle. *Science*, **173**, 233–235

Beske, H.E., Gijbels, R., Hurrle, A. and Jochum, K.P. (1981) Part IV Review and evaluation of spark source mass spectrometry as an analytical method. *Fresenius' Z. Anal. Chem.*, **309**, 329–341

Bethell, F.V. (1963) Progress Review No. 55: The distribution and origin of minor elements in coal. *J. Inst. Fuel*, **36**, 478–492; *Br. Coal Util. Res., Assoc. Mon. Bull.*, **26**, 401–430 (1962)

Bettinelli, M. (1983a) Determination of fluorine in environmental standard reference materials with a fluoride ion-selective electrode. *Analyst (London)*, **108**, 404–407

Bettinelli, M. (1983b) Fusion procedure for the trace metal analysis of coal by atomic absorption. *At. Spectrosc.*, **4**(1), 5–9

Bettinelli, M., Baroni, U. and Pastorelli, N. (1987) Determination of scandium in coal fly ash and geological materials by graphite furnace atomic absorption spectrometry and inductively coupled plasma atomic emission spectrometry. *Analyst (London)*, **112**, 23–26

Bewers, J.M., Barry, P.J. and MacGregor, D.J. (1987) Distribution and cycling of cadmium in the environment. In *Cadmium in the Environment* (eds. J.O. Nriagu and J.B. Sprague), pp. 1–18. New York: Wiley

Binns, G.J. and Harrow, G. (1897) On the occurrence of certain minerals at Netherseal colliery, Leicestershire. *Trans. Inst. Min. Eng.*, **13**, 252–255

Bishop, M. and Ward, D.L. (1958) The direct determination of mineral matter in coal. *Fuel*, **37**, 191–200

Black, P.M. (1981) Taranaki coalfields: coal quality. *NZ Energy Res. Dev. Comm.*, Publ. No. P53, 25 pp

Blackburn, R. (1985) Radiochemical pollution due to combustion of coal and coal derived fuels. *Anal. Proc.*, **22**, 272–273

Blackburn, R. and Gueran, J. (1979) Annual rate of release of radon to the atmospher as a result of coal combustion in the United Kingdom. *Radiat. Phys. Chem.*, **13**, 145–147

Block, C. (1975) Determination of lead in coal and coal ashes by flameless atomic absorption spectrometry. *Anal. Chim. Acta*, **80**, 369–373

Block, C. and Dams, R. (1973) Determination of trace elements in coal by instrumental neutron activation analysis. *Anal. Chim. Acta*, **68**, 11–24

Block, C. and Dams, R. (1975) Inorganic composition of Belgian coals and coal ashes. *Environ. Sci. Technol.*, **9**, 146–150

Block, C. and Dams, R. (1976) Study of fly ash emission during combustion of coal. *Environ. Sci. Technol.*, **10**, 1011–1017

Blomqvist, G. (1983) Coal analyses. *KHM Tekn. Rapp.*, No. 41, 85 pp

Boar, P.L. and Ingram, L.K. (1970) The comprehensive analysis of coal ash and silicate rocks by atomic-absorption spectrophotometry by a fusion technique. *Analyst (London)*, **95**, 124–130

Bock, R. (1979) *A Handbook of Decomposition Methods in Analytical Chemistry*. Glasgow: International Textbook Company, 444 pp

Bogvad, R.R. and Nielsen, A.H. (1945) Vanadium content in a series of Danish rocks. *Medd. Dan. Geol. Foren.*, **10**, 532–540

Bohor, B.F. and Gluskoter, H.J. (1973) Boron in illite as an indicator of paleosalinity of Illinois coals. *J. Sediment. Petrol.*, **43**, 945–956

Bolger, M. (1989) Trace elements in new coal fields. *SECV Rep.*, No. SO/89/157, 6 pp

Bone, K.M. and Schaap, H.A. (1980) Determination of trace elements in brown coal. *SECV Rep.*, No. SO/80/2, 21 pp

Bone, K.M. and Schaap, H.A. (1981a) Elemental mass balance at a brown coal-fired power station. *SECV Rep.*, No. SO/81/37, 78 pp; *Rep. NERDDP*, No. EG82/041, 281 pp

Bone, K.M. and Schaap, H.A. (1981b) Trace elements in Latrobe Valley brown coal: environmental and geochemical implications. *SECV Rep.*, No. SO/80/15, 67 pp

Bone, K.M., Schaap, H.A. and Hughes, T.C. (1981) Analysis of brown coal and ash by neutron activation analysis. *SECV Rep.*, No. SO/81/36, 15 pp

Bonnett, R. and Cousins, R.P.C. (1987) On the metal content and metal ion uptake of botanically specific peats and the derived humic acids. *Org. Geochem.*, **11**, 497–503

Bonnett, R. and Czechowski, F. (1980) Gallium porphyrins in bituminous coal. *Nature (London)*, **283**, 465–467

Bonnett, R. and Czechowski, F. (1981) Metals and metal complexes in coal. *Phil. Trans. R. Soc. London Ser. A.*, **300**, 51–63

Bonnett, R. and Czechowski, F. (1984) Metalloporphyrins in coal. I. Gallium in bituminous coals. *J. Chem. Soc. Perkin Trans.*, **1** (1), 125–131

Bonnett, R. and Czechowski, F. (1987) Metalloporphyrins in coal. 3. Porphyrins and metalloporphyrins in petrographic components of a subbituminous coal. *Fuel*, **66**, 1079–1083

Bonnett, R., Burke, P.J. and Czechowski, F. (1987) Metalloporphyrins in lignite, coal, and calcite. *ACS Symp. Ser.*, No. 344, 173–185

Borisova, T.F., Guren, G.F., Komissarova, L.N. and Shatskii, V.M. (1974a) A study of the distribution of gallium in the coal substance. *Solid Fuel Chem.*, **8**(5), 27–29

Borisova, T.F., Guren, G.F., Komissarova, L.N. and Shatskii, V.M. (1974b) Distribution of scandium in the coal substance. *Solid Fuel Chem.*, **8** (5), 5–7

Borisova, T.F., Guren, G.F., Komissarova, L.N. and Shatskii, V.M. (1976) Distribution of scandium in coal and brown coal. *Solid Fuel Chem.*, **10** (3), 53–56

Borovik, S.A. and Ratynskii, V.M. (1944) Tin in the coals of the Kuznetsk Basin. *CR Acad. Sci. URSS*, **45**, 120–121; *CA*, **39**, 5069 (1945)

Borrowdale, J., Jenkins, R.H. and Shanahan, C.E.A. (1959) The determination of boron in plain-carbon and alloy steels. *Analyst (London)*, **84**, 426–433

Bosserman, R.W., Auble, G.T. and Hamilton, D.B. (1984) Cation exchange characteristics of *Sphagnum* from Okefenokee swamp. In Cohen *et al.* (1984), pp. 333–342

Botto, R.I. (1980) Coal ash element analysis by ICPES using an automatic fusion device. *ICP Inf. Newsl.*, **6**(3), 126

Boumans, P.W.J.M. (ed.) (1987a) *Inductively Coupled Plasma Emission. Part 1. Methodology, Instrumentation, and Performance*. New York: Wiley, 584 pp

Boumans, P.W.J.M. (ed.) (1987b) *Inductively Coupled Plasma Emission Spectroscopy. Part 2. Applications and Fundamentals*. New York: Wiley, 486 pp

Bouska, V. (1981) *Geochemistry of Coal*. Amsterdam: Elsevier, 284 pp

Bouska, V. and Havlena, V. (1959) The coal layers of the Jan Sverma Mine (at Lampertice u Zaclere) and a geochemical investigation of their trace elements. *Rozpr. Cesk. Akad. Ved, Rada Mat. Prir. Ved*, **69** (3), 1–64

Bouska, V. and Stehlik, E. (1980) Time parameter of the enrichment of coal with trace elements. *Acta Univ. Carol., Geol.* (1–2), 91–106

Bowen, H.J.M. (1966) *Trace Elements in Biochemistry*. London: Academic, 241 pp

Bowen, H.J.M. (1979) *Environmental Chemistry of the Elements*. London: Academic, 333 pp

Boyle, R.W. (1968) The geochemistry of silver and its deposits. *Geol. Surv. Can. Bull.*, No. 160, 264 pp

Boyle, R.W. (1977) Cupriferous bogs in the Sackville area, New Brunswick, Canada. *J. Geochem. Explor.*, **8**, 495–527

Bradburn, E. (1958) Discussion. *Proc. Conf. Science in Use of Coal*, Sheffield, Inst. Fuel, London, E47–E48

Branagan, D.F. (1960) Structure and sedimentation in the Western Coalfield of New South Wales. *Australas. Inst. Min. Metall. Proc.*, No. 196, 79–116

Branagan, D.F. (1961) Structure and sedimentation in the Western Coalfield of New South Wales. *Australas. Inst. Min. Metall. Proc.*, No. 199, 157–160

Brandenstein, M., Janda, I. and Schroll, E. (1960) Rare elements in Austrian coals and bituminous rocks. *Tschermaks Mineral. Petrogr. Mitt.*, **7**, 260–285

Breger, I.A. (1958) Geochemistry of coal. *Econ. Geol.*, **53**, 823–841

Breger, I.A. (1974) The role of organic matter in the accumulation of uranium. *Rep. IAEA-SM-183/29*, 99–124

Breger, I.A. and Schopf, J.M. (1955) Germanium and uranium in coalified wood from Upper Devonian black shale. *Geochim. Cosmochim. Acta*, **7**, 287–293

Breger, I.A., Deul, M. and Meyrowitz, R. (1955) Geochemistry and mineralogy of a uraniferous subbituminous coal. *Econ. Geol.*, **50**, 610–624

Breger, I.A., Deul, M. and Rubinstein, S. (1955) Geochemistry and mineralogy of a uraniferous lignite. *Econ. Geol.*, **50**, 206–226

Briggs, H. (1929) A note on the mineralogy of coal as suggested by X-ray examination. *Colliery Guardian*, **138**, 638–640

Briggs, H. (1934) Metals in coal. *Colliery Eng. (London)*, **11**, 303–308

Brinkmann, K. (1977) Mineralogy and geochemistry of iron sulfides in the overburden of the Frechen open cast mine (Rhine brown coal area). *Neues Jahrb. Mineral. Abh.*, **129**, 333–352

Britten, R.A., Smyth, M., Bennett, A.J.R. and Shibaoka, M. (1975) Environmental interpretations of Gondwana coal measure sequences in the Sydney Basin of New South Wales. In *Gondwana Geology* (ed. K.S.W. Campbell), pp. 233–247. Canberra: Australian National University Press

Brockway, D.J., Aylmer, J.A. and Holmes, R.J. (1984) Natural radioactivity of Latrobe Valley coals and clays. *Proc. Aust. Coal Sci. Conf.* Gippsland Inst. Advanced Educ., Churchill, Vic., pp. 152–158

Bronshtein, A.N., Sendulskaya, T.I. and Shpirt, M.Y. (1960) Spetrographic determination of germanium and gallium in coal. *Zavod. Lab.*, **26**, 973–974; *CA*, **55**, 8813 (1961)

Brooks, J.D. (1956) Organic sulphur in coal. *J. Inst. Fuel*, **29**, 82–85

Brooks, J.D. and Sternhell, S. (1957) Chemistry of brown coals. I. Oxygen-containing functional groups in Victorian brown coals. *Aust. J. Appl. Sci.*, **8**, 206–221

Brown, H.R. and Swaine, D.J. (1964) Inorganic constituents of Australian coals. *J. Inst. Fuel*, **37**, 422–440

Brown, H.R. and Taylor, G.H. (1961) Some remarkable Antarctic coals. *Fuel*, **40**, 211–224

Brown, H.R., Durie, R.A. and Schafer, H.N.S. (1959) The inorganic constituents in Australian coals. I. The direct determination of the total mineral-matter content. *Fuel*, **38**, 295–308

Brown, H.R., Durie, R.A. and Schafer, H.N.S. (1960) The inorganic constituents in Australian coals. II. Combined acid-digestion-low-temperature oxidation procedure for determination of total mineral-matter content, water of hydration of silicate minerals and composition of carbonate minerals. *Fuel*, **39**, 59–70

Brown, J.K. (1987) Does a little radiation do you good? *Nucl. Spectrum*, **3**(1), 8–10

Brown, N.A., Belcher, C.B. and Callcott, T.G. (1965) Mineral matter in NSW coke-making coals: composition, determination and effects. *J. Inst. Fuel*, **38**, 198–206

Brown, R.L. and Carman, E.P. (1954) Report of research and technologic work on coal and related investigations. *US Bur. Mines Inform. Circ.*, No. 7699, 102 pp

Brownfield, M.E., Affolter, R.H. and Stricker, G.D. (1986) Crandallite group minerals in the Capps and Q coal beds, Tyonek formation, Beluga Energy Resource area, south-central Alaska. *Proc. Conf. Focus on Alaskas' Coal '86*, Anchorage, University of Alaska, Fairbanks, pp. 142–149

Brumsack, H., Heinrichs, H. and Lange, H. (1984) West German coal power plants as sources of potentially toxic emissions. *Environ. Technol. Lett.*, **5**, 7–22

BS (1971) Methods for the analysis and testing of coal and coke. Part 9 Phosphorus in coal and coke. *Br. Stand. Inst.*, BS1016: Part 9, 13 pp

BS (1977) Methods for the analysis and testing of coal and coke. Part 8 Chlorine in coal and coke. *Br. Stand. Inst.*, BS1016: Part 8, 1977, 8 pp

Bujok, J., Jarczyk, L., Rokita, E., Slominska, D. and Strzalkowski, A. (1980) Applications of PIXE method to coal analysis. *Nucl. Phys. Methods Mater. Res., Proc. Div. Conf. 7th.*, pp. 395–397; *CA*, **95**, 64837 (1981)

Buljan, M. (1949) The occurrence of elements in the sea water and in the earth's crust in relation to the Periodic System of elements. *Acta Adriat.*, **4** (6), 171–253

Bunzl, K., Rosner, G. and Schmidt, W. (1983) Distribution of lead, cobalt and nickel in the soil around a coal-fired power plant. *Z. Pflanzenernaehr. Bodenkd.*, **146**, 705–713

Burkser, E.S., Shapiro, M.J. and Kondoguri, V.V. (1931) Radioactivity of the coal and anthracite of the Donets Basin. *Biochem. Z.*, **237**, 276–281

Burkser, E.S., Kondoguri, V.V., Kapustin, N.P. and Potapov, P.P. (1934) Radioactivity of Kuznetsk coal. *Ukr. Khem. Zh.*, **9**, 441–445; *CA*, **29**, 6138 (1935)

Burmistrov, V.R. and Chekanov, V.N. (1984) Instrumental neutron-activation determination of rhenium in coal. *Izv. Akad. Nauk Kaz. SSR, Ser. Fiz.-Mat.* (2), 65–68; *CA*, **101**, 57434 (1984)

Burns, M.S., Durie, R.A. and Swaine, D.J. (1962) Significance of chemical evidence for the presence of carbonate minerals in brown coals and lignites. *Fuel*, **41**, 373–383

Burns, R.G. and Fyfe, W.S. (1967) Trace element distribution rules and their significance. *Chem. Geol.*, **2**, 89–104

Bussy, A. (1839) The presence of iodine in the combustion products from coal mines. *J. Pharm.*, **25**, 718–721; CIEB (1956), 129

Butler, J., Marsh, H. and Goodarzi, F. (1988) World coals: genesis of the world's major coalfields in relation to plate tectonics. *Fuel*, **67**, 269–274

Butler, J.R. (1953) Geochemical affinities of some coals from Svalbard. *Nor. Polarinst. Skr.*, No. 96, 1–26

Butler, L.R.P. (ed.) (1986) *Analytical Chemistry in the Exploration, Mining and Processing of Materials.* Oxford: Blackwell, 254 pp

Cadle, R.D., Wartburg, A.P., Pollock, W.H., Gandrud, B.W. and Shedlovsky, J.P. (1973) Trace constituents emitted to the atmosphere by Hawaiian volcanoes. *Chemosphere*, No. 6, 231–234

Cahill, R.A. and Mills, J.C. (1983) New data for trace element concentrations in Australian bituminous coals. *Proc. Australas. Inst. Min. Eng.*, No. 285, 39–43

Cahill, R.A. and Shiley, R.H. (1981) Forms of trace elements in coal. *Proc. Int. Conf. Coal Sci.*, Verlag Glückauf GmbH, Essen, pp. 751–755

Cahill, R.A., Kuhn, J.K., Dreher, G.B., Ruch, R.R., Gluskoter, H.J. and Miller, W.G. (1976) Occurrence and distribution of trace elements in coal. *Am. Chem. Soc. Div., Fuel Chem. Prepr.*, **21**(7), 90–93

Cairncross, B. and Cadle, A.B. (1988) Palaeoenvironmental control on coal formation, distribution and quality in the Permian Vryheid Formation, East Witbank coalfield, South Africa. *Int. J. Coal Geol.*, **9**, 343–370

Cambel, B. and Jarkovsky, J. (1967) The geochemistry of pyrite from some Czechoslovakian deposits. Quoted by Finkelman (1981a), p.215

Cameron, C.C. (1975) Some peat deposits in Washington and southeastern Aroostook Counties, Maine. *US Geol. Surv. Bull.*, No. 1317-C, 40 pp

Cameron, C.C. and Schruben, P. (1983) Variations in mineral matter content of a peat deposit in Maine resting on glacio-marine sediments. In Raymond and Andrejko (1983), pp. 63–85

Campbell, A.D. (1987) Determination of fluoride in various matrices. *Pure Appl. Chem.*, **59**, 695–702

Campbell, F. and McCord, C. (1988) Chemical composition of the coals from the San Juan Basin, New Mexico. *J. Coal Qual.*, **7**(2), 71–76

Campbell, W.J., Carl, H.F. and White, C.E. (1957) Quantitative analyses by fluorescent X-ray spectrography. Determination of germanium in coal and coal ash. *Anal. Chem.*, **29**, 1009–1017

Camplin, W.C. (1980) Coal-fired power stations – the radiological impact of effluent discharges to atmosphere. *Nat. Radiol. Prot. Board Rep.*, No. NRPB-R107, 51 pp

Cannon, H.L., Harms, T.F. and Hamilton, J.C. (1975) Lithium in unconsolidated sediments and plants of the Basin and Range Province, Southern California and Nevada. *US Geol. Surv. Prof. Pap.*, No. 918, 23 pp

Capes, C.E., McIlhinney, A.E., Russell, D.S. and Sirianni, A.F. (1974) Rejection of trace metals from coal during beneficiation by agglomeration. *Environ. Sci. Technol.*, **8**, 35–38

Carlson, R. (1985) Peat-health-environment. *Proc. Int. Conf. Coal Sci.*, Pergamon, Sydney, NSW, pp. 433–436

Carr, P.F. and Fardy, J.J. (1983) REE geochemistry of Late Permian shoshonitic lavas from the Sydney Basin, New South Wales, Australia. *Chem. Geol.*, **43**, 187–201

Carter, J.A., Donohue, D.L. and Franklin, J.C. (1977) Trace metal analysis in coal by multielement isotope dilution spark source mass spectrometry. *Am. Chem. Soc. Div., Fuel Chem. Prepr.*, **22**(5), 60–63

Carter, J.A., Walker, R.L. and Sites, J.R. (1975) Trace impurities in fuels by isotope dilution mass spectrometry. In Babu (1975), pp. 74–83

Carter, J.A., Donohue, D.L., Franklin, J.C. and Walker, R.L. (1978) Trace impurities in coal and fly ash by isotope dilution mass spectrometry. In Karr (1978a), pp. 403–420

Casagrande, D.J. (1984) Organic geochemistry of Okefenokee peats. In Cohen *et al.* (1984b), pp. 391–409

Caagrande, D.J. (1987) Sulphur in peat and coal. In Scott (1987), pp. 87–105

Casagrande, D.J. and Erchull, L.D. (1977) Metals in plants and waters in the Okefenokee swamp and their relationship to constituents found in coal. *Geochim. Cosmochim. Acta*, **41**, 1391–1394

Casagrande, D.J. and Given, P.H. (1980) Geochemistry of amino acids in some Florida peat accumulations. II Amino acid distributions. *Geochim. Cosmochim. Acta*, **44**, 1493–1507

Casagrande, D. and Siefert, K. (1977) Origin of sulfur in coal: importance of the ester sulfate content of coal. *Science*, **195**, 675–676

Casagrande, D.J., Gronli, K. and Sutton, N. (1980) The distribution of sulfur and organic matter in various fractions of peat: origins of sulfur in coal. *Geochim. Cosmochim. Acta*, **44**, 25–32

Casagrande, D.J., Siefert, K., Berschinski, C. and Sutton, N. (1977) Sulfur in peat-forming systems of the Okefenokee Swamp and Florida Everglades: origins of sulfur in coal. *Geochim. Cosmochim. Acta*, **41**, 161–167

Casella, V.R., Bishop, C.T., Glosby, A.A. and Phillips, C.A. (1981) Anion exchange method for the sequential determination of uranium, thorium and lead-210 in coal and coal ash. *J. Radioanal. Chem.*, **62**, 257–266

Casswell, J.D. (1961) *A Lance for Liberty*. London: Harrap, 325 pp

Castillo, J.R., Lanaja, J. and Aznarez, J. (1982) Determination of germanium in coal ashes by hydride generation and flame atomic-absorption spectrophotometry. *Analyst (London)*, **107**, 89–95

Caswell, S.A., Holmes, I.F. and Spears, D.A. (1984a) Total chlorine in coal seam profiles from the South Staffordshire (Cannock) coalfield. *Fuel*, **63**, 782–787

Caswell, S.A., Holmes, I.F. and Spears, D.A. (1984b) Water-soluble chlorine and associated major cations from the coal and mudrocks of the Cannock and North Staffordshire coalfields. *Fuel*, **63**, 774–781

Cavallaro, J.A., Deurbrouck, A.W., Gibbon, G.A., Hattman, E.A. and Schultz, H. (1978) A washability and analytical evaluation of potential pollution from trace elements in coal. In Karr (1978a), pp. 435–464

Cebulak, S. and Rozkowska, A. (1983) Correlation of coal seams in the central coal region of the Lublin Coal Basin, on the basis of geochemical data. *Kwart. Geol.*, **27**, 25–39; *CA*, **100**, 36751 (1984)

Cech, F. and Petrik, F. (1972) Classification and description of mineral admixtures in coal seams of the Handova-Novaky area. *Miner. Slovaca*, **4**, 257–266; *CA*, **79**, 33611 (1973)

Cecil, C.B., Stanton, R.W. and Dulong, F.T. (1981) Geology of contaminants in coal: Phase I report of investigations. *US Geol. Surv. Open-File Rep.*, No. 81-953-A, 92 pp

Cecil, C.B., Stanton, R.W., Dulong, F.T. and Renton, J.J. (1982) Geologic factors that control mineral matter in coal. In *Atomic and Nuclear Methods in Fossil Energy Research* (eds. R.H. Filby, B.S. Carpenter and R.C. Ragaini). New York: Plenum, 323–335

Cecil, C.B., Stanton, R.W., Allshouse, S.D., Finkelman, R.B. and Greenland, L.P. (1979) Geologic controls on element concentrations in the Upper Freeport coal bed. *Am. Chem. Soc. Div., Fuel Chem. Prepr.*, **24**(1), 230–235

Cecil, C.B., Stanton, R.W., Neuzil, S.G., Dulong, F.T. and Ruppert, L.F. (1983)

Paleoclimatic controls on the occurrence and quality of coal. In Cronin, Cannon and Poore (1983), pp. 40–42
Cecil, C.B., Stanton, R.W., Neuzil, S.G., Dulong, F.T., Ruppert, L.F. and Pierce, B.S. (1985) Paleoclimate controls on late Paleozoic sedimentation and peat formation in the central Appalachian Basin. *Int. J. Coal Geol.*, 5, 195–230
Chadwick, M.J. and Lindman, N. (1982) *Environmental Implications of Expanded Coal Utilization*. Oxford: Pergamon, 283 pp
Chakarvarti, S.K. and Nagpaul, K.K. (1980) Determination of the uranium content in some Indian coal and flyash samples. *Health Phys.*, 39, 358–361
Chakrabarti, J.N. (1978) Methods of determining chlorine in different states of combination in coal. In Karr (1978a), pp. 323–345
Chakrabarti, J.N. (1982) Determination of chlorine in organic combination in the coal substance. *ACS Symp. Ser.*, 205, 185–190
Chance, H.M. (1899) The discovery of new gold districts. *Trans. Am. Inst. Min. Eng.*, 29, 224–230; Gibson and Selvig (1944)
Chao, E.C.T., Minkin, J.A. and Thompson, C.L. (1982) Recommended procedures and techniques for the petrographic description of bituminous coals. *Int. J. Coal Geol.*, 2, 151–179
Chapman, A.C. (1901) Arsenic in coal and coke. *Analyst (London)*, 26, 253–256
Chatterjee, P.K. and Pooley, F.D. (1977) An examination of some trace elements in South Wales coals. *Australas. Inst. Min. Metall. Proc.*, No. 263, 19–30
Chaudhary, M.S., Ahmad, S., Mannan, A. and Qureshi, I.H. (1984) INAA of toxic elements in coal and their transfer into environments. *J. Radioanal. Nucl. Chem.*, 83, 387–396
Chaudhuri, M.K., Gupta, H.S. Das and Khathing, D.T. (1979) North-east Indian coal: an assessment of its potential as a germanium source. *Chem. Ind. (London)*, 21 July, 475–476
Chen, B.R., Qian, Q.F., Yang, Y.N. and Yang, S.J. (1986) Determination of trace elements in samples from 110 coal mines in China by INAA. *Proc. 7th Int. Conf. Modern Trends Activation Anal.*, Copenhagen, pp. 1169–1174
Chen, J.R., Kneis, H., Martin, B., Nobiling, R., Traxel, K., Chao, E.C.T. and Minkin, J.A. (1981) Trace elemental analysis of bituminous coals using the Heidelberg proton microprobe. *Nucl. Instrum. and Methods*, 181, 151–157
Chen, J.R., Martys, N., Chao, E.C.T., Minkin, J.A., Thompson, C.L., Hanson, A.L., Kraner, H.W., Jones, K.W., Gordon, B.M. and Mills, R.E. (1984) Synchrotron radiation determination of elemental concentrations in coal. *Nucl. Instrum. Methods Phys. Res. Sect. B*, 231(1–3), 241–245
Chiou, K.Y. and Manuel, O.K. (1984) Determination of tellurium and selenium in atmospheric aerosol samples by graphite furnace atomic absorption spectrometry. *Anal. Chem.*, 56, 2721–2723
Chirnside, R.C. (1950) Extraction metallurgy of germanium. *Times Rev. Ind.*, 4, 20–22
Chirnside, R.C. and Cluley, H.J. (1952) Germanium from coal. A British source of germanium for use in crystal valves. *GEC J.*, 19, 94–100
Chou, C.-L. (1984) Relationship between geochemistry of coal and the nature of strata overlying the Herrin coal in the Illinois Basin, USA. *Geol. Soc. China Mem.* (6), 269–280
Chou, C.-L. and Harvey, R.D. (1983) Composition and distribution of mineral matter in the Herrin coal of the Illinois Basin. *Proc. Int. Conf. Coal Sci.*, Pittsburgh, pp. 373–376
Chu, S-T., Wang, Y-C., Chien, C-F., Cheng, P-J. and Sun, C-H. (1981) Determination of thirty elements in the environmental standard reference material (NBS SRM 1632a) by instrumental neutron activation analysis. *K'o Hsueh Tung Pao*, 26, 256; *CA*, 94, 211213 (1981)
Chyi, L.L. (1982) The distribution of gold and platinum in bituminous coal. *Econ. Geol.*, 77, 1592–1597
CIEB (1956) *Geochemistry of Iodine*. London: Chilean Iodine Educational Bureau, 150 pp
Clark, M.C. and Swaine, D.J. (1962a) Trace elements in coal. I New South Wales Coals. *CSIRO Div. Coal Res. Tech. Commun.*, No. 45, 1–21
Clark, M.C. and Swaine, D.J. (1962b) The contents of several trace elements in the standard rocks G-1 and W-1. *Geochim. Cosmochim. Acta*, 26, 511–514
Clark, M.C. and Swaine, D.J. (1963a) Estimation of silica ratios from semi-quantitative spectrochemical analyses of coal ash. *Fuel*, 42, 315–318
Clark, M.C. and Swaine, D.J. (1963b) Trace-element contents of the National Bureau of Standards reference samples numbers 1a, 98 and 99. *Geochim. Cosmochim. Acta*, 27, 1139–1142

Clark, P.J., Zingaro, R.A., Irgolic, K.J. and McGinley, A. (1980) Arsenic and selenium in Texas lignite. *Int. J. Environ. Anal. Chem.*, **7**, 295-314

Clarke, F.W. (1916) The data of geochemistry. *US Geol. Surv. Bull.*, No. 616, 821 pp

Clayton, E. and Dale, L.S. (1985) Determination of fluorine in NBS coal and coal fly ash by proton induced gamma ray emission and spark source mass spectrometry. *Anal. Lett.*, **18**(A12), 1533-1538

Cluley, H.J. (1951) Determination of germanium. II Absorptiometric determination with phenylfluorone. III Determination of germanium in flue dust, coal and coke. *Analyst (London)*, **76**, 523-536

Clymo, R.S. (1987a) The ecology of peatlands. *Sci. Prog. (London)*, **71**, 593-614

Clymo, R.S. (1987b) Rainwater-fed peat as a precursor of coal. In Scott (1987), pp. 17-23

Cobb, J.C., Masters, J.M., Treworgy, C.G. and Helfinstine, R.J. (1979) Abundance and recovery of sphalerite and fine coal from mine waste in Illinois. *Ill. State Geol. Surv. Ill. Miner. Note*, No. 71, 1-11

Cobb, J.C., Steele, J.D., Treworgy, C.G. and Ashby, J.F. (1980) The abundance of zinc and cadmium in sphalerite-bearing coals in Illinois. *Ill. State Geol. Surv. Ill. Miner. Note*, No. 74, 1-128

Cohen, A.C. (1959) Simplified estimators for the normal distribution when samples are singly censored or truncated. *Technometrics*, **1**, 217-237

Cohen, A.D., Spackman, W. and Raymond, R. (1987) Interpreting the characteristics of coal seams from chemical, physical and petrographic studies of peat deposits. In Scott (1987). pp. 107-125

Cohen, A.D., Andrejko, M.J., Spackman, W. and Corvinus, D. (1984a) Peat deposits of the Okefenokee Swamp. In Cohen et al. (1984b), pp. 493-553

Cohen, A.D., Casagrande, D.J., Andrejko, M.J. and Best, G.R. (1984b) *The Okefenokee Swamp: Its Natural History, Geology and Geochemistry.* Los Alamos: Wetland Surveys, 709 pp

Cohen, B.L. (1979) Role of radon in comparisons of effects of radioactivity releases from nuclear power, coal burning and phosphate mining. *Rep. AECL*-6958, 237-243; *Energy Res. Abstr.*, 7(2), 2502 (1982)

Cole, D.A., Herman, R.G., Simmons, G.W. and Klier, K. (1985) Generation of free radicals in partial oxidation of coal. *Fuel*, **64**, 303-306

Cole, D.A., Simmons, G.W., Herman, R.G., Klier, K. and Czako-Nagy, I. (1987) Transformations of iron minerals during coal oxidation. *Fuel*, **66**, 1240-1248

Coleman, S.L., Finkelman, R.B. and Bragg, L.J. (1987) Selenium concentration and mode of occurrence in coals of the United States. *Geol. Soc. Am. Abstr. Program*, **19**(7), 624

Coles, D.G., Ragaini, R.C. and Ondov, J.M. (1978) Behavior of natural radionuclides in western coal-fired power plants. *Environ. Sci. Technol.*, **12**, 442-446

Collao, S., Oyarzun, R., Palma, S. and Pineda, V. (1987) Stratigraphy, palynology and geochemistry of the Lower Eocene coals of Arauco, Chile. *Int. J. Coal Geol.*, **7**, 195-208

Conrad, V.B. and Brownlee, W.D. (1988) Hydropyrolitic-ion chromatographic determination of fluoride in coal and geological materials. *Anal. Chem.*, **60**, 365-369

Conzemius, R.J., Chriswell, C.D. and Junk, G.A. (1988) The partitioning of elements during physical cleaning of coals. *Fuel Process. Technol.*, **19**, 95-106

Cook, A.C. (1962) Fluorapatite petrifactions in a Queensland coal. *Aust. J. Sci.*, **25**, 94-95

Cook, A.C. (ed.) (1975) *Australian Black Coal, Its Occurrence, Mining, Preparation and Use.* Wollongong: Australas. Inst. Min. Metall. Illawarra Branch, 291 pp

Cook, A.C. and Taylor, G.H. (1963) The petrography of some Triassic Ipswich coals. *Australas. Inst. Min. Metall. Proc.*, No. 205, 35-55

Cooke, T.D. and Bruland, K.W. (1987) Aquatic chemistry of selenium: evidence of biomethylation. *Environ. Sci. Technol.*, **21**, 1214-1219

Cooke, W.T. (1938) Occurrence of gallium and germanium in some local coal ashes. *Trans. R. Soc. South Aust.*, **62**, 318-319

Cooper, J.A., Wheeler, B.D., Wolfe, G.J., Bartell, D.M. and Schlafke, D.B. (1977) Determination of sulfur, ash and trace element content of coal, coke, and fly ash using multielement tube excited X-ray fluorescence analysis. In *Advances in X-ray Analysis* (eds. H.F. McMurdie, C.S. Barrett, J.B. Newkirk and C.O. Ruud), pp. 431-436. New York: Plenum

Cooper, M.B. and Leith, I.S. (1983) Radiological impact of the operation of coal-fired power stations within Australia. *Annu. Rev. Res. Projects*, No. ARL/TR050, 142-143

Corbett, J.O. (1983) The radiation dose from coal burning: a review of pathways and data. *Radiat. Prot. Dosim.*, **4**, 5–19

Corcoran, J.F. (1979) Mineral matter in Australian steaming coals – a survey. *Colloq. Combustion of Pulverised Coal: The Effect of Mineral Matter*, University of Newcastle, Newcastle, NSW, L4-1-L4-16

Corey, R.C., Myers, J.W., Schwartz, C.H., Gibson, F.H. and Colbassani, P.J. (1959) Occurrence and determination of germanium in coal ash from power plants. *US Bur. Mines Bull.*, No. 575, 68 pp

Correa, da Silva Filho, B. (1982) Preliminary study on the geochemical behaviour of some trace elements of drill holes of Santa Rita coal basin, Canoas district, RS (Brazil). *Acta Geol. Leopold.*, **6**(12), 189–206; *CA*, **98**, 146233 (1983)

Correa da Silva, Z.C., Bortoluzzi, C.A., Cassulo-Klepzig, M., Dias Fabricio, M.E., Guerra-Sommer, M., Marques-Toigo, M., Paim, P.S.G., Piccoli, A.E.M. and Silva Filho, B.C. (1984) Geology of Santa Rita Coal Basin, Rio Grande do Sul, Brazil. *Int. J. Coal Geol.*, **3**, 383–400

Correns, C.W. (1956) The geochemistry of the halogens. In *Physics and Chemistry of the Earth* (eds. L.H. Ahrens, K. Rankama and S.K. Runcorn), pp. 181–233. London: Pergamon

Costick, R.J. and Schafer, H.N.S. (1959) The determination of phosphorus in coal ash. *Fuel*, **38**, 277–281

Cox, J.A., Larson, A.E. and Carlson, R.H. (1984) Estimation of the inorganic-to-organic chlorine ratio in coal. *Fuel*, **63**, 1334–1335

Cressey, B.A. and Cressey, G. (1988) Preliminary mineralogical investigation of Leicestershire low-rank coal. *Int. J. Coal Geol.*, **10**, 177–191

Cronin, T.M., Cannon, W.F. and Poore, R.Z. (eds.) (1983) Paleoclimate and mineral deposits. *US Geol. Surv. Circ.*, No. 822, 59 pp

Crook, T. (1913) On the frequent occurrence of ankerite in coal. *Mineral Mag.*, **16**, 219–223

Crossley, H.E. (1944a) Fluorine in coal. III The manner of occurrence of fluorine in coals. *J. Soc. Chem. Ind., London, Trans. Commun.*, **63**, 289–292

Crossley, H.E. (1944b) Fluorine in coal. II The determination of fluorine in coal. *J. Soc. Chem. Ind., London*, **63**, 284–288

Crossley, H.E. (1946) The inorganic constituents of coal. *Inst. Fuel, London, Bull.*, Dec., 57–60, 67

Crumley, P.H., Fletcher, A.W. and Wilson, D.S. (1955) The formation of bonded deposits in pulverized-fuel-fired boilers. *J. Inst. Fuel*, **28**, 117–120

CSIRO (1957) Distribution of mineral matter in coal. *Coal Res. CSIRO*, No. 1, 15

CSIRO (1960) Studies of the characteristics of the Big Seam, Blair Athol, Queensland. *CSIRO Div. Min. Chem. Tech. Commun.*, No. 39, 94 pp

CSIRO (1962) Characteristics of coals from the Ipswich Coalfield, Southern Queensland. *CSIRO Div. Coal Res. Tech. Commun.*, No. 46, 128 pp

CSIRO (1963) Characteristics of coals from Lobe D, Leigh Creek Coalfield, South Australia. *CSIRO Div. Coal Res. Loc. Rep.*, No. 324, 65 pp

CSIRO (1964) Characteristics of coals from Lobe C, Leigh Creek Coalfield, South Australia. *CSIRO Div. Coal Res. Loc. Rep.*, No. 329, 47 pp

CSIRO (1965a) Minerals in Australian coals. *Coal Res. CSIRO*, No. 25, 7–10

CSIRO (1965b) Characteristics of further coals from Lobe D, Leigh Creek Coalfield, South Australia. *CSIRO Div. Coal Res. Loc. Rep.*, No. 336, 37 pp

CSIRO (1966) Characteristics of further coals from Lobe C, Leigh Creek Coalfield, South Australia. *CSIRO Div. Coal Res. Loc. Rep.*, No. 341, 39 pp

Cuttita, F. and Warr, J.J. (1958) Retention of lead during oxidation ashing of selected naturally occurring carbonaceous substances. *Geochim. Cosmochim. Acta*, **13**, 256–259

Daci, N.M., Berisha, M. and Gashi, S.T. (1983) Trace elements in Kosova's Basen coal. *Erdoel. Kohle, Erdgas, Petrochem. Brennst.-Chem.*, **36**, 428

Daci, N.M., Hoxha, E.M. and Vujicic, G. (1985) Trace metal analysis of fractions of a DMSO extract of Kosova Basin coal. *Glas. Hem. Drus. Beograd*, **49**(11), 723–727; *CA*, **103**, 217933 (1985)

Dai, G., Wang, Z., Tao, Y., Zhang, Z., Zhang, D., Zhang, G. and Wang, G. (1986) Investigation of fluorosis caused by burning coal. *Zhonghua Yufangyixue Zazhi*, **20**, 217–219

Dale, L.S. (1979) The emission spectrographic determination of boron in silicate materials. *Appl. Spectrosc.*, **33**, 404–406

Dale, L.S., Fardy, J.J. and Clayton, E.J. (1985) Trace element abundance data for coals using instrumental methods of analysis. *Proc. Int. Conf. Coal Sci.*, Pergamon, Sydney, NSW, pp. 857–860
Dalton, I.M. and Pringle, W.J.S. (1962) The gallium content of some Midland coals. *Fuel*, **41**, 41–48
Darwin, G.E. and Buddery, J.H. (1960) *Beryllium*. London: Butterworths, 392 pp
Das Gupta, H.N. and Chakrabarti, J.N. (1951) Estimation of chlorine in organic combination in the coal substance. *J. Indian Chem. Soc.*, **28**, 646–666
Date, A.R. and Gray, A.L. (eds.) (1989) *Applications of Inductively Coupled Plasma Mass Spectrometry*. Glasgow: Blackie, 254 pp
Daubrée, A. (1858) *C.R. Acad. Sci.*, **47**, 259; Duck and Himus (1951)
Davidson, C.F. and Ponsford, D.R.A. (1954) On the occurrence of uranium in coals. *Min. Mag.*, **91**, 265–273
Davies, I.J. (1977) Medical significance of the essential biological metals. *J. Radioanal. Chem.*, **37**, 39–64
Davis, A. (ed.) (1971) Proceedings of the Second Bowen Basin Symposium. *Geol. Surv. Queensl. Rep.*, No. 62, 210 pp
Davis, A., Russell, S.J., Rimmer, S.M. and Yeakel, J.D. (1984) Some genetic implications of silica and aluminosilicates in peat and coal. *Int. J. Coal Geol.*, **3**, 293–314
Davison, R.L., Natusch, D.F.S., Wallace, J.R. and Evans, C.A. (1974) Trace elements in fly ash – dependence of concentration on particle size. *Environ. Sci. Technol.*, **8**, 1107–1113
Davy, R. and Wilson, A.C. (1984) An orientation study of the trace- and other-element composition of some Collie coals. *Geol. Surv. West Aust. Rec.*, 1984/3, 48 pp
Day, J.C., Jones, R.H. and Belcher, C.B. (1979) Evaluation of coals for the metallurgical and power industries. *BHP Tech. Bull.*, **23**(2), 11–18
Daybell, G.N. and Pringle, W.J.S. (1958) The mode of occurrence of chlorine in coal. *Fuel*, **37**, 283–292
De Brito, A.C. (1955) Spectrographic study of the ash of Portuguese anthracites and lignites. *Rep. Port. Estudos, Notas Trab. Serv. Fom. Min.*, **10**, 236–262; *CA*, **50**, 11001 (1956)
Degenhardt, H. (1957) Investigations of the geochemical distribution of zirconium in the lithosphere. *Geochim. Cosmochim. Acta*, **11**, 279–309
Degens, E.T. (1965) *Geochemistry of Sediments – A Brief Survey*. New Jersey: Prentice-Hall, 342 pp
De Merre, M. (1956) Recovery of germanium. *Brit. Pat.*, 745,505,29 Feb.; *CA*, **50**, 13709 (1956)
Denson, N.M. (1959) Uranium in coal in the western United States. Introduction. *US Geol. Surv. Bull.*, No. 1055, 1–10
Des Cloizeaux, A. (1880) *Bull. Soc. Fr. Mineral.*, **3**, 170–171; Clarke (1916)
Deul, M. (1955) The origin of ash-forming ingredients in coal. *Econ. Geol.*, **50**, 103–104
Deul, M. and Annell, C.S. (1956) The occurrence of minor elements in ash of low-rank coal from Texas, Colorado, North Dakota and South Dakota. *US Geol. Surv. Bull.*, No. 1036-H, 155–172
Dewison, M.G. and Kanaris-Sotiriou, R. (1986) The determination of trace elements in whole coal by rhodium tube X-ray fluorescence spectrometry: a rock-based synthetic coal calibration. *Int. J. Coal Geol.*, **6**, 327–341
Diehl, R.C., Hattman, E.A., Schultz, H. and Haren, R.J. (1972) Fate of trace mercury in the combustion of coal. *US Bur. Mines, Tech. Prog. Rep.*, No. 54, 9 pp
Diessel, C.F.K. (1980) Newcastle and Tomago coal measures. *Geol. Surv. N.S.W. Bull.*, No. 26, 100–114
Dimitrakakis, P., Rankin, D. and Schaap, H. (1976) The determination of minor and trace elements in brown coal ash by X-ray emission methods. *SECV Rep.*, No. 321, 18 pp
Dissanayake, C.B. (1984) Geochemistry of the Muthurajawela peat deposit of Sri Lanka. *Fuel*, **63**, 1494–1503
Dixon, W.A. (1881) On the inorganic constituents of some epiphytic ferns. *J. Proc. R. Soc. NSW*, **15**, 175–183
Dolezal, J., Povondra, P. and Sulcek, Z. (1966) *Decomposition Techniques in Inorganic Analysis*. London: Iliffe Books, 224 pp
Donaldson, A.C., Renton, J.J. and Presley, M.W. (1985) Pennsylvanian deposystems and paleoclimates of the Appalachians. *Int. J. Coal Geol.*, **5**, 167–193
Doolan, K.J. (1982) The determination of traces of mercury in solid fuels by high-

temperature combustion and cold-vapor atomic absorption spectrometry. *Anal. Chim. Acta*, **140**, 187–195

Doolan, K.J. (1987) A pyrohydrolytic method for the determination of low fluorine concentrations in coal and minerals. *Anal. Chim. Acta*, **202**, 61–73

Doolan, K.J., Mills, J.C. and Belcher, C.B. (1979) The analysis and characterisation of mineral matter and ash. *Colloq. Combustion of Pulverised Coal: The Effect of Mineral Matter*, University of Newcastle, Newcastle, NSW, L3-1-L3-11

Doolan, K.J., Mills, J.C. and Turner, K.E. (1980) Environmental significance and analysis of trace elements in coal and coal products. *BHP Tech. Bull.*, **24**(2), 17–22

Doolan, K.J., Turner, K.E., Mills, J.C., Knott, A.C. and Ruch, R.R. (1984a) Volatilities of inorganic elements in coals during ashing. *Am. Chem. Soc. Div., Fuel Chem. Prepr.*, **29**(1), 127–134

Doolan, K.J., Turner, K.E., Knott, A.C. and Warbrooke, P. (1984b) A study of the feasibility of elemental analyses as an aid in coal seam discrimination. *BHP Rep.*, No. CRL/R/17/84, 116 pp; *Rep. NERDDP*, No. EG84/317, 123 pp

Doolan, K.J., Turner, K.E., Knott, A.C. and Warbrooke, P.R. (1985a) Coal seam discrimination using inorganic constituents. *Proc. Int. Conf. Coal Sci.*, Fuel Chem. Div. ACS, 642

Doolan, K.J., Turner, K.E., Mills, J.C., Knott, A.C. and Ruch, R.R. (1985b) Determination of trace elements in coal and coal products. Part 6 Volatilities of inorganic elements in coals during ashing. *Rep. NERDDP*, No. EG85/393, 88 pp

Dove, L.P. (1921) Sphalerite in coal pyrite. *Am. Mineral.*, **6**, 61

Drakeley, T.J. and Smith, F.W. (1922) The ultimate composition of British coals; the radium content. *J. Chem. Soc.*, **121**, 237–238

Dreher, G.B. and Schleicher, J.A. (1975) Trace elements in coal by optical emission spectroscopy. In Babu (1975), pp. 35–47

Drever, J.I., Murphy, J.W. and Surdam, R.C. (1977) The distribution of As, Be, Cd, Cu, Hg, Mo, Pb, and U associated with the Wyodak coal seam, Powder River Basin, Wyoming. *Contrib. Geol. Univ. Wyo.*, **15**, 93–101

DSIR (1949) The analyses and fusion temperatures of New Zealand coal ashes. *Dep. Sci. Ind. Res. NZ, Coal Rep.*, No. 249, 17 pp

Dubansky, A. (1983) Beryllium in coals from the North Bohemian brown coal region. *Uhli*, **31**(2), 59–63

Duck, N.W. and Himus, G.W. (1951) On arsenic in coal and its mode of occurrence. *Fuel*, **30**, 267–271

Duff, P. McL. D. (1967) Cyclic sedimentation in the Permian coal measures of New South Wales. *J. Geol. Soc. Aust.*, **14**, 293–308

Duflos, – (1847) Iodine and bromine content of Silesian coal. *Arch. Pharm.*, Hanover, **99**, 29–30; CIEB (1956), 129

Duran, J.E., Mahasay, S.R. and Stock, L.M. (1986) The occurrence of elemental sulphur in coals. *Fuel*, **65**, 1167–1168

Durie, R.A. (1961) The inorganic constituents in Australian coals. *Fuel*, **40**, 407–422

Durie, R.A. and Schafer, H.N.S. (1958) Germanium in Australian coals – a preliminary survey. *CSIRO Coal Res. Sect. Ref. Tech. Commun.*, No. 27, 25 pp

Durie, R.A. and Schafer, H.N.S. (1964) The inorganic constituents in Australian coals. IV Phosphorus and fluorine – their probable mode of occurrence. *Fuel*, **43**, 31–41

Durie, R.A. and Swaine, D.J. (1971) Inorganic constituents in coal. *Coal Res. CSIRO*, No. 45, 9–19

Durie, R.A., Schafer, H.N.S. and Swaine, D.J. (1963) Application of atomic absorption spectrometry to analysis of inorganic constituents in Victorian brown coals. *CSIRO Div. Coal Res. Invest. Rep.*, No. IR36, 11 pp

Durie, R.A., Schafer, H.N.S. and Swaine, D.J. (1965) The analysis of bituminous and brown coal ashes and related materials. *CSIRO Div. Coal Res. Tech. Commun.*, No. 47, 61 pp

Dvornikov, A.G. (1967a) Certain data on the distribution of mercury in iron disulfides from coal beds of the central Donets Basin. *Dokl. Akad. Nauk SSSR*, **172**, 1419–1422

Dvornikov, A.G. (1967b) Mercury distribution in disulfides of iron from coal seams of the central Donets Basin. *Dokl. Akad. Nauk SSSR*, **172**, 211–213

Dvornikov, A.G. (1967c) Some characteristics of iron disulfides in the Donets Basin coals. *Dopov. Akad. Nauk Ukr. RSR Ser. B*, **29**, 397–399; *CA*, **67**, 101735 (1967)

Dvornikov, A.G. (1967d) Some mercuriferous characteristics of coals of the East Donets Basin (Rostov region). *Dokl. Akad. Nauk SSSR*, **172**, 199–202; *CA*, **66**, 57653 (1967)

Dvornikov, A.G. (1967e) Mercury distribution in gas coal of the northern small folding area (Donbas). *Dopov. Akad. Nauk Ukr. RSR Ser. B*, **29**, 828–830; *CA*, **67** 110396 (1967)

Dvornikov, A.G. (1971) Increased mercury concentrations in certain coal fractions. *Coke Chem. USSR*, No. 9, 6–7

Dvornikov, A.G. (1981a) Forms of mercury occurrence in coals of the Donets Basin. *Dokl. Akad. Nauk SSSR*, **257**, 1214–1216; *CA*, **95**, 135456 (1981)

Dvornikov, A.G. (1981b) Forms of mercury in Donets Basin coals. *Geol. Zh.*, **41**, 96–104; *CA*, **95**, 135454 (1981)

Dvornikov, A.G. (1985) Mercury content of coals of different genetic types from coal basins. *Geol. Zh.*, **45**(6), 73–79; *CA*, **104**, 91777 (1986)

Dvornikov, A.G. and Tikhonenkova, E.G. (1968) Distribution and composition of iron disulfides from coals of the Donets Basin. *Zap. Vses. Mineral. Ova.*, **97**, 309–320; *CA*, **69**, 53469 (1974)

Dvornikov, A.G. and Tikhonenkova, E.G. (1974) Cinnabar find in coal mines of the Donbas. *Dokl. Akad. Sci. USSR, Earth Sci. Sect.*, **214**, 134–135

Dybczynski, R. (1980) Comparison of the effectiveness of various procedures for the rejection of outlying results and assigning consensus values in interlaboratory programs involving determination of trace elements or radionuclides. *Anal. Chim. Acta*, **117**, 53–70

Dzubay, T.G. (ed.) (1977) *X-ray Fluorescence Analysis of Environmental Samples*. Ann Arbor: Ann Arbor Science, 310 pp

Dzyuba, S.M. and Lepkii, S.D. (1981) Gold content of terrigenous formations and coal seams of the Donets Basin. *Sostav. Proiskhozhd. Razmeshchenie Osad. Porod Rud, Mater. Resp. Litol. Soves* 3rd, 164–168; *CA*, **97**, 200921 (1982) Izd. Naukova Dumka, Kiev

Ebdon, L. and Parry, H.G.M. (1987) Direct atomic spectrometric analysis by slurry atomisation. Part 2 Elimination of interferences in the determination of arsenic in whole coal by electrothermal atomisation atomic absorption spectrometry. *J. Anal. At. Spectrom.*, **2**, 131–134

Ebdon, L. and Pearce, W.C. (1982) Direct determination of arsenic in coal by atomic-absorption spectroscopy using solid sampling and electrothermal atomisation. *Analyst (London)*, **107**, 942–950

Ebdon, L. and Wilkinson, J.R. (1987) Direct atomic spectrometric analysis by slurry atomisation. Part 3 Whole coal analysis by inductively coupled plasma atomic emission spectrometry. *J. Anal. At. Spectrom.*, **2**, 325–328

Ebdon, L., Wilkinson, J.R. and Jackson, K.W. (1982) Determination of mercury in coal by non-oxidative pyrolysis and cold vapour atomic-fluorescence spectrometry. *Analyst (London)*, **107**, 269–275

Ebersbach, S. (1960) Investigations of the trace-element contents of brown coal ashes from the Menselwitz district. *Geologie*, **8**, 941–942

Edgcombe, L.J. (1956) State of combination of chlorine in coal. I Extraction of coal with water. *Fuel*, **35**, 38–48

Edgcombe, L.J. and Gold, H.K. (1955) Determination of arsenic in coal. *Analyst (London)*, **80**, 155–157

Egorov, A.P., Laktionova, N.V. and Borts, N.M. (1981) Trace elements in coals of the Neryungrin deposit. *Ugol* (6), 51–52; *CA*, **95**, 153413 (1981)

Egorov, A.P., Laktionova, N.V. and Popinako, N.V. (1979) A study of the loss of trace elements in the ashing of coals. *Solid Fuel Chem.*, **13**(2), 24–27

Egorov, A.P., Laktionova, N.V., Titova, T.A. and Tsedevsuren, T. (1983) Microelement composition of coal hydrogenation products. *Solid Fuel Chem.*, **17**(6), 80–82

Eisenbud, M. and Petrow, H.G. (1964) Radioactivity in the atmospheric effluents of power plants that use fossil fuels. *Science*, **144**, 288–289

Elliott, M.A. (ed.) (1981) *Chemistry of Coal Utilization: Second Supplementary Volume*. New York: Wiley, 2374 pp

Elseewi, A.A., Grimm, S.R., Page, A.L. and Straughan, I.R. (1981) Boron enrichment of plants and soils treated with coal ash. *J. Plant Nutr.*, **3**, 409–427

Endell, J. (1958) Fortuna brown coal fly ash as a calcium fertilizer. *Braunkohle*, **10**, 326–335; *CA*, **52**, 20829 (1958)

Erdman, J.A., Ebens, R.J. and Case, A.A. (1978) Molybdenosis: a potential problem in ruminants grazing on coal mine spoils. *J. Range Manage.*, **31**, 34–36

Ericsson, S.O. (1983) Natural radioactivity in coal. *Rep. KHM*-TR-86, 89 pp

Ershov, V.M. (1958) The character of the association of germanium with organic constituents of coal. *Geochemistry (USSR)* (6), 763–765

Ershov, V.M. (1961) Rare-earth elements in coals of the Kizel coal basin. *Geochemistry (USSR)* (3), 306–308
Eshleman, A., Siegel, S.M. and Siegel, B.Z. (1971) Is mercury from Hawaiian volcanoes a natural source of pollution? *Nature (London)*, **233**, 471–472
Eskenazi, G. (1967) Adsorption of gallium on peat and humic acids. *Fuel*, **46**, 187–191
Eskenazi, G. (1970) Adsorption of beryllium on peat and coals. *Fuel*, **49**, 61–67
Eskenazi, G. (1972) Adsorption of titanium on peat and coals. *Fuel*, **51**, 221–223
Eskenazi, G. (1972–73) Geochemistry of titanium during coal formation. *God. Sofii, Univ., Geol.-Geogr. Fak.*, **65**, 177–199
Eskenazi, G. (1977) On the binding form of tungsten in coals. *Chem. Geol.*, **19**, 153–159
Eskenazi, G.M. (1980) On the geochemistry of indium in the coal-forming process. *Geochim. Cosmochim. Acta*, **44**, 1023–1027
Eskenazi, G. (1982a) Indium content of the Marica East coals. *God. Sofii, Univ., Geol. - Geogr. Fak.*, **72**, 219–228
Eskenazi, G.M. (1982b) The geochemistry of tungsten in Bulgarian coals. *Int. J. Coal Geol.*, **2**, 99–111
Eskenazi, G.M. (1987a) Zirconium and hafnium in Bulgarian coals. *Fuel*, **66**, 1652–1657
Eskenazi, G.M. (1987b) Rare earth elements and yttrium in lithotypes of Bulgarian coals. *Org. Geochem.*, **11**, 83–89
Eskenazi, G.M. (1987c) Rare earth elements in a sampled coal from the Pirin Deposit, Bulgaria. *Int. J. Coal Geol.*, **7**, 301–314
Eskenazi, G. and Chubriev, Z. (1984) Trace elements in coal from the Pirin deposit. *Spis. Bulg. Geol. Druzh.*, **45**, 56–72; *CA*, **101**, 133627 (1984)
Eskenazi, G. and Minceva, E. (1989) On the geochemistry of strontium in Bulgarian coals. *Chem. Geol.*, **74**, 265–276
Eskenazi, G., Mincheva, E. and Ruseva, D. (1986) Trace elements in lignite lithotypes from the Elhovo coal basin. *Dokl. Bolg. Akad. Nauk*, **39**(10), 99–101; *CA*, **106**, 122749 (1987)
Estep, P.A., Kovach, J.J. and Karr, C. (1968) Quantitative infrared multicomponent determination of minerals occurring in coal. *Anal. Chem.*, **40**, 358–363
Esterle, J.S. and Ferm, J.C. (1986) Relationship between petrographic and chemical properties and coal seam geometry, Hance seam, Breathitt Formation, southeastern Kentucky. *Int. J. Coal Geol.*, **6**,199–214
Evans, J.C., Abel, K.H., Olsen, K.B., Lepel, E.A., Sanders, R.W., Wilkerson, C.L. and Hayes, D.J. (1985) Characterization of trace constituents at Canadian coal-fired power plants. *Rep. Can. Elec. Assoc.*, Vol. 1, Phase I, 172 pp
Faanhof, A. (1986) Neutron activation. In Butler (1986), pp. 67–85
Facer, R.A., Cook, A.C. and Beck, A.E. (1980) Thermal properties and coal rank in rocks and coal seams of the southern Sydney Basin, New South Wales: a palaeogeothermal explanation of coalification. *Int. J. Coal Geol.*, **1**, 1–17
Failey, M.P., Anderson, D.L., Zoller, W.H. and Gordon, G.E. (1979) Neutron-capture prompt γ-ray activation analysis for multielement determination in complex samples. *Anal. Chem.*, **51**, 2209–2221
Falcon, R.M.S. (1978) Coal in South Africa. Part II The application of petrography to the characterisation of coal. *Miner. Sci. Eng.*, **10**(1), 28–52
Falcon, R.M.S. and Snyman, C.P. (1986) An introduction to coal petrography: atlas of petrographic constituents in the bituminous coals of Southern Africa. *Geol. Soc. S. Afr. Rev. Pap.*, No. 2, 27 pp
Fardy, J.J. and Swaine, D.J. (1985) Use of neutron activation analysis to measure the variation in trace element concentrations in a coal seam. *Proc. Fourth Aust. Conf. Nucl. Techniques Anal., AINSI*, Lucas Heights, NSW, pp. 152–154
Fardy, J.J., Hügi, T. and Swaine, D.J. (1987) Trace elements in Swiss coals. *Proc. 5th Aust. Conf. Nucl. Techniques Anal., AINSI*, Lucas Heights, NSW, pp. 166–168
Fardy, J.J., McOrist, G.D. and Farrar, Y.J. (1984) The analysis of coals and fly ash for trace elements and natural radioactivity. *Proc. Aust. Coal Sci. Conf.*, Gippsland Inst. Advanced Educ., Churchill, Vic., pp. 159–166
Farne, G., Randi, G., Grimaldi, R., Mariani, E., Paris, V. and Riccio, M. (1983) Direct determination of inorganic components in coal by X-ray fluorescence spectrometry. *Riv. Combust.*, **37**, 105–118; *CA*, **100**, 36757 (1984)
Faurschou, D.K., Bonnell, G.W. and Janke, L.C. (1982) Analysis directory of Canadian commercial coals – Supplement No. 4. *CANMET Rep.*, No. 82-13E, 192 pp

Felgueroso, J., Martinez-Alonso, A., Martinez-Tarazona, M.R. and Tascon, J.M.D. (1988) The determination of mineral matter content of low-rank coals. *J. Coal Qual.,* **7**, 127-131

Fersman, A. (1958) *Geochemistry for Everyone.* Moscow: Foreign Language Publishing House, 454 pp

Fieldner, A.C., Hall, A.E. and Field, A.L. (1918) The fusibility of coal ash and the determination of the softening temperature. *US Bur. Mines Bull.,* No. 129, 146 pp

Filby, R.H. and Swaine, D.J. (1983) Trace elements in coal, fly-ash and moss by neutron activation analysis. *Proc. Third Aust. Conf. Nucl. Techniques Anal., AINSI,* Lucas Heights, NSW, pp. 148-149

Filby, R.H. and van Berkel, G.J. (1987) Geochemistry of metal complexes in petroleum, source rocks and coals: an overview. *ACS Symp. Ser.,* No. 344, 2-39

Filby, R.H., Shah, K.R. and Sautter, C.A. (1977) A study of trace element distribution in the solvent refined coal (SRC) process using neutron activation analysis. *J. Radioanal. Chem.,* **37**, 693-704

Filby, R.H., Nguyen, S., Grimm, C.A., Markowskii, G.R., Ekambaran, V., Tanaka, T. and Grossman, L. (1985) Evolution of geochemical standard reference materials for mic roanalysis. *Anal. Chem.,* **57**, 551-555

Filip, Z., Cheshire, M.V., Goodman, B.A. and McPhail, D.B. (1985) The occurrence of copper, iron, zinc and other elements and the nature of some copper and iron complexes in humic substances from municipal refuse disposed of in a landfill. *Sci. Total Environ.,* **44**, 1-16

Finkelman, R.B. (1978) Determination of trace element sites in the Waynesburg coal by SEM analysis of accessory minerals. *Scanning Electron Microsc.,* **1**, 143-148

Finkelman, R.B. (1981a) Modes of occurrence of trace elements in coal. *US Geol. Surv. Open-File Rep.,* No. OFR-81-99, 301 pp.; also PhD Thesis, University of Maryland, 1980

Finkelman, R.B. (1981b) Recognition of authigenic and detrital minerals in coal. *Geol. Soc. Amer. Abstr. Program,* **13**(7), 451

Finkelman, R.B. (1982a) The origin, occurrence, and the distribution of the inorganic constituents in low-rank coals. In Schobert (1982), pp. 70-89

Finkelman, R.B. (1982b) Modes of occurrence of trace elements and minerals in coal: an analytical approach. In *Atomic and Nuclear Methods in Fossil Energy Research* (eds. R.H. Filby, B.S. Carpenter and R.C. Ragaini). New York: Plenum, 141-149

Finkelman, R.B. (1984) Chemical analysis. In Finkelman *et al.* (1984b), pp. 21-24

Finkelman, R.B. (1986) Characterisation of the inorganic constituents in coal. *Proc. Mater. Res. Soc. Symp.,* **65**, 71-76

Finkelman, R.B. (1988) The inorganic geochemistry of coal: a scanning electron microscopy view. *Scanning Micros.,* **2**(1), 97-105

Finkelman, R.B. and Aruscavage, P.J. (1981) Concentration of some platinum-group metals in coal. *Int. J. Coal Geol.,* **1**, 95-99

Finkelman, R.B. and Bhuyan, K. (1986) Inorganic geochemistry of a Texas lignite. In Finkelman and Casagrande (1986), pp. 187-197

Finkelman, R.B. and Casagrande, D.J. (eds.) (1986) *Geology of Gulf Coast Lignites.* Houston: Environmental and Coal Associates, 219 pp

Finkelman, R.B. and Gluskoter, H.J. (1983) Characterization of minerals in coal: problems and promises. In *Fouling and Slagging Resulting from Impurities in Combustion Gases* (ed. R.W. Bryers), Proc. 1981 Eng. Foundation Conf., New York, pp. 299-318

Finkelman, R.B. and Simon, F.O. (1984) Mineralogy of 'demineralized' coal. In Finkelman *et al.* (1984b), pp. 33-34

Finkelman, R.B. and Stanton, R.W. (1978) Identification and significance of accessory minerals from a bituminous coal. *Fuel,* **57**, 763-768

Finkelman, R.B., Fiene, F.L., Miller, R.L. and Simon, F.O. (eds.) (1984b) Interlaboratory comparison of mineral constituents in a sample from the Herrin (No. 6) coal bed from Illinois. *US Geol. Surv. Circ.,* No. 932, 42 pp

Finkelman, R.B., Simons, D.S., Dulong, F.T. and Steel, E.B. (1984a) Semi-quantitative ion microprobe mass analyses of mineral-rich particles from the Upper Freeport coal. *Int. J. Coal Geol.,* **3**, 279-289

Finkelman, R.B., Stanton, R.W., Cecil, C.B. and Minkin, J.A. (1979) Modes of occurrence of selected trace elements in several Appalachian coals. *Am. Chem. Soc. Div., Fuel Chem. Prepr.,* **24**(1), 236-241

Finn, C.P. (1930) An occurrence of barytes in the Parkgate (South Yorkshire) seam. *Trans. Inst. Min. Eng.*, **80**, 25-26

Firth, J.N.M. (1973) Mineralogical distributions and disease patterns in the South Wales coalfield. In *Trace Substances in Environmental Health-VI* (ed. D.D. Hemphill), pp. 325-331, University of Missouri, Columbia

Fishbein, L. (1981) Sources, transport and alterations of metal compounds: an overview. I Arsenic, beryllium, cadmium, chromium and nickel. *Environ. Health Perspect.*, **40**, 43-64

Flaig, W. (1968) Biochemical factors in coal formation. In Murchison and Westoll (1968), pp. 197-232

Fleet, M.E.L. (1965) Preliminary investigations into the sorption of boron by clay minerals. *Clay Miner.*, **6**, 3-16

Fletcher, J.D. and Golightly, D.W. (1985) The determination of 28 elements in whole coal by direct-current arc spectrography. *US Geol. Surv. Open-File Rep.*, No. 85-204, 10 pp

Flood, P.G. (1985) Facies study of the Callide seam, central Queensland: implications for mine planning and design. *Aust. Coal Geol.*, **5**, 13-24

Flum, Z. (1957) Colorimetric method of determining arsenic in coal and tar. *Paliva*, **37**, 33-36

Folkmann, F. (1975) Analytical use of ion-induced X rays. *J. Phys. E*, **8**, 429-444

Fowkes, W.W. (1978) Separation and identification of minerals from lignites. In Karr (1978b), pp. 293-314

Francis, W. (1961) *Coal – Its Formation and Composition*. London: Edward Arnold, 806 pp

Francois, R. (1987) A study of sulphur enrichment in the humic fraction of marine sediments during early diagenesis. *Geochim. Cosmochim. Acta*, **51**, 17-27

Fraser, D.C. (1961) Organic sequestration of copper. *Econ. Geol.*, **56**, 1063-1078

Frazer, F.W. and Belcher, C.B. (1973) Quantitative determination of the mineral-matter content of coal by a radiofrequency-oxidation technique. *Fuel*, **52**, 41-46

FRISA (1960) Annual Report. *Fuel Res. Inst. S. Afr.*, 30th, 68 pp

Frost, J.K., Santoliquido, P.M., Camp, L.R. and Ruch, R.R. (1975) Trace elements in coal by neutron activation analysis with radiochemical separations. In Babu (1975), pp. 84-97

Fudagawa, N. and Kawase, A. (1985) Determination of cadmium in coal by metal furnace atomic absorption spectrometry. *Bunseki Kogaku*, **34**, 228-233; *CA*, **103**, 39514 (1985)

Fuge, R. (1988) Sources of halogens in the environment, influences on human and animal health. *Environ. Geochem. Health*, **10**, 51-61

Fuge, R. and Johnson, C.C. (1986) The geochemistry of iodine – a review. *Environ. Geochem. Health*, **8**, 31-54

Fukushima, M. and Nakaoka, A. (1984) A method for the determination of trace uranium in environmental specimens. *Denryoku Chuo Kenkyuyo Enerugi, Kankyo Gijutsu Kenkyuko Kenkyu Hokoku*, Rep. No. 28035, 20 pp

Fukuzaki, N., Tamura, R., Hirano, Y. and Mizushima, Y. (1986) Mercury emission from a cement factory and its influence on the environment. *Atmos. Environ.*, **20**, 2291-2299

Furr, A.K., Parkinson, T.F., Hinrichs, R.A., van Campen, D.R., Bache, C.A., Gutten mann, W.H., St John, L.E., Pakkala, I.S. and Lisk, D.J. (1977) National survey of elements and radioactivity in fly ashes. *Environ. Sci. Technol.*, **11**, 1194-1201

Furst, A. (1987) Biological testing of germanium. *Toxicol. Ind. Health*, **3**, 167-204

Fyfe, W.S., Kronberg, B.I. and Brown, J.R. (1982) Variations in the inorganic chemistry of coal. *Am. Chem. Soc. Div., Fuel Chem. Prepr.*, **27**(1), 116-123

Gainsford, A.R. (1985) A comparison of the inorganic content of some Waikato and Mokau coals. *Proc. Coal Res. Conf. NZ*, Pap. No. 13.2, 9 pp

Galatanu, V. and Engelmann, C. (1981) Determination of some trace elements in coal non-destructively by photonuclear activation. *J. Radioanal. Chem.*, **67**, 143-163

Gao, G., Yan, B. and Yang, L. (1984) Determination of total fluorine in coal by the combustion–hydrolysis/fluoride ion selective eletrode method. *Fuel*, **63**, 1552-1555

Garbauskas, M.F. and Wong, J. (1983) XRF analysis of trace titanium in coal using fundamental parameters. *X-ray Spectrom.*, **12**, 118-120

Gardner, D. (1977) The determination of total mercury in coal and organic matter with minimal risk of external contamination. *Anal. Chim. Acta*, **93**, 291-295

Gauger, A.W., Barrett, E.P. and Williams, F.J. (1934) Mineral matter in coal – a preliminary report. *Trans. Am. Inst. Min. Metall.*, **108**, 226-236

Geer, M.R., Davis, F.T. and Yancey, H.F. (1943) Occurrence of phosphorus in Washington coal, and its removal. *Am. Inst. Min. Metall. Eng. Tech. Publ.*, No. 1586, 8 pp; *CA*, **38**, 468 (1944)

Gehrs, C.W., Shriner, D.S., Herbes, S.E., Salmon, E.J. and Perry, H. (1981) Environmen-

tal, health and safety implications of increased coal utilization. In Elliott (1981), pp. 2159–2223
Gerhard, L., Kautz, K., Pickhardt, W., Scholz, A. and Zimmermeyer, G. (1985) A study of trace element distribution during combustion of bituminous coal in three power plants. *VGB Kraftwerksteknik*, **65**, 676–686
Germani, M.S. and Zoller, W.H. (1988) Vapor-phase concentrations of arsenic, selenium, bromine, iodine, and mercury in the stack of a coal-fired power plant. *Environ. Sci. Technol.*, **22**, 1079–1085
Germani, M.S., Gokmen, I., Sigleo, A.C., Kowalczyk, G.S., Olmez, I., Small, A.M., Anderson, D.L., Falley, M.P., Gulovali, M.C., Choquette, C.E., Lepel, E.A., Gordon, G.E. and Zoller, W.H. (1980) Concentrations of elements in the National Bureau of Standard's bituminous and subbituminous coal standard reference materials. *Anal. Chem.*, **52**, 240–245
Ghosh, B., Biswas, D. and Banerjee, N.N. (1979) Gallium in Indian coals. *Indian J. Technol.*, **17**, 61–64
Ghosh, B., Das, M.C., Ghosh, S.B. and Banerjee, N.N. (1987) Vanadium in Indian coals. *Indian J. Technol.*, **25**, 467–470
Giauque, R.D., Garrett, R.B. and Goda, L.Y. (1977) Determination of forty elements in geochemical samples and coal fly ash by X-ray fluorescence spectrometry. *Anal. Chem.*, **49**, 1012–1017
Giauque, R.D., Garrett, R.B. and Goda, L.Y. (1979) Determination of trace elements in light element matrices by X-ray fluorescence spectrometry with incoherent scattered radiation as an internal standard. *Anal. Chem.*, **51**, 511–516
Gibb, W.H. (1983) The nature of chlorine in coal and its behaviour during combustion. In Meadowcroft and Manning (1983), pp. 25–45
Gibb, W.H. and Mayne, G.F. (1980) An investigation into the British Standard high temperature method for the determination of chlorine in coal. *J. Inst. Energy*, **53**, 47–51
Gibson, F.H. and Selvig, W.A. (1944) Rare and uncommon chemical elements in coal. *US Bur. Mines Tech. Pap.*, No. 669, 23 pp
Gibson, J. and Kennedy, G.F. (1980) Coal and the environment. *World Energy Conf.*, London, 11th, **3**, 325–346
Gillis, J., Hoste, J. and Claeys, A. (1947) A specific reagent for germanium: phenylfluorone. *Anal. Chim. Acta*, **1**, 302–308
Given, P.H. (1960) The distribution of hydrogen in coals and its relation to coal structure. *Fuel*, **39**, 147–153
Given, P.H. (1961) Towards an understanding of the chemical structure of coal. In Francis (1961), pp. 717–753
Given, P.H. (1984) An essay on the organic geochemistry of coal. In *Coal Science* (eds. M.L. Gorbaty, J.W. Larsen and I. Wender), Vol. 3, pp. 63–252. Orlando: Academic Press
Given, P.H. and Cohen, A.D. (eds.) (1977) Interdisciplinary studies of peat and coal origins. *Geol. Soc. Am. Microform Publ.*, No. 7
Given, P.H. and Miller, R.N. (1985) Distribution of forms of sulfur in peats from saline environments in the Florida Everglades. *Int. J. Coal Geol.*, **5**, 397–409
Given, P.H. and Miller, R.N. (1987) The association of major, minor and trace inorganic elements with lignites. III Trace elements in four lignites and general discussion of all data from this study. *Geochim. Cosmochim. Acta*, **51**, 1843–1853
Given, P.H. and Yarzab, R.F. (1978) Analysis of the organic substance of coals: problems posed by the presence of mineral matter. In Karr (1978b), pp. 3–41
Given, P.H., Spackman, W., Davis, A. and Jenkins, R.G. (1980) Some proved and unproved effects of coal geochemistry on liquefaction behaviour with emphasis on U.S. coals. *ACS Symp. Ser.*, No. 139, 3–34
Given, P.H., Marzec, A., Barton, W.A., Lynch, L.J. and Gerstein, B.C. (1986) The concept of a mobile or molecular phase within the macro-molecular network of coals: a debate. *Fuel*, **65**, 155–164
Given, P.H., Rhoads, C., Painter, P.C., Spackman, W. and Ryan, N.J. (1983a) Aspects of the origins of some coal macerals. *Proc. Int. Conf. Coal Sci.*, pp. 389–392
Given, P.H., Spackman, W., Imbalzano, J.R., Casagrande, D.J., Lucas, A.J., Cooper, W. and Exarchos, C. (1983b) Physicochemical characteristics and levels of microbial activity in some Florida peat swamps. *Int. J. Coal Geol.*, **3**, 77–99
Gladney, E.S. (1972) Direct determination of beryllium in NBS SRM 1632 coal by flameless atomic absorption. *At. Absorpt. Newsl.*, **16**, 42–43

Gladney, E.S. (1980a) Elemental concentrations in NBS biological and environmental standard reference materials. *Anal. Chim. Acta,* **118**, 385-396
Gladney, E.S. (1980b) Compilation of elemental concentration data for NBS biological and environmental standard reference materials. *Los Alamos Rep.,* No. LA-8438-MS, 119 pp
Gladney, E.S. and Burns, C.E. (1983) 1982 compilation of elemental concentrations in eleven United States Geological survey rock standards. *Geostand. Newsl.,* 7, 3-226
Gladney, E.S. and Owens, J.W. (1976) Beryllium emissions from a coal-fired power plant. *J. Environ. Sci. Health,* **A11**, 297-311
Gladney, E.S. and Roelandts, I. (1987) Compilation of boron concentration data for NBS, USGS and CCRMP reference materials. *Geostand. Newsl.,* 11(2), 167-185
Gladney, E.S., Garcia, S.R. and Newlin, J.S. (1986) Determination of elemental composition of NBS SRM coals via automated neutron activation analysis. *Geostand. Newsl.,* **10**, 77-80
Gladney, E.S., Jurney, E.T. and Curtis, D.B. (1976) Nondestructive determination of boron and cadmium in environmental materials by thermal neutron-prompt γ-ray spectrometry. *Anal. Chem.,* **48**, 2139-2142
Gladney, E.S., Goode, W.E., Perrin, D.R. and Burns, C.E. (1981) Quality assurance for environmental analytical chemistry: 1980. *Los Alamos Rep.,* No. LA-8966-MS, 176 pp
Gladney, E.S., Perrin, D.R., Burns, C.E. and Robinson, R.D. (1982) Quality assurance for environmental analytical chemistry: 1981. *Los Alamos Rep.,* No. LA-9579-MS, 128 pp
Gladney, E.S., Small, J.A., Gordon, G.E. and Zoller, W.H. (1976) Composition and size distribution of in-stack particulate material at a coal-fired power plant. *Atmos. Environ.,* **10**, 1071-1077
Gleit, C.E. and Holland, W.D. (1962) Use of electrically excited oxygen for the low temperature decomposition of organic substances. *Anal. Chem.,* **34**, 1454-1457
Glick, D.C. and Davis, A. (1987) Variability in the inorganic element content of U.S. coals including results of cluster analysis. *Org. Geochem.,* **11**, 331-342
Glowiak, B.J. and Pacyna, J.M. (1980) Radiation dose due to atmospheric releases from coal/fired power stations. *Int. J. Environ. Stud.,* **16**, 23-28
Gluskoter, H.J. (1965) Electronic low-temperature ashing of bituminous coal. *Fuel,* **44**, 285-291
Glukoter, H.J. (1975) Mineral matter and trace elements in coal. In Babu (1975), pp. 1-22
Gluskoter, H.J. and Lindahl, P.C. (1973) Cadmium – mode of occurrence in Illinois coals. *Science,* **188**, 264-266
Gluskoter, H.J. and Rees, O.W. (1964) Chlorine in Illinois coal. *Ill. State Geol. Surv. Circ.,* No. 372, 23 pp
Gluskoter, H.J. and Ruch, R.R. (1971) Chlorine and sodium in Illinois coals as determined by neutron activation analyses. *Fuel,* **50**, 65-76
Gluskoter, H.J., Shimp, N.F. and Ruch, R.R. (1981) Coal analyses, trace elements and mineral matter. In Elliott (1981), pp. 369-424
Gluskoter, H.J., Ruch, R.R., Miller, W.G., Cahill, R.A., Dreher, G.B. and Kuhn, J.K. (1977) Trace elements in coal: occurrence and distribution. *Ill. State Geol. Surv. Circ.,* No. 499, 154 pp
Godbeer, W.C. (1987) Results for fluorine in coals and other reference materials. *Geostand. Newsl.,* **11**, 143-145
Godbeer, W.C. and Swaine, D.J. (1979) Cadmium in coal and fly ash. In *Trace Substances in Environmental Health-XIII* (ed. D.D. Hemphill), pp. 254-261, University of Missouri, Columbia
Godbeer, W.C. and Swaine, D.J. (1987) Fluorine in Australian coals. *Fuel,* **66**, 794-798
Godbeer, W.C., Morgan, N.C. and Swaine, D.J. (1981) The use of moss to monitor trace elements. *Proc. 7th Int. Clean Air Conf.,* Adelaide, Ann Arbor Science, Ann Arbor, pp. 789-798
Godbeer, W.C., Morgan, N.C. and Swaine, D.J. (1984) The accession of trace elements to the environs of a power station. *Proc. Eighth Int. Clean Air Conf.,* Melbourne, Clean Air Soc. Aust. and N.Z., Sydney, NSW, pp. 883-890
Godden, R.G. and Thomerson, D.R. (1980) Generation of covalent hydrides in atomic-absorption spectroscopy. *Analyst (London),* **105**, 1137-1156
Goldschmidt, V.M. (1930) The presence of germanium in coals and products. *Nachr. Ges. Wiss. Goettingen, Math.-Phys. Kl., Fachgruppe,* **3**, 398-401; *CA,* **25**, 3935 (1931)
Goldschmidt, V.M. (1935) Rare elements in coal ashes. *Ind. Eng. Chem.,* **27**, 1100-1102
Goldschmidt, V.M. (1937) The principles of distribution of chemical elements in minerals and rocks. *J. Chem. Soc.,* 655-673

Goldschmidt, V.M. (ed. A. Muir) (1954) *Geochemistry*. Oxford: Clarendon Press, 730 pp
Goldschmidt, V.M. and Hefter, O. (1933) The geochemistry of selenium. *Nachr. Ges. Wiss. Goettingen, Math.-Phys. Kl., Fachgruppe IV*, pp. 245-252
Goldschmidt, V.M. and Paters, C. (1931) On the geochemistry of gallium. *Nachr. Ges. Wiss. Goettingen, Math.-Phys. Kl., Fachgruppe IV*, pp. 165-183
Goldschmidt, V.M. and Peters, C. (1932a) On the geochemistry of beryllium. *Nachr. Ges. Wiss. Goettingen, Math.-Phys. Kl., Fachgruppe IV*, pp. 360-376
Goldschmidt, V.M. and Peters, C. (1932b) On the geochemistry of boron. *Nachr. Ges. Wiss. Goettingen, Math.-Phys. Kl. V*, pp. 528-545
Goldschmift, V.M. and Peters, C. (1933a) On the geochemistry of germanium. *Nachr. Ges. Wiss. Goettingen, Math.-Phys. Kl., Fachgruppe IV*, pp. 141-166
Goldschmidt, V.M. and Peters, C. (1933b) The enrichment of rare elements in hard coals. *Nachr. Ges. Wiss. Goettingen, Math.-Phys. Kl., Fachgruppe IV*, pp. 371-386
Goldschmidt, V.M. and Peters, C. (1934) The geochemistry of arsenic. *Nachr. Ges. Wiss. Goettingen, Math.-Phys. Kl., NF Fachgruppe I*, pp. 11-22
Golightly, D.W. and Simon, F.O. (1989) Methods for sampling and inorganic analysis of coal. *US Geol. Surv. Bull.*, 1823, pp 72-239
Golovko, V.A. (1960) The distribution of minor elements in Carboniferous deposits of the central districts of the USSR. *Dokl. Akad. Nauk SSSR*, **132**, 911-914; *CA*, **55**, 12191 (1961)
Goodarzi, F. (1987a) Comparison of elemental distribution in fresh and weathered samples of selected coals in the Jurassic-Cretaceous Kootenay Group, British Columbia, Canada. *Chem. Geol.*, **63**, 21-28
Goodarzi, F. (1987b) Concentration of elements in lacustrine coals from zone A Hat Creek deposit No. 1, British Columbia, Canada. *Int. J. Coal Geol.*, **8**, 247-268
Goodarzi, F. (1987c) Elemental concentrations in Canadian coals. 2 Byron Creek collieries, British Columbia. *Fuel*, **66**, 250-254
Goodarzi, F. (1988) Elemental distribution in coal seams at the Fording coal mine, British Columbia, Canada. *Chem. Geol.*, **68**, 129-154
Goodarzi, F. and Cameron, A.R. (1987) Distribution of major, minor and trace elements in coals of the Kootenay Group, Mount Allan, Alberta. *Can. Mineral.*, **25**, 555-565
Goodarzi, F. and van der Flier-Keller, E. (1988) Distribution of major, minor and trace elements in Hat Creek Deposit No.2, British Columbia, Canada. *Chem. Geol.*, **70**, 313-333
Goodarzi, F. and van der Flier-Keller, E. (1989) Organic petrology and geochemistry of intermontane coals from British Columbia 3. The Blakeburn opencast mine near Tulameen, British Columbia, Canada. *Chem. Geol.*, **75**, 227-247
Goodarzi, F., Foscolos, A.E. and Cameron, A.R. (1985) Mineral matter and elemental concentrations in selected western Canadian coals. *Fuel*, **64**, 1599-1605
Goodman, B.A. and Cheshire, M.V. (1973) Electron paramagnetic resonance evidence that copper is complexed in humic acid by the nitrogen of porphyrin groups. *Nature (London) New Biol.*, **244**, 158-159
Goodman, G.T. and Roberts, T.M. (1971) Plants and soils as indicators of metals in the air. *Nature (London)*, **231**, 287-292
Goodman, G.T., Inskip, M.J., Smith, S., Parry, G.D.R. and Burton, M.A.S. (1979) The use of moss-bags in aerosol monitoring. *Scope Rep. No. 14*, 'The Saharan Dust, Mobilization, Transport, Deposition', (ed. C. Morales). Sussex: Wiley, pp. 211-232
Gordon, G.E. (1988) Receptor models. *Environ. Sci. Technol.*, **22**, 1132-1142
Gordon, S.A. (1959a) Adsorption as a means of accumulating germanium in coals. *Nauchn. Tr. Mosk. Gorn. Inst. Sb.*, **27**, 47-56; *CA*, **55**, 11804 (1961)
Gordon, S.A. (1959b) Carbonyl compounds of germanium. *Nauchn. Tr. Mosk. Gorn. Inst. Sb.*, **27**, 129-139; *CA*, **55**, 19204 (1961)
Gordon, S.A. and Motina, A.G. (1959) Composition of the mineral portion of coal as a factor characterising the accumulation of germanium. *Nauchn. Tr. Mosk. Gorn. Inst.*, **27**, 57-63; *CA*, **55**, 7800 (1961)
Gordon, S.A., Volkov, K.Y. and Mendovskii, M.A. (1958) On the forms of germanium in coal. *Geochemistry (USSR)* (4), 484-489
Gorham, E. (1957) The development of peat lands. *Q. Rev. Biol.*, **32**, 145-166
Gorham, E. (1967) Some chemical aspects of wetland ecology. *Tech. Mem. NRC Assoc. Comm. Geotech. Res.*, No. 90, 20-38
Gorham, E., Santelmann, M.V. and McAllister, J.E. (1984) A peatland bibliography, chiefly with reference to the ecology, hydrology and biochemistry of Sphagnum bogs. *Publ. Dept. Ecology Behavioural Biology, University of Minnesota*, 153 pp

Gorsuch, T.T. (1962) Losses of trace elements during oxidation of organic materials – the formation of volatile chlorides during dry ashing in presence of inorganic chlorides. *Analyst (London)*, **87**, 112–115

Gould, R.F. (ed.) (1966) *Coal Science; Advances in Chem. Ser. No. 55.* Washington, DC: ACS, 743 pp

Gracia, I. and Martin, A. (1968) Volatilization of rhenium during combustion or gasification of rhenium-enriched lignites. *Ensayos Invest.*, **3**(10), 3–8; *CA*, **70**, 79820 (1969)

Graedel, T.E. (1982) Aqueous chemistry in the atmosphere, group report. In *Atmospheric Chemistry* (ed. E.D. Goldberg), pp. 93–118. Berlin: Springer Verlag

Graf, – (1852) Iodine in hard coal. *Arch. Pharm.*, *Hanover*, **120**, 136; CIEB (1956), p. 87

Gray, V.R. (1986) The chemical properties and composition of Mokau coal. *NZ J. Geol. Geophys.*, **29**, 447–461

Green, T., Kovac, J., Brenner, D. and Larsen, J.W. (1982) The macro-molecular structure of coals. In Meyers (1982), pp. 199–282

Greenfield, S., Jones, I.L. and Berry, C.T. (1964) High pressure plasmas as spectroscopic emission sources. *Analyst (London)*, **89**, 713–720

Greenland, L.P. and Aruscavage, E.P. (1986) Volcanic emission of selenium, tellurium and arsenic from Kilauea volcano, Hawaii. *J. Volcanology Geothermal Res.*, **27**, 195–201

Gregorowicz, Z. (1957) Germanium in Polish brown coals. *Przem. Chem.*, **13**, 700–701

Griepink, B., Scholz, A. and Wilkinson, H.C. (1988) Preparation and certification of reference samples of coal. *Fuel*, **67**, 1580–1581

Grosjean, A. (1943) Occurrences of millerite in the Carboniferous of Belgium. *Bull. Soc. Belg. Geol.*, **52**, 34–50

Gueven, N. and Lee, L.J. (1983) Characterisation of mineral matter in East Texas lignites. *Rep. TENRAC/EDF-103*, 70 pp.; *CA*, **101**, 57501 (1984)

Guidoboni, R.J. (1973) Determination of trace elements in coal and coal ash by spark source mass spectrometry. *Anal. Chem.*, **45**, 1275–1277

Guidoboni, R.J. (1978) Spark source mass spectrometry and atomic absorption spectrophotometry for the determination of trace elements in coal. In Karr (1978a), pp. 421–434

Gulyayeva, L.A. and Itkina, E.S. (1962) The halogens and vanadium, nickel and copper in coals. *Geochemistry (USSR)*, No. 4, 395–407

Gumz, W., Kirsch, H. and Mackowsky, M.-Th. (1958) *Schlackenkunde.* Berlin: Springer, 422 pp

Guó, Q., Cao, J. and Zhuo, Z. (1986) Determination of trace selenium in coal using derivative fluorimetry with oxygen bomb combustion pre-treatment. *Fenxi Huaxue*, **14**, 829–832; *CA*, **106**, 69952 (1987)

Gurevich, A.B. and Shishlov, S.B. (1987) Igneous intrusions in coal-bearing sequences. *Int. Geol. Rev.*, **29**, 951–958

Guseva, Z.L. (1959) Chemical determination of scandium in coal ash. *Nauchn. Tr. Moskov. Gorn. Inst. Sb.*, **27**, 81–88; *CA*, **55**, 16960 (1961)

Haberlandt, H. (1944) Concentration of rare elements in mineral formations due to additions of organic origin. *Forsch. Fortschr.*, **20**, 154–155; *CA*, **43**, 2137 (1949)

Hak, J. and Babcan, J. (1967) Geochemistry of germanium and beryllium in coals of the Sokolov Basin. *Geochem. Czech. Trans. 1st Conf. Geochem.*, Ostrava, pp. 163–170; *CA*, **70**, 98688 (1969)

Hall, G.E.M., MacLaurin, A.I. and Vaive, J. (1986) The analysis of geological materials for fluorine, chlorine and sulphur using pyrohydrolysis and ion chromatography. *J. Geochem. Explor.*, **26**, 177–186

Hallam, A. and Payne, K.W. (1958) Germanium enrichment in lignites from the lower Lias of Dorset. *Nature (London)*, **181**, 1008–1009

Halstead, W.D. (1981) Flue gas behaviour of trace elements in coal-fired power stations. *Proc. Int. Conf. Coal Sci.*, Verlag Glückauf GmbH, Essen, pp. 737–744

Hamilton, E.I. (1974) The chemical elements and human morbidity – water, air and places – a study of natural variability. *Sci. Total Environ.*, **3**, 3–85

Hamilton, E.I., Minski, M.J. and Cleary, J.J. (1972) Problems concerning multi-element assay in biological materials. *Sci. Total Environ.*, **1**, 1–14

Hamrla, M. (1959) Conditions of origin of the coal beds in the Karst region (northwest Yugoslavia). *Geol. Razprave Porocila*, **5**, 180–264; *CA*, **55**, 24421 (1961)

Hannan, A.H.M.A., Kehinde, L.O., Oluwole, A.F., Borishade, A.B., Oshin, O., Aladekomo, J.B. and Jervis, R.E. (1982) Determination of trace elements in Nigerian coals by neutron activation analysis. *Radiochem. Radioanal. Lett.*, **53**, 155–162

Hannan, J.C. (ed.) (1980) Environmental Controls for Coal Mining. *Proc. First Nat. Semin.,* *Univ. Sydney,* 376 pp
Hannan, J.C. (ed.) (1982) Coal and Society. *Proc. Nat. Semin., Univ. Sydney,* Australian Coal Assoc. Sydney, NSW, 396 pp
Haraguchi, H., Kurosawa, M. and Iawata, Y. (1985) Simultaneous multielement analysis of coal and fly ashes by inductively coupled plasma atomic emission spectrometry. *Bunseki Kagaku,* **34,** 252-257; *CA,* **103,** 39518 (1985)
Harder, H. (1961) Incorporation of boron into detrital clay minerals. *Geochim. Cosmochim. Acta,* **21,** 284-294
Harnly, J.M., Miller-Ihli, N.J. and O'Haver, T.C. (1982) Computer software for a simultaneous multielement atomic-absorption spectrometer. *J. Automat.Chem.,* **4**(2),54-60
Harris, L.A. and Yust, C.S. (1981) The ultrafine structure of coal determined by electron microscopy. *ACS Adv. Chem. Ser.,* **192,** 321-336
Harris, L.A., Barrett, H.E. and Kopp, O.C. (1981) Elemental concentrations and their distribution in two bituminous coals of different palaeoenvironments. *Int. J. Coal Geol.,* **1,** 175-193
Harrison, G.R. (1939) *M.I.T. Wavelength Tables.* New York: Wiley, 429 pp
Hart, R.J. and Leahy, R.M. (1983) The geochemical characterisation of coal seams from the Witbank Basin. *Spec. Publ. Geol. Soc. S. Afr.,* **7,** 169-174
Hart, R.J., Leahy, R. and Falcon, R.M. (1982) Geochemical investigation of the Witbank coalfield using instrumental neutron activation analysis. *J. Radioanal. Chem.,* **71,** 285-297
Harter, P. (1982) Trace elements from coal combustion – atmospheric emissions – a bibliography. *IEA Coal Res. Rep.,* No. ICTIS/BIB/02, 33 pp
Hartstein, A.M., Freedman, R.W. and Platter, D.W. (1973) Novel wet-digestion procedure for trace-metal analysis of coal by atomic absorption. *Anal. Chem.,* **45,** 611-614
Harvey, R.D. and De Maris, P.J. (1987) Size and maceral association of pyrite in Illinois coals and their float-sink fractions. *Org. Geochem.,* **11,** 343-349
Harvey, R.D. and Kruse, C.W. (1988) The Illinois Basin coal sample program: status and sample characterization. *J. Coal Qual.,* **7,** 109-113
Harvey, R.D. and Ruch, R.R. (1984) Overview of mineral matter in U.S. coals. *Am. Chem. Soc. Div., Fuel Chem. Prepr.,* **29**(4), 2-8
Harvey, R.D. and Ruch, R.R. (1986) Mineral matter in Illinois and other U.S. coals. In Vorres (1986), pp. 10-40
Harvey, R.D., Cahill, R.A., Chou, C.L. and Steele, J.D. (1983) Mineral matter and trace elements in the Herrin and Springfield coals, Illinois Basin coal field. *Ill. State Geol. Surv. Contract/Grant Rep.:* 1983-84, 162 pp
Hashimoto, S. (1953) Minor elements in coals from the Abrato coal mine, Yamagata Prefecture. *Misc. Repts. Res. Inst. Nat. Resources (Japan),* No. 31, 27-34
Hatch, J.R. (1983) Geochemical processes that control minor and trace element composition of United States coals. In *Cameron Volume on Unconventional Mineral Deposits* (ed. W.C. Shanks), pp. 89-98. New York: Soc. Econ. Geol.
Hatch, J.R. and Swanson, V.E. (1976) Trace elements in Rocky Mountain coals. *Symp. Geol. Rocky Mountain Coal,* Colorado Geological Survey, Denvir, CO, pp. 143-163
Hatch, J.R., Avcin, M.J. and van Dorpe, P.E. (1984) Element geochemistry of Cherokee Group coals (Middle Pennsylvanian) from south-central and southeastern Iowa. *Iowa Geol. Surv. Tech. Pap.,* No. 5, 108 pp
Hatch, J.R., Gluskoter, H.J. and Lindahl, P.C. (1976) Sphalerite in coals from the Illinois Basin. *Econ. Geol.,* **71,** 613-624
Hatcher, P.G. (1985) Origin of sedimentary humic acids, potential carriers of ore-forming elements. *US Geol. Surv. Circ.,* No. 949, 21-22
Hatcher, P.G., Breger, I.A., Maciel, G.E. and Szeverenyi, N.M. (1983) Chemical structure in coal: geochemical evidence for the presence of mixed structural components. *Proc. Int. Conf. Coal Sci.,* pp. 310-313
Hatcher, P.G., Breger, I.A., Maciel, G.E. and Szeverenyi, N.M. (1985) Geochemistry of humin. In Aiken *et al.* (1985), pp. 275-302
Hauck, R.D. (1975) The genesis and stability of nitrogen in peat and coal. *Am. Chem. Soc. Div., Fuel Chem. Prepr.,* **20**(2), 85-93
Haught, O.L. (1954) On the occurrence and distribution of minor elements in coal. *Mo. Univ. Sch. Mines Metall. Tech. Ser.,* No. 85, 17-24
Hawley, J.E. (1955) Germanium content of some Nova Scotian coals. *Econ. Geol.,* **50,** 517-532

Hawley, J.E. and Rimsaite, Y. (1954) Spectrographic studies of Nova Scotia coals. *Anal. Chem.*, **26**, 1663–1664
Hayatsu, R., Scott, R.G., Moore, L.P. and Studier, M.H. (1975) Aromatic units in coal. *Nature (London)*, **257**, 378–380
Hayes, K.F., Roe, A.L., Brown, G.E., Hodgson, K.O., Leckie, J.O. and Parks, G.A. (1987) In situ X-ray absorption study of surface complexes: selenium oxyanions on α-FeOOH. *Science*, **238**, 783–786
Haynes, B.S., Neville, M., Quann, R.J. and Sarofim, A.F. (1982) Factors governing the surface enrichment of fly ash in volatile trace species. *J. Colloid Interface Sci.*, **87**, 266–278
Haynes, B.W. (1978) Electrothermal atomic absorption determination of arsenic and antimony in combustible municipal solid waste. *At. Absorpt. Newsl.*, **17**, 49–52
Headlee, A.J.W. and Hunter, R.G. (1951) Germanium in coals of West Virginia. *W. Va. Geol. Econ. Surv. Rep. Invest.*, No. 8, 1–15
Headlee, A.J.W. and Hunter, R.G. (1953) Elements in coal ash and their industrial significance. *Ind. Eng. Chem.*, **45**, 548–551
Headlee, A.J.W. and Hunter, R.G. (1955a) Changes in the concentration of the inorganic elements during coal utilisation. *W. Va. Geol. Surv.*, 13(A), 150–159
Headlee, A.J.W. and Hunter, R.G. (1955b) Characteristics of minable coals of West Virginia. The inorganic elements in the coals. *W. Va. Geol. Surv.*, **13A**, 36–122
Hegemann, F., Giesen, K. and Kostyra, H. (1959) Spectrographic determination of secondary and trace elements in coal without previous ashing. *Ber. Dtsch. Keram. Ges.*, **36**, 145–149
Heinrich, A.G. and Foscolos, A.E. (1984) A comparison between theoretical and experimental corrections for interfering spectral lines from target elements in X-ray fluorescence spectroscopy. *Geol. Surv. Can. Pap.*, No. 83-27, 1–12
Heinrichs, H. (1975a) Determination of mercury in water, rocks, coal, and petroleum with flameless atomic absorption spectrophotometry. *Fresenius' Z. Anal. Chem.*, **273**, 197–201
Heinrichs, H. (1975b) Investigation of rocks and waters for Cd, Sb, Hg, Tl, Pb and Bi using flameless atomic absorption spectrophotometry. *Doctoral Dissertation*, University of Göttingen, West Germany, 97 pp
Heinrichs, H. (1982) Trace element discharge from a brown coal fired power plant. *Environ. Technol. Lett.*, **3**, 127–136
Heinrichs, H., Brumsack, H.J. and Lange, H. (1984) Emissions from hard and brown coal power stations in West Germany. *Fortschr. Mineral.*, **62**, 79–105
Heinrichs, H., Schulz-Dobrick, B. and Wedepohl, K.H. (1980) Terrestrial geochemistry of Cd, Bi. Tl, Pb, Zn and Rb. *Geochim. Cosmochim. Acta*, **44**, 1519–1533
Helz, A.W. (1964) A gas jet for dc arc spectroscopy. *US Geol. Surv. Prof. Pap.*, No. 475-D, D176–D178
Henderson, J.A., Wilkes, G.P., Bragg, L.J. and Oman, C.L. (1985) Analyses of Virginia coal samples collected 1978–1980. *Va. Div. Miner. Resour. Publ.*, No. 63, 56 pp
Henderson, P. (ed.) (1984) *Rare Earth Element Geochemistry*. Amsterdam: Elsevier, 510 pp
Hess, F.L. (1932) Vanadium. *US Bur. Mines Inf. Circ.*, No. 6572, 8 pp
Hewett, D.F. (1909) Vanadium deposits in Peru. *Trans. Am. Inst. Min. Eng.*, **40**, 274–299
Hickman, D.A., Rooke, J.M. and Thompson, M. (1987) Atomic spectrometry update – minerals and refractories. *J. Anal. At. Spectrom.*, **2**, 211R–230R
Hieftje, G.M. and Vickers, G.H. (1989) Developments in plasma source/mass spectrometry. *Anal. Chim. Acta*, **216**, 1–24
Hildebrand, R.T., Clardy, B.F. and Holbrook, D.F. (1981) Chemical analyses of lignite from the Wilcox and Claiborne Groups (Eocene) Southern and Eastern Arkansas. *Arkansas Geol. Comm. Inf. Circ.*, No. 28, 57 pp
Hinds, H. (1912) The coal deposits of Missouri. *Mo Bur. Geol. Mines* Ser. 2, **11**, 503 pp.; Wedge, Bhatia and Rueff (1976)
Hislop, J.S., Fisher, E.M.R., Morton, A.G., Salmon, L. and Wood, D.A. (1978) Multielement analysis of British coals. *22nd Annu. Conf. Anal. Chem. Energy Technol.* Gatlinburgh, TN, 10 pp
Hobday, D.K. (1987) Gondwana coal basins of Australia and South Africa: tectonic setting, depositional systems and resources. In Scott (1987), pp. 219–233
Hodges, N.J., Ladner, W.R. and Martin, T.G. (1983) Chlorine in coal: a review of its origin and mode of occurrence. *J. Inst. Energy*, **56**, pp. 158–169
Hoehne, K. (1957) Zircon crystals in coal beds. *Chem. Erde*, **19**, 38–50

Hokr, Z. (1977) Arsenic in the coal of the north Bohemian brown coal basin. *Vest. Ustred. Ustavu Geol.*, **52**, 267–273

Holliday, R., Hodgson, D.R., Townsend, W.N. and Wood, J.W. (1958) Plant growth on 'fly ash'. *Nature (London)*, **181**, 1079–1080

Hopke, P.K. (1988) Target transformation factor analysis as an aerosol mass apportionment method: a review and sensitivity study. *Atmos. Environ.*, **22**, 1777–1792

Horne, J.C., Ferm, J.C., Caruccio, F.T. and Baganz, B.P. (1978) Depositional models in coal exploration and mine planning in Appalachian region. *Am. Assoc. Pet. Geol. Bull.*, **62**, 2379–2411

Horton, L. (1950) Progress review No. 6: the constitution of coal. *J. Inst. Fuel*, **23**, 91–95

Horton, L. (1958) The chemical constitution of coal. *Conf. Sci. Use Coal*, Sheffield Inst. Fuel, London, A24–A28

Horton, L. and Aubrey, K.V. (1950) Distribution of minor elements in vitrain – three vitrains from the Barnsley seam. *J. Soc. Chem. Ind. London*, **69** (Suppl. No. 1), 541–548

Horwitz, W., Kamps, L.R. and Boyer, K.W. (1980) Quality assurance in the analysis of foods for trace constituents. *J. Assoc. Off. Anal. Chem.*, **63**, 1344–1354

Howarth, W.E. (1928) The occurrence of linnéite in the coal measures of South Wales. *Geol. Mag.*, **65**, 517–518

Hsieh, K.C. and Wert, C.A. (1981) Sulfide crystals in coal. *Mater. Sci. Eng.*, **50**, 117–125

Hsieh, K.C. and Wert, C.A. (1983) Examination of fine scale minerals in coal and coal products. *Proc. Int. Conf. Coal Sci.*, pp. 357–360

Hsieh, K.C. and Wert, C.A. (1985) Direct measurement of organic saulphur in coal. *Fuel*, **64**, 255–262

Huffman, C. (1960) Water-soluble boron in sample containers. *US Geol. Surv. Prof. Pap.*, No. 400-B, 493–494

Huffman, C., Rahill, R.L., Shaw, V.E. and Norton, D.R. (1972) Determination of mercury in geologic materials by flameless atomic absorption spectrometry. *US Geol. Surv. Prof. Pap.*, No. 800-C, C203–C207

Huffman, G.P. and Huggins, F.E. (1984) Analysis of the inorganic constituents in low-rank coals. In Schobert (1984), pp. 159–174

Huggins, F.E. and Huffman, G.P. (1979) Mössbauer analysis of iron-bearing phases in coal, coke, and ash. In Karr (1979), pp. 371–423

Huggins, F.E. and Huffman, G.P. (1983) An EXAFS investigation of calcium in coal. *Proc. Int. Conf. Coal Sci.*, pp. 679–682

Hügi, T., Fardy, J.J., Morgan, N.C. and Swaine, D.J. (1989) Trace elements in some Swiss coals. (In press)

Hulett, L.D. and Weinberger, A.J. (1980) Some etching studies of the microstructure and composition of large aluminosilicate particles in flyash from coal-burning power plants. *Environ. Sci. Technol.*, **14**, 965–970

Hull, D.R. and Horlick, G. (1984) Electrothermal vaporization sample introduction system for the inductively coupled plasma. *Spectrochim. Acta, Part B*, **39B**, 843–850

Hume, D.N. (1973) Pitfalls in the determination of environmental trace metals. In *Chemical Analysis of the Environment and Other Modern Techniques* (eds. S. Ahuja, E.M. Cohen, T.J. Kneip, J.L. Lambert and G. Zweig), pp. 3–16. New York: Plenum

Hunt, J.W. (1982) Relationship between microlithotype and maceral composition of coals and geological setting of coal measures in the Permian Basins of Eastern Australia. In Mallett (1982), pp. 484–502

Hunt, J.W. and Smith, J.W. (1985) $^{34}S/^{32}S$ ratios of low-sulfur Permian Australian coals in relation to depositional environments. *Chem. Geol. (Isotope Geosci. Sect.)*, **58**, 137–144

Hutcheon, J.M. (1953) The fuel industries and atomic energy – some common interests. *J. Inst. Fuel*, **26**, 306–311

Hutton, M. (1983) A prospective atmospheric emission inventory for cadmium – the European Community as a study area. *Sci. Total Environ.*, **29**, 29–47

Ibarra, J.V. and Orduna, P. (1986) Variation of the metal complexing ability of humic acids with coal rank. *Fuel*, **65**, 1012–1016

Idiz, E.F., Carlisle, D. and Kaplan, I.R. (1986) Interaction between organic matter and trace elements in a uranium rich bog, Kern County, California, U.S.A. *Appl. Geochem.*, **1**, 573–590

Idzikowski, A. (1960) Occurrence of certain trace elements in the ashes of Upper Silesian bituminous coals. II. *Bull. Acad. Pol. Sci., Ser. Sci., Chim., Geol. geogr.*, **8**, 235–243; *CA*, **58**, 15245 (1961)

Idzikowski, A. and Trzebiatowski, W (1960) Occurrence of certain trace elements in the ashes of Upper Silesian bituminous coals. I. *Bull. Acad. Pol. Sci., Ser. Sci., Chim., Geol. geogr.*, **8**, 225–233

IEA (1983) *Concise Guide to World Coalfields*. London: IEA Coal Research

Ihnat, M. (1977) Data validity in analytical chemistry. *Proc. Tenth Nat. Shellfish Sanitation Workshop*, US Dept. Health, Education, Welfare, pp. 95–105

Ihnat, M. (1984) Atomic absorption and plasma atomic emission spectrometry. In *Modern Methods of Food Analysis* (eds. K.K. Stewart and J.R. Whitaker), pp. 129–166. Westport, CT: AVI Publishing

Ilger, J.D., Ilger, W.A., Zingaro, R.A. and Mohan, M.S. (1987) Modes of occurrence of uranium in carbonaceous uranium deposits: characterisation of uranium in a South Texas (U.S.A.) lignite. *Chem. Geol.*, **63**, 197–216

Imai, N., Terashima, S. and Ando, A. (1984) Determination of selenium in geological materials by automated hydride generation and electrothermal atomic absorption spectrometry. *Bunseki Kagaku*, **33**, 288–290; *CA*, **101**, 47809 (1984)

Inagaki, M. (1951a) Spectroscopic analysis of inorganic matters in coal I. *J. Coal Res. Inst. (Japan)*, **2**, 229–234; *CA*, **49**, 7221 (1955)

Inagaki, M. (1951b) Spectroscopic analysis of inorganic matters in coal. III. *J. Coal Res. Inst. (Japan)*, **2**, 379–388; *CA*, **49**, 7221 (1955)

Inagaki, M. (1952a) Germanium in Japanese coals. *J. Coal Res. Inst. (Japan)*, **3**, 261–267; *CA*, **49**, 7219 (1955)

Inagaki, M. (1952b) Germanium – a rare element in coal. *J. Coal Res. Inst. (Japan)*, **3**, 72–80; *CA*, **49**, 7219 (1955)

Inagaki, M. (1955) Germanium compounds from coal. *Jap. Pat.*, 1875(55), Mar. 22; *CA*, **51**, 1589 (1957)

Inagaki, M. (1956) Metallic gallium. *J. Coal Res. Inst. (Japan)*, **7**, 129–138; *CA*, **50**, 13635 (1956)

Inagaki, M. (1967) Some considerations of germanium and gallium in Japanese coals. *J. Fuel Soc. Jpn.*, **46**, 684–688

Inagaki, M. and Tokuga, U. (1959) Uranium in columns of Ube coal. *Tanken*, **10**, 97–107; *CA*, **53**, 18440 (1959)

Inagaki, M., Oikawa, H. and Moriya, S. (1955) Germanium in coal. III. Recovery of germanium from gas liquor. *J. Coal Res. Inst. (Japan)*, **6**, 129–132; *CA*, **49**, 16390 (1955)

Ingram, H.A.P. (1982) Size and shape in raised mire ecosystems: a geophysical model. *Nature (London)*, **297**, 300–303

Inoue, S., Hoshi, S. and Matsubara, M. (1986) Extraction–spectrophotometric determination of vanadium with N-m-tolyl-N-phenylhydroxylamine and its application to coal and coal fly-ash. *Talanta*, **33**, 611–613

Isles, P.T. (1986) Australia's Bowen Basin. *Eng. Min. J.*, **187**(4), 24–30

ISO (1975) Sampling of hard coal. *Int. Stand. Organ.*, 1988. 90 pp

Iwasaki, I. and Ukimoto, I. (1942) Coal. I Rare elements in coal. 1 Vanadium and chromium contents of coal. *J. Chem. Soc. Jpn.*, **63**, 1678–1684; *CA*, **41**, 3408 (1947)

Iyengar, V. (1982) Presampling factors in the elemental composition of biological systems. *Anal. Chem.*, **54**, 554A–560A

Jacobi, W., Schmier, H. and Schwibach, J. (1982) Comparison of radiation exposure from coal-fired and nuclear power plants in the Federal Republic of Germany. *Proc. Int. Symp. Health Impacts Differ. Sources Energy*, IAEA, Vienna, pp. 215–227; *CA*, **98**, 97558 (1983)

Janda, I. and Schroll, E. (1959) Boron content of some East Alpine coals and other bioliths. *Tschermaks Mineral. Petrogr. Mitt.*, **7**, 118–129

Jarvis, K.E. (1988) Inductively coupled plasma mass spectrometry: a new technique for the rapid or ultra-trace level determination of the rare-earth elements in geological materials. *Chem. Geol.*, **68**, 31–39

Jaworowski, Z. and Bilkiewicz, J. (1982) Are heavy metals and radium in man related to emissions from coal burning? *Rep. IAEA-SM-254/57*, pp. 159–165

Jaworowski, Z., Bysiek, M. and Kownacka, L. (1981) Flow of metals into the global atmosphere. *Geochim. Cosmochim. Acta*, **45**, 2185–2199

JCB (1988) *Black Coal in Australia 1986–87*. Sydney: Joint Coal Board, 150 pp

Jeczalik, A. (1970) The geochemistry of uranium in uraniferous hard coals from Poland. *Biul., Inst. Geol., Warsaw*, No. 224, 103–204

Jedwab, J. (1960) Coal as a source of beryllium. *Bull. Soc. Belge Geol. Paleontol. Hydrol.*, **69**, 67–77

Jenkins, R.G. and Walker, P.L. (1978) Analysis of mineral matter in coal. In Karr (1978b), pp. 265–292
Jenny, W.P. (1903) The chemistry of ore-deposition. *Trans. Am. Inst. Min. Eng.*, **33**, 445–498; Gibson and Selvig (1944)
Jensch, E. (1887) On the metal content of Upper Silesian coals. *Chem. Industr. Berl.*, **10**, 54
Jervis, R.E., Ho, K.R. and Tiefenbach, B. (1982) Trace impurities in Canadian oil-sands, coals and petroleum products and their fate during extraction, up-grading and combustion. *J. Radioanal. Chem.*, **71**, 225–241
Joensuu, O.I. (1971) Fossil fuels as a source of mercury pollution. *Science*, **172**, 1027–1028
Johnson, E.B. (1947) Discussion. *J. Inst. Fuel*, **20**, 94
Jones, C.R. and Dawson, E.C. (1945) The arsenic content of grain dried directly with flue gas. *Analyst (London)*, **70**, 256–257
Jones, J.H. and Miller, J.M. (1939) Occurrence of titanium and nickel in the ash from some special coals. *Chem. Ind. (London)*, **58**, 237–245
Jones, L.H. and Lewis, A.V. (1960) Weathering of fly ash. *Nature (London)*, **185**, 404–405
Jorissen, A. (1896) Molybdenum, selenium, etc., in coal from Liege. *Ann. Soc. Geol. Belg.*, **23**, 101–105; Gibson and Selvig (1944)
Jorissen, A. (1905) Presence of chromium and vanadium in coal from Liege. *Acad. R. Belg. Bull.*, 178–181; Gibson and Selvig (1944)
Jorissen, A. (1913) Distribution of molybdenum in the Liege coal fields. *Bull. Soc. Chim. Belg.*, **27**, 21–25; *CA*, **7**, 3584 (1913)
Josefsson, A. (1954) Rare metal impurities in steel originating from coke ash. *Jernkontorets Ann.*, **138**, 744–749
Kaakinen, J.W., Jorden, R.M., Lawasani, M.H. and West, R.E. (1975) Trace element behaviour in a coal-fired power plant. *Environ. Sci. Technol.*, **9**, 862–869
Kaiser, G. and Tölg, G. (1986) Reliable determination of elemental traces in the ng/g range in biotic materials and in coal by inverse voltametry and atomic absorption spectrometry after combustion of the sample in a stream of oxygen. *Fresenius' Z. Anal. Chem.*, **325**, 32–40
Kakimi, T., Hirayama, J., Sekine, S. and Ikeda, K. (1969) Possibility of syngenetic concentration of uranium in a coal field. *Chishitsu Chosasho Hokoku*, No. 232, 659–675; *CA*, **71**, 115156 (1969)
Kama, W. and Siegel, S.M. (1980) Volatile mercury release from vascular plants. *Org. Geochem.*, **2**, 99–101
Kamada, E., Nakajima, R., Goto, K. and Shibata, S. (1982) Determination of twelve elements in coal by atomic absorption analysis (including main and trace elements). *Jpn. Anal.*, **31**, 551–556; *IEA Coal Abstr.*, **7**(9), 7110 (1983)
Kaplan, B.Y. and Olshevskaya, I.V. (1963) Determination of scandium in coal ash samples after paper chromatographic separation. *Zavod. Lab.*, **29**(1), 26–27; *CA*, **58**, 13661 (1963)
Karacki, S.S. and Corcoran, F.L. (1973) Coal ash analysis with an argon plasma emission excitation source. *Appl. Spectrosc.*, **27**, 41–42
Karasik, M.A., Dvornikov, A.G., Talalaev, G.K. and Zemblevskii, K.K. (1967) The mercury content of Donbas coals and their products of carbonization. *Coke Chem., USSR* (9), 14–15
Karner, F.R., Benson, S.A., Schobert, H.H. and Roaldson, R.G. (1984) Geochemical variation of inorganic constituents in a North Dakota lignite. In Schobert (1984), 175–193
Karner, F.R., Schobert, H.H., Falcone, S.K. and Benson, S.A. (1986) Elemental distribution and association with inorganic and organic components in North Dakota lignites. *ACS Symp. Ser.*, **301**, 70–89
Karr, C. (ed.) (1978a) *Analytical Methods for Coal and Coal Products*. New York: Academic, Vol. I, 580 pp
Karr, C. (ed.) (1978b) *Analytical Methods for Coal and Coal Products*. New York: Academic, Vol. II, 669 pp
Karr, C. (ed.) (1979) *Analytical Methods for Coal and Coal Products*. New York: Academic, Vol. III, 641 pp
Kautz, K., Kirsch, H. and Laufhutte, D.W. (1975) On trace-element contents in hard coals and flue-gas solids formed from them. *VGB Kraftwerkstechnik*, **10**, 672–676
Kautz, K., Pickhardt, W., Riepe, W., Schaaf, R., Scholz, A. and Zimmermeyer, G. (1984) Trace elements in hard coals, their distribution during combustion and their biological action. *Glueckauf-Forschungsh.*, **45**, 228–237

Kawamura, K. and Kaplan, I.R. (1987) Dicarboxylic acids generated by thermal alteration of kerogen and humic acids. *Geochim. Cosmochim. Acta,* **51**, 3201-3207
Kear, D. and Ross, J.B. (1961) Boron in New Zealand coal ashes. *NZ J. Sci.,* **4**, 360-380
Kear, R.W. and Menzies, H.M. (1956) 'Chlorine' in coal: its occurrence and behaviour during combustion and carbonisation. *Br. Coal Util. Res. Assoc. Mon. Bull.,* **20**(2), 53-65
Kehn, T.M. (1957) Selected annotated bibliography of the geology of uranium-bearing coal and carbonaceous shale in the United States. *US Geol. Surv. Bull.,* No. 1059A, 28 pp
Keil, F. and Gille, F. (1950) Destruction of an annular kiln stack by waste gases containing fluorine. *Tonind. Ztg.,* **74**, 226
Kekin, N.A. and Marincheva, V.E. (1970) Spectrographic method for determining arsenic in coals and cokes. *Zavod. Lab.,* **36**, 1061-1063; *CA,* **74**, 33360 (1971)
Kellerman, S.P., Haines, J. and Robert, R.V.D. (1983) The determination of trace elements in coal by atomic-absorption spectroscopy. *Mintek Rep.,* M121, 14 pp
Kemezys, M. and Taylor, G.H. (1964) Occurrence and distribution of minerals in some Australian coals. *J. Inst. Fuel,* **37**, 389-397
Kendrick, D.T., Kyle, P.R. and Kuellmer, F.J. (1988) Analysis of NBS coal standard reference materials 1632a and 1635 by instrumental neutron activation analysis. *Geostand. Newsl.,* **12**, 375-377
Kessler, F.M. and Dockalova, L. (1955) The determination of managanese in coal ashes. *Paliva,* **35**, 178-181; *CA,* **50**, 3732 (1956)
Kessler, M.F., Malan, O. and Valeska, F. (1965) The relevance of trace elements and minor elements for the correlation and identification of seams and the complex utilization of coal. *Rozpr. Cesk. Akad. Ved, Rada Mat. Prir. Ved,* **75**, 1-123
Kessler, T., Sharkey, A.G. and Friedel, R.A. (1973) Analysis of trace elements in coal by spark-source mass spectrometry. *US Bur. Mines Rep. Invest.,* No. 7714, 8 pp
Khan, I.A. and Sen, D. (1959) Studies on the separation of boron from coke. *J. Sci. Ind. Res. India,* **18B**, 434-436
Khanamirova, A.A. (1963) A study of bonds between germanium and organic part of coal. *Tr. Inst. Goryuch. Iskop., Akad. Nauk SSSR,* **21**, 88-98; *CA,* **60**, 329 (1964)
Khizhnyak, N.D., Tsebrii, L.S. and Lenkevich, Z.K. (1971) Determination of germanium in coals by derivative polarography. *Ukr. Khim. Zh.,* **37**, 359-362; *CA,* **75**, 65936 (1971)
KHM (1983) *The Health and Environmental Effects of Coal. Final Rep. Projekt KHM,* 241 pp
Kingston, H. and Pella, P.A. (1981) Preconcentration of trace metals in environmental and biological samples by cation exchange resin filters for X-ray spectometry. *Anal. Chem.,* **53**, 223-227
Kingston, H.M. and Jassie, L.B. (eds.) (1988) Introduction to microwave sample preparation: theory and practice. *ACS Professional Ref. Book,* Washington DC, 263 pp
Kirkby, W.A. (1950) Minor elements in coal. *Fuel,* **29**, 72
Kirsch, H., Schirmer, U. and Schwarz, G. (1980) The origin of the trace elements zinc, cadmium and vanadium in bituminous coals and their behaviour during combustion. *VGB Kraftwerkstechnik,* **60**, 734-744
Kiss, L.T. (1966) X-ray fluorescence determination of brown coal inorganics. *Anal. Chem.,* **38**, 1731-1735
Kiss, L.T. (1982) Chemistry of Victorian brown coals. *Aust. Coal Geol.,* **4**(1), 153-168
Kiss, L.T. and King, T.N. (1977) The expression of results of coal analysis: the case for brown coals. *Fuel,* **56**, 340-341
Kiss, L.T. and King, T.N. (1979) Reporting of low-rank analysis – the distinction between minerals and inorganics. *Fuel,* **58**, 547-549
Kitaev, I.V. (1970) Trace elements in Jurassic-Cretaceous coals and carbonaceous rocks of the Bureya and Tyrminsk synclines. *Vop. Geol., Geokhim. Metallogen. Sev.-Zapad. Sekt. Tikhookean. Poyasa, Mater. Nauch. Sess.* 1969, pp. 207-208; *CA,* **75**, 90192 (1971)
Kitaev, I.V. (1971) Distribution of some rare elements in coals of the Bureinsk and Tyrminsk basins. *Vop. Litol. Geokhim. Vulkanogenno-Osad. Yuga Dal'nego Vostoka,* pp. 193-207; *CA,* **81**, 108286 (1974)
Kivinen, E. (1936) Information on the occurrence of iron carbonates in Finnish bogs. *Z. Pflanzenernaehr. Dueng. Bodenkd.,* **45**, 96-104
Kizilshteyn, L.Y. and Syunyakova, N.N. (1964) Analysis of the mode of accumulation of germanium in coal by the methods of mathematical statistics. *Dokl. Acad. Sci. USSR (Earth Sci. Sect.),* **151**, 169-170
Knapp, G. (1984) Decomposition methods in elemental trace analysis. *Trends Anal. Chem.,* **3**, 182-185

Knott, A.C. and Warbrooke, P. (1983) Determination of trace elements in coal and coal products. Part 5 Characterisation of a range of Australian raw coals. *Rep. NERDDP*, No. EG85/392, 55 pp
Knott, A.C., Mills, J.C. and Belcher, C.B. (1978) Synthetic calibration standards for optical emission and X-ray spectrometry. *Can. J. Spectrosc.*, **23**, 105–111
Knott, A.C., Thompson, S.C. and Lee, J.B. (1985) Characterization, treatment and disposal of coal wastes in an environmentally acceptable manner. *BHP Rep. of NERDDP Project* (Grant No. 81/1201), 76 pp
Knott, A.C., Thompson, S.C. and Ruch, R.R. (1985) The effects of coal cleaning procedures on inorganic and trace elements in coal products. Part 2 of *Rep. NERDDP*, No. EG/85/433, 51 pp
Knott, A.C., Turner, K.E., Lee, J.B., Dubrawski, J.V. and Warbrooke, P. (1984) Characterisation of a range of Australian 'as sold' coals. *Rep. BHP*, No. CRL/R/20/84, 88 pp
Knox, W.F.J. (1980) Metal emissions from fossil fuel utilization and their effects on the Canadian environment – outline and bibliography. Ottawa: NRC, 55 pp
Kochenov, A.V. and Kreshtapova, V.N. (1967) Rare and dispersed elements in the peats of the northern part of the Russian Platform. *Geochem. Int.*, **4**, 286–294
Koglin, E., Schenk, H.J. and Schwochau, K. (1978) Spectroscopic studies on the binding of uranium by brown coal. *Appl. Spectrosc.*, **32**, 486–488
Köhler, R. (1931) Methods of determining iodine in coal. *Mitt. Lab. Preuss. Geol. Landesanst.*, (13), 1–9; CIEB (1956), 87
Kohno, Y., Takanashi, S. and Fujiwara, T. (1982) Effects of trace elements from coal-fired power station on vegetation: a review. *Rep. CRIEPI*, No. 481017, 80 pp
Koirtyohann, S.R. (1980) Current status and future needs in atomic absorption instrumentation. *Anal. Chem.*, **52**, 736A–744A
Kojima, T. and Furusawa, T. (1986) Behaviour of elements in coal ash with sink-float separation of coal and organic affinity of the elements. *Nenryo Kyokai-Shi*, **65**, 143–149
Konieczynski, J. (1960) Spectrographic method for determination of trace amounts of boron in coals, cokes, pitch and tars. *Pr. Gl. Inst. Gorn. Komun. Ser. B*, No. 254, 1–8; *CA*, **55**, 7801 (1961)
Konieczynski, J. (1970) Investigations on the occurrence of boron in hard coals. *Pr. Gl. Inst. Gorn. Komun.*, No. 482, 25 pp
Kononova, M.M. (1966) *Soil Organic Matter: Its Nature, Its Role in Soil Formation and in Soil Fertility*. New York: Pergamon, 2nd edn, 544 pp
Koppenaal, D.W., Lett, R.G., Brown, F.R. and Manahan, S.E. (1980) Determination of nickel, copper, selenium, cadmium, thallium and lead in coal gasification products by isotope dilution spark source mass spectrometry. *Anal. Chem.*, **52**, 44–49
Korkisch, J., Farag, A. and Hecht, F. (1958) Determination of uranium in phosphates, coal ash, and bauxite by ion exchange. *Z. Anal. Chem.*, **161**, 92–100
Korolev, D.F. (1957) Some peculiarities of molybdenum distribution in rocks of the Bylymsk coal deposits. *Geochemistry (USSR)* (5), 493–499
Korotev, R.L. (1987) Chemical homogeneity of National Bureau of Standards coal fly ash. *J. Radioanal. Nucl. Chem.*, **110**, 179–189
Korotev, R.L. and Lindstrom, D.J. (1982) Small samples of NBS fly ash as an INAA standard. *Trans. Am. Nucl. Soc.*, **41**, 187
Kortenski, I. (1986) Trace elements in coal ashes from Sofia Pliocene Basin. *Spis. Bulg. Geol. Druzh.*, **47**, 165–172; *CA*, **106**, 122744 (1987)
Kosasa, K. and Nakajima, S. (1987) Determination of manganese and chromium in coal by atomic emission excited in a shock tube. *Spectrochim. Acta, Part B*, **42B**, 501–503
Koseki, K., Ogawa, K. and Hasuda, T. (1986) Determination of trace elements in coals by spark source mass spectrometry. *Nippon Kogyo Kaishi*, **102**, 722–725; *IEA Coal Abstr.*, **11**(10), 08273 (1987)
Kostadinov, K. and Djingova, R. (1979) Determination of mercury in coal by destructive neutron activation analysis. *God. Sofii. Univ., Khim. Fak.*, **70** (Part 2), 47–51; *CA*, **93**, 75332 (1980)
Kostadinov, K.N. and Djingova, R.G. (1980) Trace element investigation of coal samples by thermal and epithermal neutron activation analysis. *Radiochem. Radioanal. Lett.*, **45**, 297–304
Krejci-Graf, K. (1983) Minor elements in coals. In Augustithis (1983), pp. 533–597

Krichko, A.A. (1984) Fiftieth anniversary of the Fossil Fuels Institute. *Coke Chem. USSR* (12), 1–5

Kronberg, B.I., Murray, F.H., Daddar, R. and Brown, J.R. (1988) Fingerprinting geological materials using SSMS. *Chem. Geol.*, **68**, 351–359

Kronberg, B.I., Brown, J.R., Fyfe, W.S., Peirce, M. and Winder, C.G. (1981) Distributions of trace elements in Western Canadian coal ashes. *Fuel*, **60**, 59–63

Kronberg, B.I., Murray, F.H., Fyfe, W.S., Winder, C.G., Brown, J.R. and Powell, M. (1987) Geochemistry and petrography of the Mattagami Formation lignites (Northern Ontario). In Volborth (1987), pp. 245–263

Kruse, C.W., Harvey, R.D. and Rapp, D.M. (1987) Illinois Basin coal sample program and access to information on Illinois Basin coals. *Am. Chem. Soc. Div., Fuel Chem. Prepr.*, **32**(4), 359–365

Kryukova, V.N., Kindeeva, V.P., Baskova, L.V. and Latyshev, V.P. (1985) Arsenic in coal of eastern Siberia. *Solid Fuel Chem.*, **19**(1), 120–123

Kube, W.R., Schobert, H.H., Benson, S.A. and Karner, F.R. (1984) The structure and reactions of Northern Great Plains lignites. In Schobert (1984), pp. 39–51

Kuhl, J. (1957) Mineral raw materials accompanying coal deposits and their utilisation. *Przegl. Geol.*, **5**, 248–255; *CA*, **52**, 992 (1958)

Kuhl, J. and Ziolkowski, J. (1954) Rare elements in Upper Silesian coal. *Przegl. Gorn.* (5),, 180; *CA*, **52**, 10541 (1958)

Kuhn, J.K. (1973) Trace elements in whole coal determined by X-ray fluorescence. *Norelco Rep.*, **20**(3), 7–10

Kuhn, J.K. and Henderson, L.R. (1977) A survey of emission X-ray analysis of coal. *Am. Chem. Soc. Div., Fuel Chem. Prepr.*, **22**(5), 68–70

Kuhn, J.K., Harfst, W.F. and Shimp, N.F. (1973) X-ray fluorescence analysis of whole coal. *Am. Chem. Soc. Div., Fuel Chem. Prepr.*, **18**(4), 72–77

Kuhn, J.K., Harfst, W.F. and Shimp, N.F. (1975) X-ray fluorescence analysis of whole coal. In Babu (1975), pp. 66–73

Kuhn, J.K., Fiene, F.L., Cahill, R.A., Gluskoter, H.J. and Shimp, N.F. (1980) Abundance of trace and minor elements in organic and mineral fractions of coal. *Environ. Geol. Notes, Ill. State Geol. Surv.*, No. 88, 67 pp

Kühn, W. (1960) Trace-element contents in brown coal ashes of the Bitterfield district and south-west Mecklenburg. *Geologie*, **8**, 942

Kulinenko, D.R. (1972) Quantitative evaluation of trace element mobilities during coal accumulation (in the Donets and Lvov-Volzn Basins). *Geol. Zh.*, **32**(2), 86–91

Kullerud, G., Steffen, R.M., Simms, P.C. and Rickey, F.A. (1979) Proton induced X-ray emission (PIXE) – a new tool in geochemistry. *Chem. Geol.*, **25**, 245–256

Kulskaya, O.A. and Vdovenko, O.F. (1957) Spectral method of quantitative determination of scandium in natural objects and coal ash. *Ukr. Khim. Zh.*, **23**, 799–804; *CA*, **52**, 13526 (1958)

Kunstmann, F.H. and Bodenstein, L.B. (1961) The arsenic content of South African coals. *J. S. Afr. Inst. Min. Metall.*, **62**, pp. 234–244

Kunstmann, F.H. and Gass, S.B. (1957) The occurrence of titanium in South African coals. *Fuel Res. Inst. S. Afr. Rep.*, No. 41, 13 pp

Kunstmann, F.H. and Hamersma, J.C. (1955) The occurrence of germanium in South African coal and derived products. *Fuel Res. Inst. S. Afr. Rep.*, No. 12, 25 pp

Kunstmann, F.H. and Harris, J.F. (1966) The occurrence of vanadium in South African coals. *Fuel Res. Inst. S. Afr. Bull.*, No. 44, 7 pp

Kunstmann, F.H. and Müller, E.F.E. (1959) The absorptiometric determination of germanium with phenylfluorone. *Analyst (London)*, **84**, 324–325

Kunstmann, F.H. and van Rensburg, M.J.J. (1967) The manganese content of South African coals. *Fuel Res. Inst. S. Afr. Rep.*, No. 21/1967, 8 pp

Kunstmann, F.H. and van Veijeren, A.M. (1964) A method for the determination of low chlorine contents in coal and the results of its application to South African coals. *Fuel Res. Inst. S. Afr. Bull.*, No. 65, 25 pp

Kunstmann, F.H., Harris, J.F., Bodenstein, L.B. and van den Berg, A.M. (1963) The occurrence of boron and fluorine in South African coals and their behaviour during combustion. *Fuel Res. Inst. S. Afr. Bull.*, No. 63, 45 pp

Kurmanov, M.I., Govor, U.S., Dobruskina, S.R., Sandler, N.I., Soloveva, G.G. and Filippova, T.F. (1957) The effect of arsenic on the properties of the high strength steels

12 KhN3A, 30KhN3A and 18KhN3A. *Byull. Nauchno-Tekh. Inf. Ukr. Nauchno-Issled. Inst. Met.*, No. 3, 59–75; *CA,* **54**, 10759 (1960)

Kusmierska, J. and Badura, A. (1983) Use of an X-ray fluorescence technique for the determination of chlorine in coal. *Przegl. Gorn.*, **39**, 335–342; *CA,* **101**, 94100 (1984)

Kuznetsova, V.V. (1961) Determination of rhenium in coal. *Zh. Anal. Khim.*, **16**, 736–737; *Anal. Abstr.*, **9**, 2774 (1962)

Kuznetsova, V.V. and Saukov, A.A. (1961) Possible forms of occurrence of molybdenum and rhenium in coals of Middle Asia. *Geochemistry (USSR)* (9), 822–829

Kvalheim, A. (1947) Spectrochemical determination of the major constituents of minerals and rocks. *J. Opt. Soc. Am.*, **37**, 585–592

Labonte, M. and Goodarzi, F. (1985) Use of the dendograph for data processing in fuel science. *Fuel*, **64**, 1177–1179

Ladner, W.R. (1984) The state of combination of chlorine in coal. *Fuel*, **63**, 726

Lag, J. and Steinnes, E. (1978) Regional distribution of selenium and arsenic in humus layers of Norwegian forest soils. *Geoderma*, **20**, 3–14

Lakin, H.W., Curtin, G.C. and Hubert, A.E. (1974) Geochemistry of gold in the weathering cycle. *US Geol. Surv. Bull.*, No. 1330, 80 pp

Laktionova, N.V., Egorov, A.P. and Popinako, N.V. (1978) Quantitative spectral determination of trace elements in coals. *Solid Fuel Chem.*, **12**(6), 91–94

Laktionova, N.V., Egorov, A.P., Belyavtseva, N.V., Ekaterinina, L.N., Alyautdinova, R.K. and Motovilova, L.V. (1987) Microelements in the humic acids of the coals of the Moscow Basin. *Solid Fuel Chem.*, **21**(4), 37–41

Landergreen, S. (1945) Contribution to the geochemistry of boron. II The distribution of boron in some Swedish sediments, rocks and iron ores. The boron cycle in the upper lithosphere. *Ark. Kemi, Mineral. Geol.*, **19A**, No. 26, 1–31

Landergreen, S. and Manheim, F.T. (1963) On the dependence of the distribution of heavy metals on facies. *Fortschr. Geol. Rheinl. Westfalen*, **10**, 173–192

Landheer, F., Dibbs, H. and Labuda, J. (1982) Trace elements in Canadian coals. *Rep. EPS*, No. 3-AP-82-6, 41 pp

Langmyhr, F.J. and Aadalen, U. (1980) Direct atomic absorption spectrometric determination of copper, nickel and vanadium in coal and petroleum coke. *Anal. Chim. Acta*, **115**, 365–368

Lantzy, R.J. and Mackenzie, F.T. (1979) Atmospheric trace metals: global cycles and assessment of man's impact. *Geochim. Cosmochim. Acta*, **43**, 511–525

Larsen, J.W. (1985) Coal: macromolecular structure and reactivity. *Am. Chem. Soc. Div., Fuel Chem. Prepr.*, **30**(4), 444–449

Larsen, J.W., Mohammadi, M., Yiginsu, I. and Kovac, J. (1984) The molecular weight distribution of coal extracts. Coalification is not a condensation polymerization. *Geochim. Cosmochim. Acta*, **48**, 135–141

Lauf, R.J. (1981) Application of materials-characterization techniques to coal and coal wastes. *Rep. ORNL/TM-7663*, NTIS Order No. DE81026087, 75 pp

Laverick, M.K. and Grice, C.R. (1980) Australian coals for international steelmaking and power generation. *Tech. Pap. 11th World Energy Conf. Div. 2*, **2**, 402–424

Lawrence, L.J., Warne, S.St J. and Booker, M. (1960) Millerite in the Bulli coalseam. *Aust. J. Sci.*, **23**(3), 87–88

Lazebnik, P.V., Grinvald, M.A. and Dolgopolov, V.M. (1967) Correlation between germanium and sulfur contents in coals from various areas of the USSR. *Geol. Gorn. Delo*, pp. 59–60; *CA*, **71**, 103849 (1969)

Leblang, G.M., Rayment, P.A. and Smyth, M. (1981) The Austinvale coal deposit – Wandoan a palaeoenvironmental analysis. *J. Coal Geol. Group Geol. Soc. Aust.*, **1**, 185–195

Leloux, M.S., Lich, N.P. and Claude, J.-R. (1987) Flame and graphite furnace atomic absorption spectroscopy methods for thallium – a review. *At. Spectrosc.*, **8**(2), 71–75

Leonhardt, J.W., Bothe, H.-K., Langrock, E.-J., Maul, E., Morgenstern, P., Müller, D. and Thümmel, H.-W. (1982) Coal analysis by means of neutron-gamma activation analysis and X-ray techniques. *J. Radioanal. Chem.*, **71**, 181–187

Lepel, E.A., Stefansson, K.M. and Zoller, W.H. (1978) The enrichment of volatile elements in the atmosphere by volcanic activity: Augustine volcano 1976. *J. Geophys. Res.*, **83**, No. C12, 6213–6220

Lerman, A (1966) Boron in clays and estimation of paleosalinities. *Sedimentology*, **6**, 267–286

Lessing, R. (1925) The inorganic constituents of coal. *J. Soc. Chem. Ind. London*, **44**, 277T–283T
Lessing, R. (1934) Fluorine in coal. *Fuel*, **13**, 347–348
Leutwein, F. (1956) Occurrence of trace metals in peats and brown coals. *Freiberg. Forschungsh. Ser. C, Angew. Naturwiss.* (3), 28–48; *CA*, **52**, 4958 (1958)
Leutwein, F. and Rösler, H.J. (1956) Geochemical investigations of Palaeozoic and Mesozoic coals from central and east G rmany. *Freiburg. Forschungsh.*, C19, 1–196; Gumz, Kirsch and Mackowsky (1958)
Levander, O.A. (1987) A global view of human selenium nutrition. *Annu. Rev. Nutr.*, **7**, 227–250
Leventhal, J.S., Briggs, P.H. and Baker, J.W. (1983) Geochemistry of the Chattanooga shale, Dekalb County, Central Tennessee. *Southeast. Geol.*, **24**, 101–116
Li, S., Guo, T. and Hu, W. (1986) Spectrometric determination of vanadium with diphenylcarbazide and cetyltrimethylammonium bromide. *Fenxi Huaxue*, **14**, 129–131; *CA*, **104**, 236498 (1986)
Lightowlers, P.J. and Cape, J.N. (1988) Sources and fate of atmospheric HCl in the U.K. and western Europe. *Atmos. Environ.*, **22**, 7–15
Lim, M.Y. (1979) Trace elements from coal combustion – atmospheric emissions. *IEA Coal Res. Rep.*, No. ICTIS/TR05, 58 pp
Lindahl, P.C. (1985a) Electrically heated quartz cell and holder for an atomic absorption hydride generation system. *At. Spectrosc.*, **6**, 123–124
Lindahl, P.C. (1985b) Determination of arsenic and selenium in coal by hydride generation/atomic absorption spectrophotometry – an interlaboratory study for the evaluation of a proposed standard test method. *Argonne Nat. Lab. Rep.*, No. ANL/ACL-85-3, 28 pp
Lindahl, P.C. (1985c) The applicability of standard test methods to the analysis of coal samples for coal research. *Am. Chem. Soc. Div., Fuel Chem. Prepr.*, **30**(4), 184–186
Lindahl, P.C. and Bishop, A.M. (1982) Determination of trace elements in coal by an oxygen bomb combustion/atomic absorption spectrophotometric method. *Fuel*, **61**, 658–662
Lindahl, P.C. and Finkelman, R.B. (1984) Factors influencing trace element variations in U.S. coals. *Am. Chem. Soc. Div., Fuel Chem. Prepr.*, **29**(4), 28–35
Lindberg, S.E. (1986) Mercury vapour in the atmosphere: three case studies on emission, deposition and plant uptake. In *Toxic Metals in the Atmosphere* (eds. J.O. Nriagu and C.I. Davidson), pp. 535–60. New York: Wiley-Interscience
Littke, R. (1987) Petrology and genesis of Upper Carboniferous seams from the Ruhr region, West Germany. *Int. J. Coal Geol.*, **7**, 147–184
Little, P. and Martin, M.H. (1974) Biological monitoring of heavy metal pollution. *Environ. Pollut.*, **6**, 1–19
Littlejohn, R.F. (1984) Emission of trace elements from coal-fired industrial boilers. A survey of relevant literature. *Energy Res.*, **8**, 375–386
Lloyd, S.J. and Cunningham, J. (1913) The radium content of some Alabama coals. *Am. Chem. J.*, **50**, 47–51
Logvinenko, N.V. (1972) Composition and origin of iron and manganese carbonates in sedimentary rocks. *Lithol. Miner. Resour. (USSR)*, **7**, 328–337
Lopez de Azcona, J.M. and Camunas Puig, A. (1947) Trace elements in Asturian coal ashes. *Bol. Inst. Geol. Min. Esp.*, **60**, 3–9
Louis, H. (1901–2) Note on a mineral vein in Wearmouth colliery. *Trans. Inst. Min. Eng.*, **22**, 127–129
Louis, H. (1927) The production of clean coal. *Chem. Ind. (London)*, **46**, 545–552
Lovering, T.G. (ed.) (1976) Lead in the environment. *US Geol. Surv. Prof. Pap.*, No. 957, 90 pp
Lowenthal, D.H. and Rahn, K.A. (1988) Tests of regional elemental tracers of pollution aerosols. 2 Sensitivity of signatures and apportionments to variations in operating parameters. *Environ. Sci. Technol.*, **22**, 420–426
Lowenthal, D.H., Wunschel, K.R. and Rahn, K.A. (1988) Tests of regional elemental tracers of pollution aerosols. 1 Distinctness of regional signatures, stability during transport, and empirical validation. *Environ. Sci. Technol.*, **22**, 413–420
Lowry, H.H. (ed.) (1945) *Chemistry of Coal Utilization*. New York: Wiley, Vol. I, pp. 1–920, Vol. II, pp. 921–1868
Lowry, H.H. (ed.) (1963) *Chemistry of Coal Utilization*. New York: Wiley, Suppl. Vol., 1142 pp
Lucas, A.J., Given, P.H. and Spackman, W. (1988) Studies of peat as the input to

coalification. I Rationale and preliminary examination of polysaccharides in peats. *Int. J. Coal Geol.*, **9**, 235-251

Luke, C.L. (1968) Determination of trace elements in inorganic and organic materials by X-ray fluorescence spectroscopy. *Anal. Chim. Acta*, **41**, 237-250

Lustigova, M. (1976) Determination of manganese content in coal ash. *Sb. Pr. UVP*, **32**, 149-178; *CA*, **87**, 138350 (1977)

Lustigova, M. and Kubant, J. (1983) Research on fluorine in coal. *Uhli*, **31**, 294-298

L'vov, B.V. (1961) The analytical use of atomic absorption spectra. *Spectrochim. Acta*, **17**, 761-770

Lynch, L.J., Sakurovs, R., Webster, D.S. and Redlich, P.J. (1988) H n.m.r. evidence to support the host/guest model of brown coals. *Fuel*, **67**, 1036-1041

Lynskey, B.J., Gainsford, A.R. and Hunt, J.L. (1984) Trace and major element analyses of 5 Waikato coals: an interlaboratory study. *NZ J. Sci.*, **27**, 443-464

Lyon, W.S. (1977) *Trace Element Measurements at the Coal-Fired Steam Plant*. Boca Raton: CRC Press, 144 pp

Lyon, W.S., Lindberg, S.E., Emery, J.F., Carter, J.A., Ferguson, N.M., van Hook, R.I. and Raridon, R.J. (1978) Analytical determination and statistical relationships of forty-one elements in coal from three coal-fired steam plants. *Rep.* IAEA-SM-227/61, pp. 615-625

Lyons, P.J. and Alpern, B. (1989) Peat and coal: origin, facies, and depositional models. *Int. J. Coal Geol.*, **12**, 1-798

MacAdam, S. (1852) On the general distribution of iodine. *Edinburgh New Philos. J.*, **53**, 315-326; CIEB (1956), 87

McBride, J.P., Moore, R.E., Witherspoon, J.P. and Blanco, R.E. (1977) Radiological impact of airborne effluents of coal-fired and nuclear power plants. *Rep.* ORNL-5315, 43 pp

McBride, J.P., Moore, R.E., Witherspoon, J.P. and Blanco, R.E. (1978) Radiological impact of airborne effluents of coal and nuclear plants. *Science*, **202**, 1045-1050

McCabe, P.J. (1984) Depositional environments of coal and coal-bearing strata. In Rahmani and Flores (1984b), pp. 13-42

McCabe, P.J. (1987) Facies studies of coal and coal-bearing strata. In Scott (1987), pp. 51-66

McCarroll, J. (1980) Health effects associated with increased use of coal. *J. Air Pollut. Control Assoc.*, **30**, 652-656

Machlan, L.A., Gramlich, J.W., Murphy, T. and Barnes, I.L. (1976) The accurate determination of lead in biological and environmentl samples by isotope dilution mass spectrometry. *Natl. Bur. Stand. (US) Spec. Publ.*, No. 422, 929-935

McClaine, L.A., Allen, R.V. and McConnell, R.K. (1968) Volcanic smoke clouds. *J. Geophys. Res.*, **73**, 5235-5246

McClendon, J.F. (1939) *Iodine and the Incidence of Goiter*. Minneapolis: University Press, 126 pp

McCurdy, D.L., Wichman, M.D. and Fry, R.C. (1985) Rapid coal analysis. Part II Slurry atomization DCP emission analysis of NBS coal. *Appl. Spectrosc.*, **39**, 984-988

McElroy, M.W., Carr, R.C., Ensor, D.S. and Markowski, G.R. (1982) Size distribution of fine particles from coal combustion. *Science*, **215**, 13-19

McGlynn, J.A. and Rice, T.D. (1982) Occurrence and distribution of fluorine in Sydney Basin coal seams. *Symp. Characteristics of Australian Coals and their Consequences for Utilization* North Ryde, 8.1-8.4

McGowan, G.E. (1960) The determination of fluorine in coal. An adaptation of spectrophotometric methods. *Fuel*, **39**, 245-252

McIntyre, N.S., Martin, R.B., Chauvin, W.J., Winder, C.G., Brown, J.R. and MacPhee, J.A. (1985) Studies of elemental distributions within discrete coal macerals. Use of secondary ion mass spectrometry and X-ray photoelectron spectroscopy. *Fuel*, **64**, 1705-1712

Mackay, A.M. and Wilson, B.I. (1978) Borate in Waikato coal. *NZ J. Sci.*, **21**, 611-614

MacKenzie, F.T., Lantzy, R.J. and Paterson, V. (1979) Global trace metal cycles and predictions. *Math. Geol.*, **11**, 99-142

McKenzie, R.M. (1962) Interference by iron in the spectrographic determination of molybdenum. *Spectrochim. Acta*, **18**, 1009-1010

Mackowsky, M.-Th. (1955) Sedimentary rhythms in coal deposits. *Neues. Jahrb. Geol. Palaeontol. Monatshe.*, 438-439

Mackowsky, M.-Th. (1968) Mineral matter in coal. In Murchison and Westoll (1968), pp. 309–321
Mackowsky, M.-Th. (1982) Minerals and trace elements occurring in coal. In Stach et al. (1982), pp. 153–171
Magee, E.M., Hall, H.J. and Varga, G.M. (1973) Potential pollutants in fossil fuels. *USEPA Rep.*, No. R2-73-249
Magyar, B. (1982) *Guide-lines to Planning Atomic Spectrometric Analysis.* Amsterdam: Elsevier, 273 pp
Mahanti, H.S. and Barnes, R.M. (1983) Determination of trace elements in coal and other energy-related materials by inductively-coupled plasma emission spectometry after collection on a poly (dithiocarbamate) resin. *Anal. Chim. Acta,* **149**, 395–400
Maksimova, M.F. and Shmariovich, E.M. (1982) Rhenium in infiltrational uranium-coal deposits. *Geol. Rudn. Mestorozhd.,* **24**(3), 71–78; *CA,* **97**, 130736 (1982)
Malberg, E. (1961) Geochemical investigations of East German brown coal deposits. *Geologie,* **10**, 728
Malcolm, R.L. (1985) Geochemistry of stream fulvic and humic substances. In Aiken et al. (1985a), pp. 181–209
Malhotra, V.M. and Graham, W.R.M. (1985) Origin of the Mn^{2+} e.p.r. spectrum in Pittsburgh No. 8 bituminous coal. *Fuel,* **64**, 270–272
Malik, A.U. and Ahmad, S. (1979) Analytical studies on Indian coals with special reference to their ash and mineral contents. *J. Mines Met. Fuels,* **27**(5), 148–152; *CA,* **93**, 75327 (1980)
Mallett, C.W. (ed.) (1982) Coal resources, origin, exploration and utilization in Australia. *Proc. Symp. Coal Group Geol. Soc. Aust.,* Melbourne, 597 pp
Manchester Brewers' Association Commission (1901) Arsenic in beer. *J. Soc. Chem. Ind., London,* **50**, 644–646; *Analyst (London),* **26**, 13–15
Manskaya, S.M. and Drozdova, T.V. (1968) *Geochemistry of Organic Substances.* Oxford: Pergamon, 345 pp
Manskaya, S.M., Drozdova, T.V. and Emelyanova, M.P. (1960) Distributions of copper in peats and peat soils of the Belorussian SSR. *Geochemistry (USSR)* (6), 630–643
Marczak, M. and Lewinska-Ochwat, L. (1987) Vanadium in organic matter of coal and vitrinite from the Niedobczyce IG-1 borehole. *Przegl. Geol.,* **35**(3), 130–133; *IEA Coal Abstr.,* **11**, 09810 (1987)
Marczak, M. and Parzentny, H. (1985) Geochemical and ecological evaluation of high-lead coal from the Chelm deposit. *Przegl. Geol.,* **33**, 680–683; *IEA Coal Abstr.,* **10**, 7521 (1986)
Margoshes, M. and Scribner, B.F. (1959) The plasma jet as a spectroscopic source. *Spectrochim. Acta,* **15**, 138–145
Marquardt, D., Luederitz, P. and Grosser, J. (1983) Multielemental analysis of brown coal and brown coal ash by atomic spectrometry in the inductive coupled plasma. *Instrum. Anal. Toxikol. (Hauptteil Vortr. Symp.)*, 184–187; *CA,* **100**, 194831 (1984)
Marshall, G.D., Robert, R.V.D. and Burden, K.J. (1987) Gallium – its determination and distribution in South African coal fly ash. *Symp. Ash a New Resource,* Pretoria **3**, 1–7
Martin, A. and Garcia-Rosell, L. (1971) Uranium and rhenium in sedimentary rocks. III Lignites of the Ebro depression. *Bol. Geol. Min.,* **82**, 178–185; *CA,* **75**, 89947 (1971)
Martin, J.E., Harward, E.D., Oakley, D.T., Smith, J.M. and Bedrosian, P.H. (1971) Radioactivity from fossil-fuel and nuclear power plants. *Rep. IAEA-SM-146/19,* pp. 325–337
Martin, M.H. and Coughtrey, P.J. (1982) *Biological Monitoring of Heavy Metal Pollution.* London: Applied Science Publishers, 475 pp
Martin, R.R., McIntyre, N.S. and MacPhee, J.A. (1985) An investigation of coal using secondary ion mass spectrometry (SIMS). *Int. Conf. Coal. Sci.,* Pergamon, Sydney, NSW, pp. 796–799
Martinez-Tarazona, R. and Cardin, J.M. (1986) The indirect determination of chlorine in coal by atomic absorption spectrophotometry. *Fuel,* **65**, 1705–1708
Martinez-Tarazona, M.R., Palacios, J.M. and Cardin, J.M. (1988) The mode of occurrence of chlorine in high volatile bituminous coals from the Asturian Central coalfield. *Fuel,* **67**, 1624–1628
Martin Perez, A. (1963) Content of germanium in Spanish mineral coals. II Experimental section. *Bol. Inf. Inst. Nac. Carbon,* No. 58, 110–149
Martins, A.F., Braganca de Moraes, A. and Baron, G.M. (1984) Application of extractive

spectrophotometric methods for the determination of trace elements of environmental relevance in coal from Candiota, R.S. (Brazil). *Cienc. Nat.*, **6**, 75–82

Mason, B. (1949) Oxidation and reduction in geochemistry. *J. Geol.*, **57**, 62–72

Massmann, H. (1968) Studies of atomic absorption and atomic fluorescence in a graphite cell. *Spectrochim. Acta, Part B*, **23B**, 215–226

Mattson, S. and Koutler-Andersson, E. (1955) Geochemistry of a raised bog. *Ann. R. Agric. Coll. Swed.*, **21**, 321–366; in Shotyk (1988)

Maylotte, D.H., Wong, J., St Peters, R.L., Lytle, F.W. and Greegor, R.B. (1981) X-ray absorption spectroscopic investigation of trace vanadium sites in coal. *Science*, **214**, 554–556

Meadowcroft, D.B. and Manning, M.I. (eds.) (1983) *Corrosion Resistant Materials for Coal Conversion Systems.* London: Applied Science, 600 pp

Mechacek, E. (1972) Microelements in beds of the Handlova-Novaky coal basin. *Geol. Zb. (Bratislava)*, **23**, 311–330; *CA*, **78**, 126937 (1973)

Mechacek, E. (1976) Microelements in coal seams of the Modry Kamen basin Dolina mine, southern Slovakia. *Acta Geol. geogr. Univ. Comenianae, Geol.*, **28**, 135–154; *CA*, **85**, 180333 (1976)

Medvedev, K.P. and Akimova, L.M. (1964) Rapid determination of germanium in coal, coal blends and coke. *Coke Chem, USSR*, No. 3, 17–19

Medvedev, K.P. and Batrakova, I.A. (1959) The content and accumulation of rare and trace elements in hard coals. *Koks Khim.*, No. 6, 13–17

Mehdi, S. and Datar, D.S. (1960) X-ray study of some Indian coals. Effect of heat on minerals in Kothagudem coal. *J. Sci. Ind. Res. (India)*, **19B**, 484–488) *CA*, **55**, 15884 (1961)

Mei, Z. (1985) The second high-selenium-containing region found in China-Ziyang County in Shaaxi Province. *Shaanxi Xinyiyao*, **14**(7), 38–40

Meij, R., Janssen, L.H.J.M. and van der Kooij, J. (1986) Air pollutant emissions from coal-fired power stations. *Kema Sci. Tech. Rep.*, **4**(6), 51–69

Meij, R., van der Kooij, J., van der Sluys, J.L.G., Siepman, F.G.C. and van der Sloot, H.A. (1984) Characteristics of emitted fly ash and trace elements from utility boilers fired with pulverized coal. *Kema Sci. Tech. Rep.*, **2**, 1–8

Meissner, C.R., Cecil, C.B. and Stricker, G.D. (eds.) (1977) Coal geology and the future – symposium abstracts and selected references. *US Geol. Surv. Circ.*, No. 757, 20 pp

Menkovskii, M.A. and Aleksandrova, A.N. (1962) Acid demineralisation under reducing conditions for characterisation of germanium compounds in coals. *Dokl. Akad. Nauk SSSR*, **146**, 868–870; *CA*, **58**, 3235 (1963)

Menkovskii, M.A. and Alexsandrova, A.N. (1963) Choice of conditions for ashing coal for the determination of germanium. *Zavod. Lab.*, **29**, 797–799

Merritt, R.D. (1988) General trends in major, minor and trace elements in coal with a comparative profile of Alaskan coal. *J. Coal Qual.*, **7**(3), 95–103

Meserole, F.B., Schwitzgebel, K., Magee, R.A. and Mann, R.M. (1979) Trace element emissions from coal-fired power plants. *J. Eng. Power*, **101**, 620–624

Meszaros, E. (1981) *Atmospheric Chemistry Fundamental Aspects.* Amsterdam: Elsevier, 201 pp

Meyberg, F., Berger, H. and Dannecker, W. (1986) Effects of some typical environmentally relevant matrixes on trace element analysis by ICP-AES. *Fortschr. Atomspektrom. Spurenanal.*, **2**, 591–605

Meyers, R.A. (ed.) (1982) *Coal Structure.* New York: Academic, 340 pp

Mikhailova, A.I. and Gladkaya, G.G. (1980) Boron in the coals of Eastern Siberia. *Solid Fuel Chem.*, **14**(3), 119–120

Mikhailova, A.I., Larina, V.A. and Vlasov, N.A. (1978) The question of the ashing of coals in the determination of trace elements. *Solid Fuel Chem.*, **12**(3), 12–13

Millancourt, B., Bresson, M.A., Masclet, P., Maffiolo, G. and Dubois, J. (1986) Emissions of fluorine and chlorine compounds and poly-aromatic hydrocarbons from coal-fired power stations. *Proc. Seventh World Clean Air Congr.*, Clean Air Soc. Aust. and NZ., Sydney, NSW, **4**, 35–41

Millard, H.T. (1977) Recent application of neutron activation analysis to coal. *Am. Chem. Soc. Div., Fuel Chem. Prepr.*, **22**(5), 64–67

Miller, H.P. (1949) The problems of coal geochemistry. *Econ. Geol.*, **44**, 649–662

Miller, R.N. (1984) The methodology of low-temperature ashing. In Finkelman *et al.* (1984b), pp. 9–15

Miller, R.N. and Given, P.H. (1986) The association of major, minor and trace inorganic elements with lignites. I Experimental approach and study of a North Dakota lignite. *Geochim. Cosmochim. Acta*, **50**, 2033–2043

Miller, R.N. and Given, P.H. (1987a) The association of major, minor and trace inorganic elements with lignites. III Trace elements in four lignites and general discussion of all data from this study. *Geochim. Cosmochim. Acta*, **51**, 1843–1853

Miller, R.N. and Given, P.H. (1987b) The association of major, minor and trace inorganic elements with lignites. II Minerals, and major and minor element profiles, in four seams. *Geochim. Cosmochim. Acta*, **51**, 1311–1322

Miller, R.N., Yarzab, R.F. and Given, P.H. (1979) Determination of the mineral-matter contents of coals by low-temperature ashing. *Fuel*, **58**, 4–10

Miller, W.H. (1842) On the specific gravity of sulphuret of nickel. *London, Edinburgh and Dublin Phil. Mag. J. Sci.*, **20**, 378–379

Mills, J.C. (1983) Determination of boron, beryllium and lithium in coal ash and geological materials by spark optical emission spectrometry. *Anal. Chim. Acta*, **154**, 227–234

Mills, J.C. (1986) An acid dissolution procedure for the determination of boron in coal ash and silicates by inductively-coupled plasma emission spectrometry with conventional glass nebulisers. *Anal. Chim. Acta*, **183**, 231–238

Mills, J.C. and Belcher, C.B. (1981) Analysis of coal, coke, ash and mineral matter by atomic spectroscopy. *Prog. Anal. At. Spectrosc.*, **4**(1–2), 49–80

Mills, J.C. and Turner, K.E. (1980) Direct determination of trace elements in coal and coal products. Part 1 Wavelength dispersive X-ray fluorescence spectrometry. *Rep. NERDDP*, No. EG/81/20, 42 pp

Mills, J.C., Doolan, K.J. and Knott, A.C. (1983) Determination of trace elements in coal and coal products. Part 4 Methods for the determination of fluorine, boron, beryllium and lithium in coal and coal products. *Rep. NERDDP*, No. EG84/358, 86 pp

Mills, J.C., Turner, K.E., Roller, P.W. and Belcher, C.B. (1981) Direct determination of trace elements in coal: wavelength-dispersive X-ray spectrometry with matrix correction using Compton scattered radiation. *X-ray Spectrom.*, **10**(3), 131–137

Minchev, D. and Eskenazi, G. (1963) Germanium and other trace elements in the ash of the Belogradcik coals. *Spis. Bulg. Geol. Druzh.*, **24**, 299–306; *CA*, **60**, 10433 (1964)

Minchev, D. and Eskenazi, G. (1963–64) Trace elements in the coals of Bulgaria. Germanium and other trace elements in the coals of the Pchslarovo deposit, East Rhodope Mountains. *Annu. Univ. Sofia*, **58**, 245–261

Minchev, D. and Eskenazi, G. (1966) Trace elements in the coals of Bulgaria. Germanium and other trace elements in the coals of the Vulche Pole deposit, East Rhodope Mountain. *God. Sofii. Univ., Geol.-Geogr. Fak.*, **59**(1), 357–372

Minchev, D. and Eskenazi, G. (1969) Trace elements in the coals of Bulgaria. Zonal distribution of germanium in erratic vitrain fragments. *Spis. Bulg. Geol. Druzh.*, **30**, 105–112; *CA*, **72**, 57694 (1970)

Minchev, D. and Eskenazi, G. (1972) Trace elements in the coal basins of Bulgaria. Trace elements in the coals from the Marica-Iztok basin. *God. Sofii. Univ., Geol.-Geogr. Fak.*, 1971–1972, **64**(1), 263–291; *Abstr. Bulg. Sci. Lit. Geol. Geog.*, **25**(1), 21–22 (1972)

Mingaye, J.C.H. (1903) Notes from the chemistry laboratory. *Rec. Dep. Mines Geol. Surv. NSW*, **7** (Part 3), 219–221

Mingaye, J.C.H. (1907) Notes on analyses of Japanese coals. *Rec. Geol. Surv. NSW*, **8**, 251–257

Minguzzi, C. and Naldoni, K.M. (1950) Supposed traces of arsenic in wood: its determination in the wood of some trees. *Atti. Soc. Toscana Sci. Nat. Pisa Mem.*, No. 57, Ser. A, 38–48; *CA*, **46**, 2634 (1952)

Minkin, J.A., Chao, E.C.T. and Thompson, C.L. (1979) Distribution of elements in coal macerals and minerals: determination by electron microprobe. *Am. Chem. Soc. Div., Fuel Chem. Prepr.*, **24**(1), 242–249

Minkin, J.A., Chao, E.C.T., Thompson, C.L., Nobiling, R. and Blank, H. (1982) Proton microprobe determination of elemental concentrations in coal macerals. *Scanning Electron Microsc.* (1), 175–184

Minkin, J.A., Chao, E.C.T., Thompson, C.L., Wandless, M.V., Dulong, F.T., Larson, R.R. and Neuzil, S.G. (1983) Submicroscopic (< 1 μm) mineral contents of vitrinites in selected bituminous coal beds. *Microbeam Anal.*, 18th, 27–30

Minkin, J.A., Finkelman, R.B., Thompson, C.L., Chao, E.C.T., Ruppert, L.F., Blank, H. and Cecil, C.B. (1984) Microcharacterisation of arsenic- and selenium-bearing pyrite in

Upper Freeport coal, Indiana County, Pennsylvania. *Scanning Electron Microsc.* (4), 1515–1524

Minkkinen, P. and Yliruokanen, I. (1978) The arsenic distribution in Finnish peat bogs. *Kem.-Kemi* (7–8), 331–335

Mishra, U.C., Lalit, B.Y. and Ramachandran, T.V. (1980) Radioactivity release to the environment by thermal power stations using coal as a fuel. *Sci. Total Environ.*, **14**, 77–83

Mishra, U.C., Lalit, B.Y. and Ramachandran, T.V. (1984) Relative radiation hazards of coal-based and nuclear power plants in India. *Radiat. Risk, Prot., Int. Congr.*, 6th, **1**, 537–540

Mitchell, B.D., Bracewell, J.M., De Endredy, A.S., McHardy, W.J. and Smith, B.F.L. (1968) Mineralogical and chemical characteristics of a gley soil from north-east Scotland. *Trans. Int. Congr. Soil Sci.*, American Elsevier, New York, **III**, 67–77

Mitchell, R.L. Trace elements in Scottish peats. *Int. Peat Symp. Sect.*, B3, 4 pp

MM (1987) Western world coal and lignite mines. *Mineral. Mag.*, **157**, 243–250

MM (1988) Western world coal and lignite mines. *Mineral. Mag.*, **159**, 188–196

Mohan, M.S., Zingaro, R.A., MacFarlane, R.D. and Irgolic, K.J. (1982) Characterisation of a uranium-rich organic material obtained from a South Texas lignite. *Fuel*, **61**, 853–858

Monkhouse, A.C. (1950) The minor constituents of coal. *Coke Gas*, **12**, 363–368

Monnot, G.A. (1953) Determination of small quantities of elements in the ash of carbon compounds and silicates. *Chim. Anal. (Paris)*, **35**, 274–276

Montano, P.A. (1979) Characterisation of iron bearing minerals in coal. *Am. Chem. Soc. Div., Fuel Chem. Prepr.*, **24**(1), 218–229

Moon, P.G.G. (1901) *J. Gas Light*, **77**, 1061; Duck and Himus (1951)

Moore, E.S. (1947) *Coal – Its Properties, Analysis, Classification, Geology, Extraction, Uses and Distribution*. New York: Wiley, 473 pp

Moore, P.D. (1987) Ecological and hydrological aspects of peat formation. In Scott (1987), pp. 7–15

Morales, G. and Gasos, P. (1985) Combustion of uraniferous lignites in a fluidised bed. *Proc. Int. Conf. Fluid. Bed. Combust.*, 8th (2), 903–911; *CA*, **104**, 189465 (1986)

Morgan, G.T. (1935) Recent researches on certain of the rarer elements. *J. Chem. Soc.*, 554–570

Morgan, G.T. and Davies, G.R. (1937) Germanium and gallium in coal ash and flue dust. *Chem. Ind. (London)*, 717–721

Morgan, M.E., Jenkins, R.G. and Walker, P.L. (1981) Inorganic constituents in American lignites. *Fuel*, **60**, 189–193

Morris, J.S. and Bobrowski, G. (1979) Determination of ^{226}Ra, ^{214}Pb and ^{214}Bi in flyash samples from eighteen (18) coal fired power plants in the United States. *Rep. METC/SP-79/10* (Pt. 1), 460–470

Morse, J.W., Millero, F.J., Cornwell, J.C. and Rickard, D. (1987) The chemistry of the hydrogen sulfide and iron sulfide systems in natural waters. *Earth-Sci. Rev.*, **24**, 1–42

Mosher, B.W. and Duce, R.A. (1983) Vapor phase and particulate selenium in the marine atmosphere. *J. Geophys. Res.*, **88**(C11), 6761–6768

Moss, K.N., Hirst, A.A. and Needham, L.W. (1929–1930) The occurrence and estimation of lead and zinc compounds in coal. *Trans. Inst. Min. Eng.*, **79**, 435–440

Mott, R.A. (1943) The origin and composition of coals. *Fuel*, **22**, 20–26

Moureu, C. and LePape, A.A. (1914) Helium in firedamp and radioactivity of coal. *C.R. Acad. Sci.*, **158**, 598–603

Mraw, S.C., de Neufville, J.P., Freund, H., Baset, Z., Gorbaty, M.L. and Wright, F.J. (1983) The science of mineral matter in coal. In *Coal Science* (eds. M.L. Gorbaty, J.W. Larsen and I. Wender), pp. 1–63. New York: Academic Press

Mroz, E.J. and Zoller, W.H. (1975) Composition of atmospheric particulate matter from the eruption of Heimaey, Iceland. *Science*, **190**, 461–464

Mukherjee, B. and Dutta, R. (1950) A note on the constituents of the ashes of Indian coals determined spectroscopically. *Fuel*, **29**, 190–192

Mukherjee, B. and Ghosh, A. (1977) Distribution and behaviour of trace elements in some Permian coals of India. *Indian Mineral.*, **17**, 23–30

Mullai, F. (1984) Heavy minerals in coals of the coal-bearing basin of Tirana. *Bul. Shkencave Gjeol.*, **3**, 125–137; *CA*, **102**, 9340 (1985)

Murchison, D. and Westoll, T.S. (eds.) (1968) *Coal and Coal-Bearing Strata*. Edinburgh: Oliver and Boyd, 418 pp

Nadkarni, R.A. (1975) Multielement analysis of coal and coal fly ash standards by instrumental neutron activation analysis. *Radiochem. Radioanal. Lett.*, **21**, 161–176

Nadkarni, R.A. (1980) Multitechnique multielemental analysis of coal and fly ash. *Anal. Chem.*, **52**, 929–935

Nadkarni, R.A. (1981) Determination of volatile elements in coal and other organic materials by oxygen bomb combustion. *Int. Lab.* (Sept.), 26–34

Nadkarni, R.A. (1982a) Comprehensive elemental analysis of coal and fly ash. *ACS Symp. Ser.*, No. 205, 147–162

Nadkarni, R.A. (1982b) Applications of hydride generation-atomic absorption spectrometry to coal analysis. *Anal. Chim. Acta*, **135**, 363–368

Nadkarni, R.A. (1984) Applications of microwave oven sample dissolution in analysis. *Anal. Chem.*, **56**, 2233–2237

Nadkarni, R.A. and Botto, R.I. (1984) Determination of germanium in coal ashes by inductively coupled plasma atomic emission spectrometry. *Appl. Spectrosc.*, **38**, 595–598

Nadkarni, R.A. and Morrison, G.H. (1978) Use of standard reference materials as multielement irradiation standards in neutron activation analysis. *J. Radioanal. Chem.*, **43**, 347–369

Nakaoka, A., Fukushima, M., Ichikawa, Y. and Takagi, S. (1982) Radiological evaluation of natural radionuclides from coal fired power plants. *Environ. Technol. Lab. CRIEPI*, Rep. No. 281036, 40 pp

Nakashima, R., Kamata, E. and Shibata, S. (1984) Atomic absorption spectrometric determination of trace elements in coal by acid digestion in a sealed polytetrafluoroethylene vessel. *Jpn. Anal.*, **33**, E343–E350; *IEA Coal Abstr.*, **9**(5), 3489 (1985)

Nakashima, R., Kamata, E., Goto, K. and Shibata, S. (1983) Comparison of coal digestion methods for atomic absorption determination of cadmium in coal. *Bunseki Kagaku*, **32**, T92–T96; *CA*, **99**, 161061 (1983)

Narayan, R., Kullerud, G. and Wood, K.V. (1988) A new perspective on the nature of 'organic' sulfur in coal. *Am. Chem. Soc. Div., Fuel Chem. Prepr.*, **33**(1), 193–199

NAS (1979) Redistribution of Accessory Elements in Mining and Mineral Processing. Part I Coal and Oil Shale. *Rep. Nat. Acad. Sci., Washington, DC*, 180 pp

NAS (1976) Selenium. *Rep. Nat. Acad. Sci., Washington, DC*, 203 pp

Ndiokwere, C.L., Guinn, V.P. and Burtner, D. (1983) Trace elemental composition of Nigerian coal measured by neutron activation analysis. *J. Radioanal. Chem.*, **79**, 123–128

Neavel, R.C. (1981) Origin, petrography and classification of coal. In Elliott (1981), pp. 91–158

Neill, P.H., Maciel, G., Given, P.H. and Weldon, D. (1987) Aromaticity of some high sulphur coals: surprising degree of heterogeneity. *Fuel*, **66**, 96–98

Nelson, J.B. (1953) Assessment of the mineral species associated with coal. *Br. Coal Util. Res. Assoc. Mon. Bull.*, **27**(2), 41–55

Nelson, W.J. (1987) Coal deposits of the United States. *Int. J. Coal Geol.*, **8**, 355–365

Newmarch, C.B. (1950) The correlation of Kootenay coal seams. *Can. Min. Metall. Bull.*, **43**, 141–148

Newmarch, C.B. (1953) Geology of the Crowsnest coal basin, with special reference to the Fernie area. *B.C. Dep. Mines Bull.*, No. 33, 107 pp

Nguyen, S., Grimm, C.A., Filby, R.H. and Swaine, D.J. (1983) Determination of trace elements in mosses as pollutant indicators by INAA. *Abstr. Pap. Am. Chem. Soc. Div. Nucl. Chem. Technol.* 185th. Nat. Meet. Seattle, American Chemical Society, Washington, DC

Nicholls, G.D. (1963) Environmental studies in sedimentary geochemistry. *Sci. Prog. (London)*, **51**, 12–31

Nicholls, G.D. (1968) The geochemistry of coal-bearing strata. In Murchison and Westoll (1968), pp. 269–307

Nicholls, G.D. and Loring, D.H. (1962) The geochemistry of some British carboniferous sediments. *Geochim. Cosmochim. Acta*, **26**, 181–223

Nichols, C.L. and D'Auria, J.M. (1981) Seam and location differentiation of coal specimens using trace element concentrations. *Analyst (London)*, **106**, 874–882

Nissenbaum, A. and Swaine, D.J. (1976) Organic matter-metal interactions in recent sediments: the role of humic substances. *Geochim. Cosmochim. Acta*, **40**, 809–816

Noll, W. (1933) The geochemistry of strontium. *Chem. Erde*, **8**, 507–600

Noller, B.N., Bloom, H. and Arnold, A.P. (1981) Sampling and analysis of metals in atmospheric particulates by graphite furnace atomic absorption spectrometry. *Prog. Anal. At. Spectrosc.*, **4**, 81–189

North, F.J. and Howarth, W.E. (1928) On the occurrence of millerite and associated minerals in the coal measures of South Wales. *Proc. S. Wales Inst. Eng.*, **44**, 325–348

Norton, G.A., Araghi, H.G. and Markuszewski, R. (1985) Removal of trace elements during chemical cleaning of coal. *Am. Chem. Soc. Div., Fuel Chem. Prepr.*, **30**(2), 58–65

Nriagu, J.O. (1979) Global inventory of natural and anthropogenic emissions of trace metals to the atmosphere. *Nature (London)*, **279**, 409–411

Nriagu, J.O., Pacyna, J.M., Milford, J.B. and Davidson, C.I. (1988) Distribution and characteristic features of chromium in the atmosphere. In *Chromium in the Natural and Human Environments* (eds. J.O. Nriagu and E. Nieboer), pp. 125–172. New York: Wiley

Nunn, R.C., Lovell, H.L. and Wright, C.C. (1953) Spectrographic analyses of trace elements in anthracite. *Trans. Annu. Anthracite Conf. Lehigh University*, **11**, 51–65

Nurminskii, N.N. (1971) Use of the Kjeldahl method for determination of germanium in coals. *Khim. Tverd. Topl.* (3), 128–130; *CA*, **75**, 65935 (1971)

Obrusnik, I. and Posta, S. (1983) Instrumental neutron activation analysis of NBS 1633a flyash and 1632a bituminous coal reference samples with the use of short irradiation. *Geostand. Newsl.*, **7**, 291–293

Ochsenkuhn-Petropulu, M. and Parissakis, G. (1985) Polarographic determination of uranium after separation and enrichment by ion-exchange/extraction. Application to seawater, lignites and fly ash. *Int. Conf. Heavy Met. Environ.*, 5th, CEP Consultants, Edinburgh, **2**, 496–500

Ode, W.H. (1963) Coal analysis and mineral matter. In Lowry (1963), pp. 202–231

Odor, L. (1967) Report on the geochemical investigation of the Eocene brown-coal sequence of the Balinka II area. *Magy. Ail. Foldt. Intez. Evi Jel.*, 315–343

Odor, L. (1969) The Be content of Transdanubian Eocene coals. *Magy. All. Foldt. Intez. Evi Jel.*, 123–131

O'Gorman, J.V. and Walker, P.L. (1971) Mineral matter characteristics of some American coals. *Fuel*, **50**, 135–151

Okamoto, K. (1979) Radioactive emission from thermal power plants and related problems. *Proc. Aust. Inst. Energy First Nat. Conf.*, **1**, P40–P41

Olczak, C. (1985) Occurrence of germanium in phenolated waste water from coking plants and in the products of waste water treatment. *Koks Smola Gaz.*, **30**(2), 30–33; IEA Coal Abstr., **10**(3), 2247 (1986)

Oman, C.L., Finkelman, R.B., Coleman, S.L. and Bragg, L.J. (1988) Selenium in coals from the Powder River Basin, Wyoming and Montana. *US Geol. Surv. Circ.*, No. 1025, 16–17

Ondov, J.M., Zoller, W.H., Olmez, I., Aras, N.K., Gordon, G.E., Rancitelli, L.A., Abel, K.H., Filby, R.H., Shah, K.R. and Ragaini, R.C. (1976) Four-laboratory comparative instrumental nucleas analysis of the NBS coal and fly ash standard reference materials. *Nat. Bur. Stand. (US) Spec. Publ.*, No. 422, 1, 211–223

Ong, H.L. and Swanson, V.E. (1966) Adsorption of copper by peat, lignite, and bituminous coal. *Econ. Geol.*, **61**, 1214–1231

Ono, K., Inada, Y. and Konno, I. (1955) Utilization of lignite containing germanium. *Bull. Res. Inst. Miner. Dress. Met. Tohuku Univ.*, **11**, 159–164; *CA*, **50**, 13405 (1956)

O'Reilly, J.E. and Hale, M.A. (1977) Direct atomic absorption analysis of powdered whole-coal slurries. *Anal. Lett.*, **10**, 1095–1104

O'Reilly, J.E. and Hicks, D.G. (1979) Slurry-injection atomic absorption spectrometry for analysis of whole coal. *Anal. Chem.*, **51**, 1905–1915

Orvini, E., Gills, T.E. and LaFleur, P.D. (1974) Method for determination of selenium, arsenic, zinc, cadmium and mercury in environmental matrices by neutron activation analysis. *Anal. Chem.*, **46**, 1294–1297

Otte, M.U. (1953) Trace elements in some German hard coals. *Chem. Erde*, **16**, 239–294

Ourisson, G., Albrecht, P. and Rohmer, M. (1984) The microbial origin of fossil fuels. *Sci. Am.*, **251**(2), 34–41

Owens, J.W., Gladney, E.S. and Knab, D. (1982) Determination of boron in geological materials by inductively-coupled plasma emission spectrometry. *Anal. Chim. Acta*, **135**, 169–172

Pacyna, J.M. (1980) Radionuclide behavior in coal-fired plants. *Ecotoxicol. Environ. Saf.*, **4**, 240–251

Pacyna, J.M. (1986) Source-receptor relationships for trace elements in Northern Europe. *Water, Air, Soil Pollut.*, **30**, 825–835
Pacyna, J.M. and Ottar, B. (1985) Transport and chemical composition of the summer aerosol in the Norwegian Arctic. *Atmos. Environ.*, **19**, 2109–2120
Pacyna, J.M., Ottar, B., Tomza, U. and Maenhaut, W. (1985) Long-range transport of trace elements to NY Alesund, Spitsbergen. *Atmos. Environ.*, **19**, 857–865
Page, A.L., Elseewi, A.A. and Straughan, I.R. (1979) Physical and chemical properties of fly-ash from coal-fired power plants with reference to environmental impacts. *Residue Rev.*, **71**, 83–120
Painter, P.C., Rimmer, S.M., Snyder, R.W. and Davis, A. (1981) A Fourier transform infrared study of mineral matter in coal; the application of a least squares curve-fitting program. *Appl. Spectrosc.*, **35**, 102–106
Painter, P.C., Coleman, M.M., Jenkins, R.G., Whang, P.W. and Walker, P.L. (1978) Fourier transform infrared study of mineral matter in coal; a novel method for quantitative mineralogical analysis. *Fuel*, **57**, 337–344
Pakalns, P. (1972) Interference of titanium in the determination of thorium in rocks with arsenazo. III *Anal. Chim. Acta*, **58**, 463–469
Palmer, C.A. and Filby, R.H. (1983) Determination of mode of occurrence of trace elements in the Upper Freeport coal bed using size and density separation procedures. *Proc. Int. Conf. Coal Sci.*, pp. 365–368
Palmer, C.A. and Filby, R.H. (1984) Distribution of trace elements in coal from the Powhatan No. 6 mine, Ohio. *Fuel*, **63**, 318–328
Palmer, C.A. and Wandless, M.V. (1985) Distribution of trace elements in coal minerals of selected eastern United States coals. *Proc. Int. Conf. Coal Sci.*, Pergamon, Sydney, NSW, pp. 792–795
Palmer, C.A., Crandell, W.B., Evans, J.R., Gillison, J.R., Moore, R., Sellers, G.A., Skeen, C.J. and Winters, L.J. (1987) Analysis of trace elements in the new U.S. Geological Survey standard coal (CLB-1) and selected premium coal reference materials. *Am. Chem. Soc. 193rd Nat. Mtg.*, GEOC-108
Pankhurst, Q.A., McCann, V.H. and Newman, N.A. (1986) Identification of the iron-bearing minerals in some bituminous coals using Mössbauer spectroscopy. *Fuel*, **65**, 880–883
Papastefanou, C. and Charalambous, S. (1979) On the radioactivity of fly ashes from coal power plants. *Z. Naturforsch.*, **34A**, 533–537; *CA*, **91**, 95889 (1979)
Papastefanou, C. and Charalambous, S. (1980) Hazards from radioactivity of fly ash of Greek coal power plants (CPP). *Proc. Congr. Int. Radiat. Prot. Soc.*, 5th, Pergamon Press, Oxford, **1**, pp. 153–158
Papp, C.S.E. and Fischer, L.B. (1987) Application of microwave digestion to the analysis of peat. *Analyst (London)*, **112**, 337–338
Papp, C.S.E. and Harms, T.F. (1985) Comparison of digestion methods for total elemental analysis of peat and separation of its organic and inorganic components. *Analyst (London)*, **110**, 237–242
Pareek, H.S. and Bardhan, B. (1985) Trace elements and their variation along seam profiles of certain coal seams of Middle and Upper Barakar formations (Lower Permian) in East Bokaro Coalfield, District Hazaribagh, Bihar, India. *Int. J. Coal Geol.*, **5**, 281–314
Parratt, R.L. and Kullerud, G. (1979) Sulfide minerals in coal bed V, Minnehaha mine, Sullivan County, Indiana. *Mineral. Deposita*, **14**, 195–206
Parry, H.G.M. and Ebdon, L. (1988) Coal analysis by analytical atomic spectrometry (ICP-AES and ICP-MS) without sample dissolution. *Anal. Proc.*, **25**, 69–71
Patterson, C.C. and Settle, D.M. (1987) Magnitude of lead flux to the atmosphere from volcanoes. *Geochim. Cosmochim. Acta*, **51**, 675–681
Paulson, L.E., Beckering, W. and Fowkes, W.W. (1972) Separation and identification of minerals from Northern Great Plains Province lignite. *Fuel*, **51**, 224–227
Pearce, W.C. and Hill, J.W.F. (1986) The mode of occurrence and combustion characteristics of chlorine in British coal. *Prog. Energy Combust. Sci.*, **12**, 117–162
Pearce, W.C., Thornewill, D. and Marston, J.H. (1985) Multi-element analysis of solutions of coal ash using inductively coupled plasma optical-emission spectrometry. *Analyst (London)*, **110**, 625–630
PECH (1980) *Trace-Element Geochemistry of Coal Resource Development Related to Environmental Quality and Health*. Washington, DC: National Academy Press, 153 pp

Pendias, H. (1964) Geochemical investigations of coals from seams of the Walbrzych and Bialy Kamien beds in the Walbrzych Basin. *Kwart. Geol.*, **8**, 769–786

Percy, J. (1875) *Metallurgy: Introduction, Refractory Materials and Fuel.* London: John Murray, 596 pp

Perricos, D.C. and Belkas, E.P. (1969) Determination of uranium in uraniferous coal. *Talanta*, **16**, 745–748

Peterson, M.J. and Zink, J.B. (1964) A semiquantitative spectrochemical method for analysis of coal ash. *US Bur. Mines Rep. Invest.*, No. 6496, 15 pp

Petrov, N.P. (1963) Molybdenum in a brown coal deposit of Uzbekistan. *Int. Geol. Rev.*, **5**, 335–338

Phillips, J.A. (1896) *A Treatise on Ore Deposits.* London: Macmillan (2nd edn, recast by H. Louis), 950 pp

Phillips, T.L. and Cecil, C.B. (1985) Paleoclimatic controls on coal resources of the Pennsylvanian System of North America – introduction and overview of contributions. *Int. J. Coal Geol.*, **5**, 1–6

Pickering, W.F. (1977) *Pollution Evaluation – The Quantitative Aspects.* New York: Dekker, 208 pp

Pickering, W.F. (1986) Metal ion speciation – soils and sediments (a review). *Ore Geol. Rev.*, **1**, 83–146

Pietzner, H. and Wolf, M. (1964) Geochemical investigations on brown coal ashes and coal-petrographical investigations of brown coals from the Niederrhein brown coal district. *Fortschr. Geol. Rheinl. Westfalen*, **12**, 517–550; *CA*, **63**, 9680 (1965)

Pike, S., Dewison, M.G. and Spears, D.A. (1989) Sources of error in low temperature plasma ashing procedures for quantitative mineral analysis of coal ash. *Fuel*, **68**, 664–668

Pilkington, E.S. (1957) A survey of some Australian sources of germanium. *Aust. J. Appl. Sci.*, **8**, 98–111

Pillay, K.K.S., Thomas, C.C. and Kaminski, J.W. (1969) Neutron activation analysis of the selenium content of fossil fuels. *Nucl. Appl. Technol.*, **7**, 478–483

Pinta, M. (1970) Review of contamination problems in trace analysis. *CNRS Rapp.*, 923, 25–39

Piperno, E. (1975) Trace element emissions: aspects of environmental toxicology. In Babu (1975), pp. 192–209

Platz, B. (1887) The presence of copper in coal and coke. *Stahl Eisen*, **7**, 258; Gibson and Selvig (1944)

Pollock, E.N. (1975) Trace impurities in coal by wet chemical methods. In Babu (1975), pp. 23–34

Porritt, R.E. and Swaine, D.J. (1976) Mercury and selenium in some Australian coals and fly-ash. *Aust. Inst. Fuel. Conf.*, Institute of Fuel, Sydney, NSW, pp. 18.1–18.9

Pougnet, M.A.B. and Orren, M.J. (1986) The determination of boron by inductively coupled plasma atomic emission spectroscopy. Part 2 Applications to South African environmental samples. *Int. J. Environ. Anal. Chem.*, **24**, 267–282

Powell, A.R. (1954) Extraction of germanium and gallium. *Chem. Ind. (London)*, **40**, 1225–1226

Powell, A.R., Lever, F.M. and Walpole, R.E. (1951) Extraction and refining of germanium and gallium. *J. Appl. Chem.*, **1**, 541–555

Prather, J.W., Guin, J.A. and Tarrer, A.R. (1979) X-ray fluorescence analysis of trace elements in coal and solvent refined coal. In Karr (1979), pp. 357–369

Prather, J.W., Tarrer, A.R. and Guin, J.A. (1977) X-ray fluorescence analysis of trace metals in solvent refined coal. *Am. Chem. Soc. Div., Fuel Chem. Prepr.*, **22**(5), 72–77

Pringle, W.J.S. and Bradburn, E. (1958) The mineral matter in coal II – the composition of the carbonate minerals. *Fuel*, **37**, 166–180

Purchase, N.G. (1985) Trace elements in New Zealand coals – a review. *Proc. Coal Res. Conf. NZ*, Coal Res. Assoc. NZ, Wellington, **1**, 13.1, 10 pp

Purchase, N.G. (1987) Major and trace element variations in Huntly power station ash streams. *Proc. Coal Res. Conf. NZ*, Coal Res. Assoc. NZ, Wellington, **1**, R10.1, 8 pp

Que Hee, S.S., Finelli, V.N., Fricke, F.L. and Wolnik, K.A. (1982) Metal content of stack emissions, coal and fly ash from some Eastern and Western power plants in the U.S.A. as obtained by ICP-AES. *Int. J. Environ. Anal. Chem.*, **13**, 1–18

Querol, X., Chinchon, S. and Lopez-Soler, A. (1989) Iron sulfide precipitation sequence in

Albian coals from the Maestrazgo Basin southeastern Iberian Range, northeastern Spain. *Int. J. Coal Geol.*, **11**, 171–189

Raask, E. (1985a) *Mineral Impurities in Coal Combustion – Behaviour, Problems, and Remedial Measures.* Washington: Hemisphere, 484 pp

Raask, E. (1985b) The mode of occurrence and concentration of trace elements in coal. *Prog. Energy Combust. Sci.*, **11**, 97–118

Radmacher, W. and Hessling, H. (1959) Spectrographic determination of trace elements in bituminous coals. *Z. Anal. Chem.*, **167**, 172–182

Radmacher, W. and Mohrhauer, P. (1955) The direct determination of the mineral matter content of hard coals. *Brennst.-Chem.*, **36**, 236–239

Radmacher, W. and Schmitz, W. (1965) Analysis of solid fuels. Determination of the chemical composition of ashes from combustibles. *Brennst.-Chem.*, **46**(1), 21–27

Rafter, T.A. (1945) Boron and strontium in New Zealand coal ashes. *Nature (London)*, **155**, 332

Rahmani, R.A. and Flores, R.M. (1984a) Sedimentology of coal and coal-bearing sequences of North America: a historical review. In Rahmani and Flores (1984b), pp. 3–10

Rahmani, R.A. and Flores, R.M. (eds.) (1984b) *Sedimentology of Coal and Coal-Bearing Sequences.* Spec. Publ. Int. Assoc. Sedimentologists No. 7, Oxford: Blackwell, 412 pp

Raistrick, A. and Marshall, C.E. (1939) *The Nature and Origin of Coal and Coal Seams.* London: English Universities Press, 282 pp

Rait, N. (1981) Multiple-element semiquantitative analysis of one-milligram geochemical samples by D.C. arc emission spectrography. *Chem. Geol.*, **32**, 317–333

Raj, S.S. (1986) PIXE – a new analytical technique for the analysis of coal and coal derived fuels. *Am. Chem. Soc. Div., Fuel Chem. Prepr.*, **31**(1), 35–42

Ramage, H. (1927) Gallium in flue dust. *Nature (London)*, **119**, 783

Rao, C., Raja, S., Datta, N.R., Pal, J.C. and Mukherjee, K.N. (1980) A note on the trace element contents in some selected tertiary and Gondwana coalfields of India. *Indian Miner.*, **34**(2), 59–61; *CA*, **95**, 153404 (1981)

Rao, C.P. and Gluskoter, H.J. (1973) Occurrence and distribution of minerals in Illinois coals. *Ill. State Geol. Surv. Circ.*, No. 476, 56 pp

Rao, S.R.R., Datta, P.B., Luthra, G.B.S. and Lahiri, A. (1951) Reduction of phosphorus in coking coal. *Trans. Min. Geol. Metall. Inst. India*, **47**, 112–127

Ratynskii, V.M. (1946) Germanium in coal. *Tr. Biogeokhim. Lab., Akad. Nauk SSSR*, **8**, 183–223; *CA*, **47**, 7961 (1953)

Ratynskii, V.M. (1975) The role of aqueous solutions in the accumulation of rare elements by fossil coals. In *Recent Contributions to Geochemistry and Analytical Chemistry* (ed. A.I. Tugarinov), pp. 530–538. New York: Wiley

Ratynskii, V.M. and Glushnev, S.V. (1967a) Relationships in the distribution of metals in coals. *Dokl. Akad. Nauk SSSR*, **177**, 236–239

Ratynskii, V.M. and Glushnev, S.V. (1967b) Principles of the distribution of rare, minor and non-ferrous metals in coals. *Khim. Tverd. Topl.* (5), 47–53; *CA*, **68**, 61569 (1968)

Ratynskii, V.M. and Zharov, Y.N. (1977) Nature of compounds of gallium with the organic part of fossil coals. *Dokl. Akad. Nauk SSSR*, **235**, 188–189; *CA*, **67**, 120355 (1977)

Ratynskii, V.M. and Zharov, Y.N. (1979) Nature of the distribution of gallium in coals. *Dokl. Akad. Nauk SSSR*, **246**, 1219–1222; *CA*, **91**, 126449 (1979)

Ratynskii, V.M., Shpirt, M.Y. and Krasnobaeva, N.V. (1980) Rhenium in fossil coals. *Dokl. Akad. Nauk SSSR*, **251**, 1489–1492; *CA*, **93**, 134827 (1980)

Ratynskii, V.M., Shpirt, M.Y., Musyal, S.A. and Beloshapko, M.A. (1982) Gold in coals. *Solid Fuel Chem.*, **16**(4), 83–85

Raymond, R. and Andrejko, M.J. (eds.) (1983) Mineral matter in peat, its occurrence, form, and distribution. *Rep. Los Alamos Nat. Lab.*, No. LA-9907-OBES, UC-11, 242 pp

Raymond, R. and Gooley, R. (1979) Electron probe microanalyzer in coal research. In Karr (1979), pp. 337–356

Raymond, R., Bish, D.L. and Gooley, R. (1983) Occurrence of szomolnokite in Kentucky No. 14 coal and possible implications concerning formation of iron sulfides in peats and coals. In Raymond and Andrejko (1983), pp. 159–167

Raymond, R., Cameron, C.C. and Cohen, A.D. (1987) Relationship between peat geochemistry and depositional environments, Cranberry Island, Maine. *Int. J. Coal Geol.*, **8**, 175–187

Rees, W.J. and Sidrak, G.H. (1956) Plant nutrition on fly ash. *Plant Soil*, **8**, 141–159

Reid, W.T. (1981) Coal ash – its effect on combustion systems. In Elliott (1981), pp. 1389–1445
Rekus, A.F. and Haberkorn, A.R. (1966) Identification of minerals in single particles of coal by the X-ray powder method. *J. Inst. Fuel*, **39**, 474–477
Renton, J.J. (1978) Systematic variability in the mineralogy of the low temperature ash of some North American coals. *Energy Sources*, **4**(2), 91–112
Renton, J.J. (1982) Mineral matter in coal. In *Coal Structure* (ed. R.A. Myers), pp. 283–326. New York: Academic Press
Renton, J.J. (1984) Semi-quantitative determination of coal minerals by X-ray diffractometry. *Am. Chem. Soc. Div., Fuel Chem. Prepr.*, **29**(4), 21–27
Reynolds, F.M. (1950) Gallium and germanium extraction from flue dust. *Chem. Prod.*, **13**, 152–153
Rhett, D.W. (1979) Mechanism of uranium retention in intractable uranium ores from northwestern New Mexico. *J. Met.*, **31**(10), 45–50
Rice, R. and Bragg, R.L. (1983) Multielement analysis of coal ash utilizing sequential ICP. *Proc. Coal Test. Conf.*, 3rd, The Conference, Charleston, WV, pp. 70–75
Rice, T.D. (1988) Determination of fluorine and chlorine in geological materials by induction furnace pyrohydrolysis and standard-addition ion-selective electrode measurement. *Talanta*, **35**, 173–178
Riddle, C., Vander Voet, A. and Doherty, W. (1988) Rock analysis using inductively coupled plasma mass spectrometry: a review. *Geostand. Newsl.*, **12**, 203–234
Riepe, W. (1986) Analytical chemistry of elementary constituents of coal. In Butler (1986), pp. 207–214
Rigin, V.I. (1981) Atomic fluorescence determination of mercury in coal, ash, plants and soil. *Zh. Anal. Khim.*, **36**, 1522–1528; *CA*, **95**, 214576 (1981)
Rigin, V.I. (1984) Extraction-atomic fluorescence determination of beryllium in coal, ash, air and water. *Zh. Anal. Khim.*, **39**, 807–812; *CA*, **101**, 65134 (1984)
Rigin, V.I. (1985) Decomposition of coal samples for atomic spectrochemical analysis. *Zh. Anal. Khim.*, **40**, 253–257; *CA*, **102**, 151815 (1985)
Rigin, V.I (1986) Group determination of trace elements by a hybrid method with nonselective atomic fluorescence recording. *Zh. Anal. Khim.*, **41**, 788–797; *CA*, **105**, 90330 (1986)
Rigin, V.I. (1987) Ion-chromatographic determination of halides, nitrogen, phosphorus and sulfur in fossil coal. *Zh. Anal. Khim.*, **42**(6), 1073–1076
Riley, K.W. (1982) Spectral interference by aluminium in the determination of arsenic using the graphite furnace: choice of resonance lines. *At. Spectrosc.*, **3**, 120–121
Riley, K.W. (1984) Significant reactions of aluminium, magnesium and fluoride during the graphite furnace atomic-absorption spectrophotometric determination of arsenic in coal. *Analyst (London)*, **109**, 181–182
Rimmer, S.M. and Davis, A. (1984a) Lateral variability in mineralogy and petrology of the Lower Kittanning seam, Western Pennsylvania and Eastern Ohio. Final Rep. Part II Rep. DOE/PC/30013-T6, 261 pp; *Energy Res. Abstr.*, **9**(23), 47318 (1984)
Rimmer, S.M. and Davis, A. (1984b) Influence of environments of deposition on the inorganic composition of coals. *Am. Chem. Soc. Div., Fuel Chem. Prepr.*, **29**(4), 9–13
Rimmer, S.M. and Davis, A. (1986) Geologic controls on the inorganic composition of Lower Kittanning coal. *ACS Symp. Ser.*, No. 301, 41–52
Rincon, J.M., Lesmes, L.E., Vargas, O. and Velazquez, J.M. (1978) The analysis of some Colombian coal ashes. *Quim. Ind. (Bogota)*, **8**(5), 23–28
Rindt, D.K., Karner, F.R., Beckering, W. and Schobert, H.H. (1980) Current research on the inorganic constituents in North Dakota lignites and some effects on utilisation. *Am. Chem. Soc. Div., Fuel Chem. Prepr.*, **25**(1), 210–218
Ring, E.J. and Hansen, R.G. (1984) The preparation of three South African coals for use as reference materials. *Mintek Rep.*, No. MI69, 130 pp
Robbat, A., Finseth, D.H. and Lett, R.G. (1984) Organic titanium in coal and the deposition of titanium on direct liquefaction catalysts. *Fuel*, **63**, 1710–1715
Roberts, J.R. and van Rensburg, T.J. (1985) SEM and EDX studies towards the association of minerals and macerals in coal. *Proc. Electron. Microsc. Soc. S. Afr.*, **15**, 11–12
Robertson, S.D., Holding, S.T., Gilmore, G.R. and Ledingham, K.W.D. (1983) Assessment of coal quality – the nuclear option. *Fuel*, **62**, 1046–1052

Robson, A. (1984) The radiological impact of electricity generation by UK coal and nuclear systems. *Sci. Total Environ.*, **35**, 417–430

Robson, A. et al. (1981) Radioactive emissions from coal-fired power stations. *CEGB Newsl.*, No. 114, 8 pp; *Fuel Energy Abstr.*, No. 83-01/24-0089 (1983)

Roelandts, I., Robaye, G., Weber, G. and Delbrouck-Habaru, J.M. (1986) The application of proton-induced gamma-ray emission (PIGE) analysis to the rapid determination of fluorine in geological materials. *Chem. Geol.*, **54**, 35–42

Roga, B., Ihnatowicz, A., Weclewska, M. and Ihnatowicz, M. (1958) Research on boron in Polish coals. *Pr. Gl. Inst. Gorn. Ser. B Komun.*, No. 212, 3–7

Ronov, A.B., Bredanova, N.V. and Migdisov, A.A. (1988) General compositional-evolutionary trends in continental-crust sedimentary and magmatic rocks. *Geochem. Int.*, **25**(9), 27–42

Rook, H.L. (1972) Rapid, quantitative separation for the determination of selenium using neutron activation analysis. *Anal. Chem.*, **44**, 1276–1278

Roscoe, B.A., Chen, C-Y. and Hopke, P.K. (1984) Comparison of the target transformation factor analysis of coal composition data with X-ray diffraction analysis. *Anal. Chim. Acta*, **160**, 121–134

Rösler, H.J., Beuge, P., Schrön, W., Hahne, K. and Bräutigam, S. (1977) Inorganic components of lignites and their significance in lignite exploration. *Freiberg. Forschungsh.*, C **331**, 53–70; *CA*, **89**, 27290 (1978)

Rosner, G., Bunzl, K., Hötzl, H. and Winkler, R. (1984) Low level measurements of natural radionuclides in soil samples around a coal-fired power plant. *Nucl. Instrum. Methods in Phys. Res.*, **223**, 585–589

Rowe, J.J. and Steinnes, E. (1976) Instrumental activation analysis of coal and fly ash with thermal and epithermal neutrons. *Proc. Int. Conf. Modern Trends in Activation Analysis*, Akad. Kiado, Budapest, **1**, 529–538

Rowe, J.J. and Steinnes, E. (1977a) Instrumental activation analysis of coal and fly ash with thermal and epithermal neutrons. *J. Radioanal. Chem.*, **37**, 849–856

Rowe, J.J. and Steinnes, E. (1977b) Determination of 30 elements in coal and fly ash by thermal and epithermal neutron-activation analysis. *Talanta*, **24**, 433–439

Rozkowska, A., Orlowska, D. and Zyczkowska, A. (1982) Determination of arsenic in the ash of hard coal by X-ray fluorescence. *Tech. Poszukiwan Geol.*, **21**(5), 39–41

Ruch, R.R., Gluskoter, H.J. and Kennedy, E.J. (1971) Mercury content of Illinois coals. *Ill. State Geol. Surv. Environ. Geol. Notes*, No. 43, 15 pp

Ruch, R.R., Gluskoter, H.J. and Shimp, N.F. (1973) Occurrence and distribution of potentially volatile trace elements in coal: an interim report. *Ill. Geol. Surv. Geol. Note*, No. 61, 43 pp

Ruch, R.R., Gluskoter, H.J. and Shimp, N.F. (1974) Occurrence and distribution of potentially volatile trace elements in coal. *Environ. Geol. Notes Ill. State Geol. Surv.*, No. 72, 96 pp

Ruch, R.R., Cahill, R.A., Frost, J.K., Camp, L.R. and Gluskoter, H.J. (1977) Survey of trace elements in coals and coal-related materials by neutron activation analysis. *J. Radioanal. Chem.*, **38**, 415–424

Rühling, A. and Tyler, G. (1970) Sorption and retention of heavy metals in the woodland moss *Hylocomium splendens* (Hedwv). *Oikos*, **21**, 92–97

Rummery, R.A. and Howes, K.M.W. (eds.) (1978) Management of lands affected by mining. *Proc. Workshop, CSIRO Div. Land Resources*, 172 pp

Rusanov, A.K. and Bodunkov, B.I. (1940) Spectral analysis of solutions and minerals. VIII Direct determination of germanium in coal ash. *Zavod. Lab.*, **9**, 183–186; *CA*, **34**, 5624 (1940)

Russell, S.J. (1977) Characterisation by scanning electron microscopy of mineral matter in residues of coal liquefaction. *Scanning Electron Microsc.*, **1**, 95–100

Russell, S.J. and Rimmer, S.M. (1979) Analysis of mineral matter in coal, coal gasification ash, and coal liquefaction residues by scanning electron microscopy and Xpray diffraction. In Karr (1979), pp. 133–162

Rust, G.W. (1931) Colloidal primary copper ores at Cornvall Mines, Southwestern Missouri. *J. Geol.*, **43**, 398–426

Ryabchenko, S.N. (1968) Behaviour of germanium during the oxidation of coals. *Protsessy Term. Prevrashch. Kamennykh Uglei* (ed. V.I. Alekhina), pp. 370–375, Izd. Nauka Sib. Otd. Novosibirsk; *CA*, **71**, 23556 (1969)

Ryabchenko, S.N. and Bochkareva, K.I. (1968) Distribution of germanium through a coal substance in relation to its petrographic composition and oxidation state. *Protsessy Term. Prevrashch. Kamennykh Uglei* (ed. V.I. Alekhina), pp. 335-344, Izd. Nauka Sib. Otd. Novosibirsk; *CA*, **71**, 23557 (1969)

Ryabchenko, S.N. and Lisin, D.M. (1965) Chemical character of germanium compounds in the Kuzbass coals and its changes as the functions of coal petrographic composition, metamorphic grade and degree of oxidation. *Mater. Soveshch. Rab. Lab. Geol. Organ.*, 9th, Kiev, pp. 110-115; *CA*, **66**, 117696 (1967)

Ryabchenko, S.N., Alekhina, V.I. and Lisin, D.M. (1968) Nature of germanium compounds in coals. *Protsessy Term. Prevrashch. Kamennykh Uglei* (ed. V.I. Alekhina), pp. 376-382, Izd. Nauka Sib. Otd. Novosibirsk; *CA*, **71**, 23549 (1969)

Ryabchenko, S.N., Grishaeva, V.N. and Lefeld, L.G. (1968) Germanium content in products of the thermal dissolution of coals. *Protsessy Term. Prevrashch. Kamennykh Uglei* (ed. V.I. Alekhina), pp. 345-350, Izd. Nauka Sib. Otd. Novosibirsk; *CA*, **71**, 23552 (1969)

Ryczek, M. et al (1970) Metallic gallium from gaseous products of burning of coal or brown coal. Pol. 60, 127 (Cl.C22b), 31 Jul. 1970, 3 pp.; *CA*, **74**, 89967 (1971)

Ryer, T.A. and Langer, A.W. (1980) Thickness change involved in the peat-to-coal transformation for a bituminous coal of Cretaceous age in Central Utah. *J. Sediment. Petrol.*, **50**, 987-992

Ryer, T.A., Phillips, R.E., Bohor, B.F. and Pollastro, R.M. (1980) Use of altered volcanic ash falls in stratigraphic studies of coal-bearing sequences: an example from the Upper Cretaceous Ferron Sandstone Member of the Mancos shale in Central Utah. *Geol. Soc. Am. Bull.*, **91** (10, Part 1), 579-586

Sabbioni, E., Pietra, R. and Gaglione, P. (1982) Long-term occupational risk of rare-earth pneumoconiosis – a case report as investigated by neutron activation analysis. *Sci. Total Environ.*, **26**, 19-32

Sabbioni, E., Goetz, L., Springer, A. and Pietra, R. (1983) Trace metals from coal-fired power plants: derivation of an average data base for assessment studies of the situation in the European Communities. *Sci. Total Environ.*, **29**, 213-227

SABS (1977a) Code of practice for the sampling of coal and preparation of a sample for analysis. Part 1 The sampling of coal. *S. Afr. Bur. Stand.*, SABS-0135-1977-Pt-1, 99 pp

SABS (1977b) Code of practice for the sampling of coal and preparation of a sample for analysis. Part 2 Preparation of a sample for analysis. *S. Afr. Bur. Stand.*, SABS-0135-1977-Pt-2, 58 pp

Safonov, S.I. (1940) The scattering method in spectral analysis. *Zavod. Lab.*, **9**, 187-188; *CA*, **36**, 6937 (1942)

Sager, M. (1984) Determination of germanium in rocks and ores by distillation and spectrophotometry. *Mikrochim. Acta*, **2**, 381-388

Sager, M. (1987) Rapid determination of fluorine in solid samples. *Monatsh. Chem.*, **118**, 25-29

Salmi, M. (1955) Prospecting for bog-covered ore by means of peat investigations. *Bull. Comm. Geol. Finl.*, No. 169, 34 pp

Salmon, L., Toureau, A.E.R. and Lally, A.E. (1984) The radioactivity content of United Kingdom coal. *Sci. Total Environ.*, **35**, 403-415

Samuel, J.O. (1933) Spectroscopy as an aid to coal seam examination and identification. *Chem. Ind. (London)*, **52**, 155-156

Sanchez Gomez, M.L. and Ramos Martin, M.C. (1987) Application of cluster analysis to identify sources of airborne particles. *Atmos. Environ.*, **21**, 1521-1527

Sanzolone, R.F. and Chao, T.T. (1981) Determination of sub-microgram amounts of selenium in geological materials by atomic absorption spectrophotometry with electrothermal atomisation after solvent extraction. *Analyst (London)*, **106**, 647-652

Saprykin, F.Y. (1965) Forms of germanium occurrence in coals. *Mater. Soveshch. Rab. Lab. Geol. Organ.*, 9th, Kiev, pp. 103-109; *CA*, **66**, 117695 (1967)

Saprykin, F.Y. and Kulachkova, A.F. (1975) Role of natural organic substances during migration and concentration of trace elements. *Tr. Vses. Nauchno-Issled. Geol. Inst.*, **241**, 77-89

Saprykin, F.Y., Kler, V.R. and Kulachkova, A.F. (1970) Geochemical characteristics of rare element concentration in various facies types of coal- and bituminous shale-bearing formations. *Uglenosn. Formatsii Ikh. Genezis*, pp. 89-93; *CA*, **75**, 79079 (1971)

Sato, K. and Sakata, M. (1985) Multielement determination of coal ash in inductively coupled plasma-atomic emission spectrometry. *Bunseki Kagaku*, **34**, 271-275; *CA*, **103**, 39539 (1985)

Satoh, K. (1984) The simultaneous analysis of many elements in coal ash. *Denryoku Chuo Kenkyujo Enerugi, Kenkyo Gijutsu Kenkyujo Kenkyu Hokoku* (283051), pp. 1–31; *IEA Coal Abstr.*, **9**(5), 3487 (1985)

Saunders, K.G. (1980) Microstructural studies of chlorine in some British coals. *J. Inst. Energy.*, **53**, 109–115

Savelev, V.F. and Timofeev, N.I. (1977) Selenium in coal fields. *Sb. Nauchn. Tr. Tashkent. Un-t* (530), pp. 101–106; *CA*, **92**, 183353 (1980)

Savul, M. and Ababi, V. (1958) The copper, zinc and lead content of several types of Romanian coal. *Acad. Repub. Pop. Rom. Fil. Iasi, Stud. Cercet, Stiint., Chim.*, **2**, 251–269

Saxby, J.D. (1969) Metal-organic chemistry of the geochemical cycle. *Rev. Pure Appl. Chem.*, **19**, 131–150

Saxby, J.D. (1973) Diagenesis of metal-organic complexes in sediments: formation of metal sulphides from cystine complexes. *Chem. Geol.*, **12**, 241–248

Schafer, H.N.S. (1984) Determination of carboxyl groups in low-rank coals. *Fuel*, **63**, 723–726

Schobert, H.H., Compiler (1982) Low-rank coal basic coal science workshop. *Proc. CONF-811268 US Dept. Energy*, DOE, Washington, DC, 284 pp

Schobert, H.H. (ed.) (1984) *The Chemistry of Low-Rank Coals.* ACS Symp. Ser. No. 264, 315 pp

Schobert, H.H. (1987) *Coal The Energy Source of the Past and Future.* Washington, DC: ACS, 298 pp

Schobert, H.H., Benson, S.A., Jones, M.L. and Karner, F.R. (1981) Studies in the characterisation of United States low-rank coals. *Proc. Int. Conf. Coal Sci.*, Verlag Glückauf GmbH, Essen, pp, 10–15

Schobert, H.H., Karner, F.R., Olson, E.S., Kleesattel, D.P. and Zygirlocke, C.J. (1987) New approaches to the characterisation of lignites: a combined geological and chemical study. In Volborth (1987), pp. 355–380

Schofield, A. and Haskin, L. (1964) Rare-earth distribution patterns in eight terrestrial materials. *Geochim. Cosmochim. Acta*, **28**, 437–446

Scholz, A. (1987) The occurrence of trace elements in bituminous coals. *Glueckauf-Forschungsh.*, **48**(2), 99–103

Scholz, A., Frigge, J., Hermann, P., Rathleff, D. and Schwarz, G. (1985) Analysis for trace elements. *Staub-Reinhalt. Luft*, **45**, 467–472

Schönbein, C.F. (1838) On the cause of colour changes which many substances undergo through the effect of heat. *Poggendorff's Ann. Phys. Chem.*, **45**, 263

Schopf, J.M. (1948) Variable coalification: the processes involved in coal formation. *Econ. Geol.*, **43**, 207–225

Schroeder, W.H., Dobson, M., Kane, D.M. and Johnson, N.D. (1987) Toxic trace elements associated with airborne particulate matter: a review. *J. Air Pollut. Control Assoc.*, **37**, 1267–1285

Schultz, H., Hattman, E.A. and Booher, W.B. (1973) The fate of some trace elements during coal pretreatment and combustion. *Am. Chem. Soc. Div., Fuel Chem. Prepr.*, **18**(4), 108–113

Scott, A.C. (ed.) (1987) *Coal and Coal-bearing Strata: Recent Advances. Geol. Soc. Spec. Publ.*, No. 32, Oxford: Blackwell Scientific, 332 pp

Scott, R.O. (1954) In Goldschmidt (1954), p. 431

Scott, R.O. and Swaine, D.J. (1959) Further comments on the vanadium-calcium line coincidence at 3185Å. *Geochim. Cosmochim. Acta*, **16**, 195–196

Scott, R.O. and Ure, A.M. (1972) Some sources of contamination in trace analysis. *Soc. Anal. Chem. Proc.*, **9**, 288–293

Seidl, J. and Stamberg, J. (1960) A new type of selective ion exchanger. *Chem. Ind. (London)*, pp. 1190–1191

Senesi, N., Sposito, G. and Martin, J.P. (1986) Copper (II) and iron (III) complexation by soil humic acids: an IR and ESR study. *Sci. Total Environ.*, **55**, 351–362

Senftle, F.E., Thorpe, A.N., Alexander, C.C. and Finkelman, R.B. (1982) Ferromagnetic and superparamagnetic contamination in pulverised coal. *Fuel*, **61**, 81–86

Sen Gupta, J.G. (1982) Determination of scandium and lanthanum in silicate rocks and coal with a simplified separation procedure and atomic absorption spectrometry. *Anal. Chim. Acta*, **138**, 295–302

Severson, R.C. and Gough, L.P. (1983) Boron in mine soils and rehabilitation plant species at selected surface coal mines in Western United States. *J. Environ. Qual.*, **12**, 142–146

Shabad, T. (1986) News notes. *Int. Geol. Rev.*, **28**, 372–375
Shacklette, H.T. and Boerngen, J.G. (1984) Element concentrations in soils and other surficial materials of the conterminous United States. *US Geol. Surv. Prof. Pap.*, No. 1270, 105 pp
Shacklette, H.T., Boerngen, J.G. and Keith, J.R. (1974) Selenium, fluorine and arsenic in surficial materials of the conterminous United States. *US Geol. Surv. Circ.*, No. 692, 14 pp
Shan, X., Yuan, S. and Ni, Z. (1985) Determination of gallium in sediment, coal, coal fly ash and botanical samples by graphite furnace atomic absorption spectrometry using nickel matrix modification. *Anal. Chem.*, **57**, 857–861
Shan, X., Yuan, S. and Ni, Z. (1986) Determination of thallium in river sediment, coal, coal fly ash and botanical samples by graphite furnace atomic absorption spectrometry. *Can. J. Spectrosc.*, **31**, 35–39
Sharkey, A.G., Kessler, T. and Friedel, R.A. (1975) Trace elements in coal dust by spark-source mass spectrometry. In Babu (1975), pp. 48–56
Sharkey, A.G., Shultz, J.L. and Friedel, R.A. (1963) Advances in coal spectrometry. Mass spectrometry. *US Bur. Mines Rep. Invest.*, No. 6318, 32 pp
Sharova, A.K., Gertman, E.M. and Semerneva, G.A. (1973) Nature of the bonding of germanium with humic acids of coals. *Khim. Tverd. Topl.* (2), 58–62; *CA*, **79**, 21475 (1973)
Shaw, D.M. (1958) A vanadium-calcium spectral line coincidence at 3185Å and its effect on vanadium abundance data. *Geochim. Cosmochim. Acta*, **15**, 159–161
Sheibley, D.W. (1975) Trace elements by instrumental neutron activation analysis for pollution monitoring. In Babu (1975), pp. 98–117
Shibaoka, M. (1971) A geochemical study of the inorganic constituents of the Bowen Basin and other Queensland coals. *Proc. Second Bowen Basin Symp., Geol. Surv. Queensl. Rep.*, Geol. Surv. Queensland, Brisbane, No. 62, 49–60
Shibaoka, M. and Bennett, A.J.R. (1975) Geological interpretation of ply structure of the Bulli seam, Sydney Basin, New South Wales. *J. Geol. Soc. Aust.*, **22**, 327–343
Shibaoka, M. and Smyth, M. (1975) Coal petrology and the formation of coal seams in some Australian sedimentary basins. *Econ. Geol.*, **70**, 1463–1473
Shinn, J.H. (1985) The structure of coal and its liquefaction products: a reactive model. *Proc. Int. Conf. Coal Sci.*, Pergamon, Sydney, NSW, pp. 738–740
Shotyk, W. (1988) Review of the inorganic geochemistry of peats and peatland waters. *Earth-Sci. Rev.*, **25**, 95–176
Shpirt, M.Y. and Sendulskaya, T.I. (1969) Distribution of germanium and types of germanium compounds in solid fuel. *Khim. Tverd. Topl.* (2), 3–11; *CA*, **71**, 41083 (1969)
Shpirt, M.Y., Ratynskii, V.M., Zharov, Y.N. and Zekel, L.A. (1984) Forms of trace element compounds and their behaviour in processing coals. *Razvit. Uglekhim. 50 Let*, pp. 224–235; *CA*, **103**, 198277 (1985)
Shpirt, M.Y., Ruban, V.A., Itkin, Y.V. and Zekel, L.A. (1984) Mineral components in the complex utilisation of coals and carbonaceous rocks. *Solid Fuel Chem.*, **18**(6), 118–123
Sim, P.G. and Lewin, J.F. (1975) Potentially toxic metals in New Zealand coals. *NZ J. Sci.*, **18**, 635–641
Simms, P.C., Rickey, P.A. and Mueller, K.A. (1977) Multielemental analysis using proton induced photon emission. *Am. Chem. Soc. Div., Fuel Chem. Prepr.*, **22**(5), 49–54
Simon, L. and Hally, J. (1984) Rapid determination of arsenic in brown coal of the North Bohemian brown coal district. *Acta Mont.*, **68**, 253–263; *CA*, **103**, 163069 (1985)
Simpson, E.S. (1912) Rare metals and their distribution in Western Australia. *J. Nat. Hist. Sci. Soc. West. Aust.*, **4**, 83–108
Singh, J.J. and Deepak, A. (eds.) (1980) *Environmental and Climatic Impact of Coal Utilization*. New York: Academic, 655 pp
Singh, M.P. (1987) On the diversity in distribution of mineral matter in the macroscopic ingredients of Lower Gondwana coals of India: a SEM study. *Proc. Int. Conf. Coal Sci.*, Elsevier, Amsterdam, pp. 93–101
Singh, R.M., Singh, M.P. and Chandra, D. (1983) Occurrence, distribution and probable source of the trace elements in Ghugas coals, Wardha Valley, districts Chandrapur and Yeotmal, Maharashtra, India. *Int. J. Coal Geol.*, **2**, 371–381
Sinnatt, F.S. and Baragwanath, G.E. (1938) The hydrogenation of Victorian brown coals. *Rep. Fuel Res. Station*, Vol. III, 349–503
Sinnatt, F.S., Grounds, A. and Bayley, F. (1921) Origin and nature of the white partings in coal seams as illustrated by the coals of Lancashire, England. *Coal Age*, **20**, 93–95
Skalska, S. and Held, S. (1956) Determination of trace amounts of boron in graphite, coal,

coke, and carbon black by the method of spectral-emission analysis. *Chem. Anal.*, **1**, 294–300; *CA*, **51**, 7688 (1957)

Skowronek, J. (1986) Natural radioactivity of coals from the Rybnik-Jastrzebie and Zabrze coal association regions. *Przegl. Gorn.*, **42**, 87–92; *CA*, **105**, 155872 (1986)

Slack, A.V. (1981) Control of pollution from combustion processes. In Elliott (1981), pp. 1447–1490

Smales, A.A. and Salmon, L. (1955) Determination by radioactivation of small amounts of rubidium and caesium in sea water and related materials of geochemical interest. *Analyst (London)*, **80**, 37–50

Smirnov, B.I. (1969a) Accumulation of accessory trace elements by coals of the Il'nitsk deposit. *Geol. Geokhim. Goryuch, Iskop.*, No. 18, 71–76

Smirnov, B.I. (1969b) Forms of chemical element occurrence in lignites of the Began deposit (Transcarpathians). *Izv. Vyssh. Uchebn. Zaved., Geol. Razved.*, **12**(9), 72–75; *CA*, **72**, 46033 (1970)

Smith, A.C. (1958) The determination of trace elements in pulverized-fuel ash. *J. Appl. Chem.*, **8**, 636–645

Smith, G.A. (1941a) Phosphorus in coal: its distribution and modes of occurrence. *J. Chem. Metall. Min. Soc. S. Afr.*, **42**, 102–112

Smith, G.A. (1941b) The minor constituents of South Africa coals. *J. Chem. Met. Mining Soc. S. Afr.*, **42**, 1–38; *CA*, **36**, 3928 (1942)

Smith, I. (1987) Trace elements from coal combustion: emissions. *IEA Coal Res. Rep.*, No. IEACR/01, 87 pp

Smith, I.C. and Carson, B.L. (1977a) *Trace Metals in the Environment*. Volume 1–Thallium. Ann Arbor: Ann Arbor Science, 394 pp

Smith, I.C. and Carson, B.L. (1977b) *Trace Metals in the Environment*. Volume 2–Silver. Ann Arbor: Ann Arbor Science, 469 pp

Smith, I.C. and Carson, B.L. (1978) *Trace Metals in the Environment*. Volume 3–Zirconium. Ann Arbor: Ann Arbor Science, 405 pp

Smith, I.C. and Carson, B.L. (eds.) (1981) *Trace Metals in the Environment*. Volume 6–Cobalt. Ann Arbor: Ann Arbor Science, 1202 pp

Smith, I.C., Carson, B.L. and Ferguson, T.L. (1978) *Trace Metals in the Environment*. Volume 4–Palladium and Osmium. Ann Arbor: Ann Arbor Science, 193 pp

Smith, I.C., Carson, B.L. and Hoffmeister, F. (1978) *Trace Metals in the Environment*. Volume 5–Indium. Ann Arbor: Ann Arbor Science, 552 pp

Smith, J.W. and Batts, B.D. (1974) The distribution and isotopic composition of sulfur in coal. *Geochim. Cosmochim. Acta*, **38**, 121–133

Smith, J.W., Gould, K.W. and Rigby, D. (1982) The stable isotope geochemistry of Australian coals. *Org. Geochem.*, **3**, 111–131

Smith, R.D. (1980) The trace element chemistry of coal during combustion and the emissions from coal-fired plants. *Prog. Energy Combust. Sci.*, **6**, 53–119

Smith-Briggs, J.L. (1984) Natural radionuclides near a coal-fired power station. *Nucl. Instrum. Methods in Phys. Res.*, **223**, 590–592

Smyth, M. (1966) The association of minerals with macerals and microlithotypes in some Australian coals. *CSIRO Div. Coal Res. Tech. Commun.*, No. 48, 35 pp

Smyth, M. (1967) Coal petrology applied to the correlation of some New South Wales Permian coals. *Australas. Inst. Min. Metall.*, **221**, 11–18

Smyth, M. (1968) The petrography of some New South Wales Permian coals from the Tomago Coal Measures. *Australas. Inst. Min. Metall. Proc.*, No. 225, 1–9

Smyth, M. (1970) Type seam sequences for some Permian Australian coals. *Australas. Inst. Min. Metall. Proc.*, No. 233, 7–16

Snelling, A.A. and Mackay, J.B. (1985) The role of volcanism in the rapid formation of coal seams. *Proc. Inst. Conf. Coal Sci.*, p. 641

Sofiyev, L.S., Gorlenko, K.A. and Semasheva, I.N. (1964) The nature of bonding of germanium in coal. *Geochem. Int.* (2), 297–300

Sokolov, N.E. and Vidishev, V.E. (1958) A few data on the mineral impurities and coal layers of the Karaganda Basin. *Tr. Karagandinst. Gorn. Inst.* (1), 230–236; *CA*, **54**, 17166 (1960)

Somer, G., Cakir, O. and Solak, A.O. (1984) Differential-pulse polarographic determination of trace heavy elements in coal samples. *Analyst (London)*, **109**, 135–137

Soong, R. and Berrow, M.L. (1979) Mineral matter in some New Zealand coals. 2. Major and trace elements in some New Zealand coal ashes. *NZ J. Sci.*, **22**, 229–233

Soong, R. and Gluskoter, H.J. (1977) Mineral matter in some New Zealand coals. 1. Low-temperature ashing and mineralogical composition of such coal ashes. *NZ J. Sci.*, **20**, 273–277
Soong, R., Godbeer, W.C. and Swaine, D.J. (1984) The determination of fluorine in some New Zealand coals by the fluorine ion selective electrode method. *NZ J. Sci.*, **27**, 151–154
Spackman, W., Ryan, N.J., Rhoads, C.A. and Given, P.H. (1988) Studies of peat as the input to coalification. II Sampling sites and preliminary fractionation. *Int. J. Coal Geol.*, **9**, 253–265
Spears, D.A. and Caswell, S.A. (1986) Mineral matter in coals: cleat minerals and their origin in some coals from the English Midlands. *Int. J. Coal Geol.*, **6**, 107–125
Spedding, P.J. (1988) Peat. *Fuel*, **67**, 883–900
Spencer, L.J. (1910) On the occurrence of alstonite and ullmanite (a species new to Britain) in a barytes-witherite vein in New Brancepeth colliery, near Durham. *Mineral. Mag.*, **15**, 302–311
Spielholtz, G.I., Toralballa, G.C. and Steinberg, R.J. (1971) Determination of arsenic in coal and in insecticides by atomic absorption spectroscopy. *Mikrochim. Acta* (6), 918–923
Spiro, C.L., Wong, J., Lytle, F.W., Greegor, R.B., Maylotte, D.H. and Lamson, S.H. (1984) X-ray absorption spectroscopic investigation of sulfur sites in coal: organic sulfur identification. *Science*, **226**, 48–50
Sprunk, G.C. and O'Donnell, H.J. (1942) Mineral matter in coal. *US Bur. Mines Tech. Pap.*, No. 648, 67 pp
Stach, E., Mackowsky, M.-Th., Teichmüller, M., Taylor, G.H., Chandra, D. and Teichmüller, R. (1982) *Stach's Textbook of Coal Petrology*. Berlin: Gebrüder Borntraeger, 3rd ed, 535 pp
Stadnichenko, T. (1953) Accumulation of minor elements in coal ash and its economic implications. *Econ. Geol.*, **48**, 332
Stadnichenko, T., Murata, K.J. and Axelrod, J.M. (1950) Germaniferous lignite from the District of Columbia and vicinity. *Science*, **112**, 109
Stadnichenko, T., Zubovic, P. and Sheffey, N.B. (1961) Beryllium content of American coals. *US Geol. Surv. Bull.*, No. 1084-K, 253–295
Stadnichenko, T., Murata, K.J., Zubovic, P. and Hufschmidt, E.L. (1953) Concentration of germanium in the ash of American coals, a progress report. *US Geol. Surv. Circ.*, No. 272, 34 pp
Stanton, R.W., Cecil, C.B., Pierce, B.S., Ruppert, L.F. and Dulong, F.T. (1986) Geologic processes affecting the quality of the Upper Freeport coal bed, west-central Pennsylvania. *US Geol. Surv. Open-File Rep.*, No. 86-173, 22 pp
Staub, J.R. and Cohen, A.D. (1978) Kaolinite enrichment beneath coals; a modern analog, Snuggedy Swamp, South Carolina. *J. Sediment. Petrol.*, **48**, 203–210
St.-Claire Deville, C. and Leblanc, F. (1857) The chemical composition of the gases emitted by the volcanic vents of southern Italy. *C.R. Acad. Sci.*, **44**, 769–773; CIEB (1956), p. 140
Steinmetz, G.L., Mohan, M.S. and Zingaro, R.A. (1988) Characterisation of titanium in United States coals. *Energy Fuels*, **2**, 684–692
Steinnes, E. (1979) Instrumental activation analysis of coal and coal ash with thermal and epithermal neutrons. In Karr (1979), pp. 279–302
Steven, W. and Balajiva, K. (1959) The influence of minor elements on the isothermal embrittlement of steels. *J. Iron Steel Inst. London*, **193** (Part 2), 141–147
Stevenson, F.J. (1983) Trace metal-organic matter interactions in geologic environments. In Augustithis (1983), pp. 671–691
Stevenson, F.J. (1985) Geochemistry of soil humic substances. In Aiken *et al.* (1985), pp. 13–52
Stock, A. and Cucuel, F. (1934) The occurrence of mercury. *Naturwissenschaften*, **22**, 390–393
Stoklasa, J. (1927) The distribution of iodine in nature and its physiological significance in plant and animal organisms. *Z. Angew. Chem.*, **40**, 20–27; CIEB (1956), 87
Stolper, W. (1956) Investigation of the iodine content of solid fuels and their products with reference to iodine extraction. *Bergakademie*, **8**, 587–588; *Fuel Abstr.*, **22**, 5986 (1957)
Stone, R.W. (1912) Coal near the Black Hills, Wyoming-South Dakota. *US Geol. Surv. Bull.*, No. 499, 66 pp
Störr, M., Adolphi, P. and Kasbaum, E. (1987) Clay minerals in low-temperature lignite ash. *Int. Geol. Rev.*, **29**, 1360–1365
Strakhov, N.M., Zalmanzon, E.S. and Glagoleva, M.A. (1956) Types of distribution of

dispersed amounts of elements in sediments of the humid zones. *Geochemistry (USSR)*, No. 6, 560-569

Straszheim, W.E., Yousling, J.G., Younkin, K.A. and Markuszewski, R. (1988) Mineralogical characterisation of lower rank coals by SEM-based automated image analysis and energy-dispersive X-ray spectrometry. *Fuel*, **67**, 1042-1047

Straughan, I.R., Elseewi, A.A., Page, A.L., Kaplan, I.R., Hurst, R.W. and Davis, T.E. (1981) Fly-ash derived strontium as an index to monitor deposition from coal-fired power plants. *Science*, **212**, 1267-1269

Strehlow, R.A., Harris, L.A. and Yust, C.S. (1978) Submicron-sized mineral component of vitrinite. *Fuel*, **57**, 185-186

Stricker, G.D. (1974) Areal and stratigraphic distribution of trace elements in the Pennsylvanian coals of the Illinois Basin. *Geol. Soc. Am. Abstr. Annu. Meet.*, pp. 974-975

Strock, L.W. (1936) *Spectrum Analysis with the Carbon Arc Cathode Layer ('Glimmschicht')*. London: Hilger, 56 pp

Strock, L.W. (1957) Emission spectrochemical analysis-basic principles and applications. In *Trace Analysis* (eds., J.H. Yoe and H.J. Koch), pp. 346-397, New York: Wiley

Stutzer, O., Revised Noé, A.C. (1940) *Geology of Coal*. Chicago: University of Chicago Press, 461 pp

Styron, C.E., Bishop, C.T., Casella, V.R., Jenkins, P.H. and Yanko, W.H. (1981) Assessment of the impact of radionuclides in coal ash. *Rep. MLM*-2829(OP), 15 pp

Suhr, N.H. and Gong. H. (1977) Comparison of atomic absorption and D.C. plasma-arc spectrometry in multi-element analysis of coal. *Am. Chem. Soc., Div. Fuel Chem., Prepr.*, **22**(5), 56-59

Suhr, N.H. and Ingamells, C.O. (1966) Solution technique for analysis of silicates. *Anal. Chem.*, **38**, 730-734

Sulcek, Z. and Povondra, P. (1989) *Methods of Decomposition in Inorganic Analysis*. Baton Rouge: CRC, 325 pp

Sulcek, Z., Povondra, P. and Dolezal, J. (1977) Decomposition procedures in inorganic analysis. *Crit. Rev. Anal. Chem.*, **6**, 255-323

Sun, J. and Jervis, R.E. (1986) Concentrations and distributions of trace and minor elements in Chinese and Canadian coals and ashes. *Proc. 7th Int. Conf. Modern Trends Activation Anal.*, pp. 1055-1062

Svoboda, J. (1958) The industrial poisoning of bees. *Int. Beekeep. Congr. (Pathol.) Rep.*, No. 17, 79-81

Swain, F.M., Blumentals, A. and Millers, R. (1959) Stratigraphic distribution of amino acids in peats from Cedar Creek bog, Minnesota, and Dismal Swamp, Virginia. *Limnol. Oceanogr.*, **4**, 119-127

Swaine, D.J. (1951) The Distribution of Trace Elements in Soils. Ph.D. Thesis, University of Aberdeen, 265 pp

Swaine, D.J. (1955) The trace-element content of soils. *Commonw. Bur. Soil Sci. Tech. Commun.*, No. 48, 157 pp

Swaine, D.J. (1957) The trace-element content of some soils and rock from Macquarie Island, South Pacific Ocean. *Aust. Nat. Antarct. Res. Exped. Sci. Rep. Ser. AIII*, 10 pp

Swaine, D.J. (1960) Trace elements. In The soils of the country around Kelso and Lauder (Sheets 25 and 26), *Mem. Soil Surv. G.B.: Scotl.*, pp. 146-148

Swaine, D.J. (1961) Occurrence of nickel in coal and dirt samples from the Coal Cliff colliery (Bulli seam). *Aust. J. Sci.*, **23**, 301-302

Swaine, D.J. (1962a) Trace elements in coal. II Origin, mode of occurrence and economic importance. *CSIRO Div. Coal Res. Tech. Commun.*, No. 45, 21-109

Swaine, D.J. (1962b) Boron in New South Wales Permian coals. *Aust. J. Sci.*, **25**, 265-266

Swaine, D.J. (1962c) The trace-element content of fertilizers. *Commonw. Bur. Soils Tech. Commun.*, No. 52, 306 pp

Swaine, D.J. (1963a) The spectrographic determination of molybdenum using the line at 3170Å. *Spectrochim. Acta.*, **19**, 841-842

Swaine, D.J. (1963b) Applications of emission spectroscopy in coal research. *Abstr. Fourth Aust. Spectros. Conf.*, Academy of Science, Canberra, A14

Swaine, D.J. (1964) Scandium in Australian coals and related materials. *Am. Chem. Soc., Div. Fuel Chem., Prepr.*, **8**(3), 172-177

Swaine, D.J. (1967) Inorganic constituents in Australian coals. *Mitt. Naturforsch. Ges. Bern*, **NF24**, 49-61

Swaine, D.J. (1968) The laser as a source of excitation in emission spectroscopy. *Hilger J.*, **11**(4), 69–71
Swaine, D.J. (1969a) Trace elements in dirt bands in coal seams. *Abstrs. Symp. Adv. Study Sydney Basin*, University of Newcastle, Newcastle, NSW, p. 46
Swaine, D.J. (1969b) The identification and estimation of carbonaceous materials by DTA. In Thermal Analysis (eds. R.F. Schwenker and P.D. Garn), pp. 1377–1386. New York: Academic
Swaine, D.J. (1971) Boron in coals of the Bowen Basin as an environmental indicator. *Proc. Second Bowen Basin Symp. Geol. Surv. Queensl. Rep.*, No. 62, 41–48
Swaine, D.J. (1975) Trace elements in coals. In *Recent Contributions to Geochemistry and Analytical Chemistry* (ed. A.I. Tugarinov), pp. 539–550, New York: Wiley
Swaine, D.J. (1977a) Trace elements in coal. In *Trace Subst. Environ. Health-XI*, (ed. D.D. Hemphill), pp. 107–116, University of Missouri, Columbia
Swaine, D.J. (1977b) Trace elements in fly ash. *NZ Dep. Sci. Ind. Res. Bull.*, No. 218, 127–131
Swaine, D.J. (1978a) Coal mining: an overview. In Rummery and Howes (1978), pp. 11–17
Swaine, D.J. (1978b) Lead in the environment. *J. Proc. R. Soc. NSW*, **111**, 41–47
Swaine, D.J. (1978c) Selenium: from magma to man. In *Trace Substances in Environ. Health-XII* (ed. D.D. Hemphill), pp. 129–134, University of Missouri, Columbia
Swaine, D.J. (1978d) The fate of trace elements in coal after combustion. *Proc. Int. Clean Air Conf.*, Brisbane, Ann Arbor Science, Ann Arbor, pp. 519–524
Swaine, D.J. (1978e) Trace elements in coal. *Symp. 'Heavy Metals in Coal' R. Swed. Acad. Eng. Sci.*, Stockholm, 20 pp
Swaine, D.J. (1979) Trace elements in Australian bituminous coals and fly-ashes. *Colloq. Combustion of Pulverised Coal: The Effect of Mineral Matter*, University of Newcastle, Newcastle, NSW, W3-14-W3-18
Swaine, D.J. (1980a) Trace-element aspects of coal mining, preparation and storage. In Hannan (1980a), pp. 264–274
Swaine, D.J. (1980b) Nickel in coal and fly ash. In *Nickel in the Environment* (ed. J.O. Nriagu), pp. 67–92. New York: Wiley
Swaine, D.J. (1981a) Trace elements in Australian coals–where, how much and why bother? *Symp. Trace Elements in Coal*, BHP Newcastle
Swaine, D.J. (1981b) Fly-ash for use–not waste. *Proc. 1st Int. Waste Recycling Symp.*, Clean Japan Center, Tokyo, pp. 405–417
Swaine, D.J. (1982) The importance of trace elements in Australian coals. ert Energy News J., **4**(3), 18–22
Swaine, D.J. (1983) Geological aspects of trace elements in coal. In Augustithis (1983), pp. 521–532
Swaine, D.J. (1984a) The fate of trace elements during combustion. *Proc. Aust. Coal Sci. Conf.*, Gippsland Inst. Advanced Educ., Churchill, Vic., pp. 1–10
Swaine, D.J. (1984b) Sulfide minerals in coal with emphasis on Australian occurrences. In *Syngenesis and Epigenesis in the Formation of Mineral Deposits* (eds. A. Wauschkuhn, C. Kluth and R.A. Zimmermann), pp. 120–129. Berlin: Springer-Verlag
Swaine, D.J. (1984c) Variations in trace-element contents in one coal seam. *Abstr. Pap. Int. Chem. Congr. Pacific Basin Soc.*, Honolulu, American Chemical Society, Washington, DC, 07C11
Swaine, D.J. (1985a) Modern methods in bituminous coal analysis: trace elements. *Crit. Rev. Anal. Chem.*, **15**(4), 315–346
Swaine, D.J. (1985b) Trace inorganic constituents in flyash. *Pap. Specialist Workshop Effects of Coal Ash Combustion Systems*, University of Newcastle, Newcastle, NSW, pp. 39–46
Swaine, D.J. (1986a) The rapid weathering of a siltstone. *J. Proc. R. Soc. NSW*, **119**, 83–88
Swaine, D.J. (1986b) Inorganic manganese in coal. *Fuel*, **65**, 1622–1623
Swaine, D.J. (1988) Victor Moritz Goldschmidt's contributions to coal science. *Fuel*, **67**, 877–879
Swaine, D.J. (1989) Trace elements in the Permian coals. *Aust. Bur. Miner. Resourc. Geol. Geophys.*, Bull., No. 231, 297–300
Swaine, D.J. and Mitchell, R.L. (1960) Trace-element distribution in soil profiles. *J. Soil Sci.*, **11**, 347–368
Swaine, D.J. and Taylor, G.F. (1970) Arsenic in phosphatic boiler deposits. *J. Inst. Fuel*, **43**, 261

Swaine, D.J., Godbeer, W.C. and Morgan, N.C. (1983) Use of moss to measure the accession of trace elements to an area around a power station. *Proc. Int. Conf. Heavy Metals Environ.*, Heidelberg 2, CEP Consultants, Edinburgh, pp. 1053–1056

Swaine, D.J., Godbeer, W.C. and Morgan, N.C. (1984a) Environmental consequences of coal combustion. *Rep. NERDDP*, No. EG84/315, 230 pp

Swaine, D.J., Godbeer, W.C. and Morgan, N.C. (1984b) Variations in contents of trace elements in coal from one seam. *Rep. NERDDP*, No. EG84/339, 103 pp

Swaine, D.J., Godbeer, W.C. and Morgan, N.C. (1989) The deposition of trace elements from the atmosphere. In *Trace Elements in New Zealand: Environmental, Human and Animal* (eds. R.G. McLaren, R.J. Haynes and G.P. Savage), pp.1–10. Lincoln: NZ Trace Elements Group

Swanson, V.E. and Huffman, C. (1976) Guidelines for sample collecting and analytical methods used in the U.S. Geological Survey for determining chemical composition of coal. *US Geol. Surv. Circ.*, No. 735, 11 pp

Swanson, V.E., Medlin, J.H., Hatch, J.R., Coleman, S.L., Wood, G.H., Woodruff, S.D. and Hildebrand, R.T. (1976) Collection, chemical analysis, and evaluation of coal samples in 1975. *US Geol. Surv. Open-file Rep.*, No. 76–468, 503 pp

Sweatman, T.R., Norrish, K. and Durie, R.A. (1963) An assessment of X-ray fluorescence spectrometry for the determination of inorganic constituents in brown coals. *CSIRO Misc. Rep.*, No. 177, 43 pp

Syabryai, V.T., Kornienko, T.G. and Kuzminskaya, I.N. (1971) Forms of germanium occurrence in the organic matter of solid caustobioliths. *Dopov. Akad. Nauk Ukr. RSR, Ser. B*, **33**, 23–26; *CA*, **74**, 144707 (1971)

Symonds, R.B., Rose, W.I. and Reed, M.H. (1988) Contribution of Cl-and F-bearing gases to the atmosphere by volcanoes. *Nature (London)*, **334**, 415–418

Symonds, R.B., Rose, W.I., Reed, M.H., Lichte, F.E. and Finnegan, D.L. (1987) Volatilization, transport and sublimation of metallic and non-metallic elements in high temperature gases at Merapi Volcano, Indonesia. *Geochim. Cosmochim. Acta*, **51**, 2083–2101

Szadeczky-Kardoss, E. and Vogl, M. (1955) Geochemical investigations on the ashes of Hungarian coals. *Fold. Kozl.*, **85**, 7–43; *CA*, **51**, 15349 (1957)

Szalay, A. (1957) The role of humus in the geochemical enrichment of uranium in coal and other bioliths. *Acta Phys. Acad. Sci. Hung.*, **8**, 25–35

Szalay, A. (1964) Cation exchange properties of humic acids and their importance in the geochemical enrichment of UO_2»fa»fa and other cations. *Geochim. Cosmochim. Acta*, **28**, 1605–1614

Szalay, A. (1969) Accumulation of uranium and other trace metals in coal and organic shales and the role of humic acids in these geochemical enrichments. *Ark. Mineral. Geol.*, **5**(3), 23–36

Szalay, A. and Szilagyi, M. (1969) Accumulation of microelements in peat humic acids and coal. In *Advances in Organic Geochemistry 1968* (eds. P.A. Schenck and I. Havenaar), pp. 567–578. Oxford: Pergamon Press

Szalay, S. (1954) Enrichment of uranium in some brown coals in Hungary. *Acta Geol. Acad. Sci. Hung.*, **2**, 299–311; *CA*, **48**, 12629 (1954)

Szalay, S. and Almassy, G. (1956) Analytical investigations on the uranium content of Hungarian coals. *Mag. Tud. Akad. Kem. Tud. Oszt. Kozl.*, **8**, 33–38; *CA*, **52**, 7657 (1958)

Szilagyi, M. (1971) The role of organic material in the distribution of Mo, V and Cr in coal fields. *Econ. Geol.*, **66**, 1075–1078

Szonntagh, J., Farady, L. and Janosi, A. (1955) Chromatographic determination of uranium in the ash of Hungarian coals. *Magyar Kem. Foly.*, **61**, 312–314; *CA*, **52**, 10541 (1958)

Takacs, P. and Horvath, A. (1959) Coal utilization by-products as raw materials for germanium and gallium production in Hungary. *Nehezvegyip. Kut. Intez. Kozl.*, **1**, 291–295

Takizawa, Y., Minagawa, K. and Fujii, M. (1981) A practical and simple method in fractional determination of ambient forms of mercury in air. *Chemosphere*, **10**, 801–809

Talmi, Y. and Andren, A.W. (1974) Determination of selenium in environmental samples using gas chromatography with a microwave emission spectrometric detection system. *Anal. Chem.*, **46**, 2122–2126

Tamura, H., Inue, T. and Fudagawa, N. (1986) Determination of trace elements in coals and fly ashes by metal furnace atomic absorption spectrometry. *Shizuoka-ken Kogyo Gijutsu Senta Kenkyu Hokoku* (30), 73–79; *CA*, **105**, 155862 (1986)

Tarpley, E.C. and Ode, W.H. (1959) Evaluation of an acid-extraction method for determining mineral matter in American coals. *US Bur. Mines Rep. Invest.*, No. 5470, 1–15

Taupitz, K. (1984) Uraniferous coals in the Neocene of NE-Macedonia (Greece). *Erzmetall*, **37**, 57–65; *CA*, **100**, 141863 (1984)

Taylor, C.J. (1973) Mercury and other potentially toxic trace elements in British coals. *NCB Yorkshire Regional Lab.*, Rep. No. NCB/YRL/Misc., 1, 21 pp

Taylor, G.F. (1968) Dolomite in New South Wales Permian coals. *Aust. J. Sci.*, **31**, 230–231

Taylor, G.H. (1967) Coals of the Bowen Basin, Eastern Queensland. *CSIRO Div. Min. Chem. Tech. Commun.*, No. 50, 51 pp

Taylor, G.H. and Liu, S.Y. (1987) Biodegradation in coals and other organic-rich rocks. *Fuel*, **66**, 1269–1273

Taylor, G.H. and Warne, S. St J. (1960) Some Australian coal petrological studies and their geological implications. *Proc. Int. Comm. Coal Petrology*, No. 3, pp. 75–87

Taylor, I.S. (1983) Analytical quality assurance in good laboratory practice. *Chem. Aust.*, **50**, 82–86

Taylor, S.R. (1965) Geochemical analysis by spark source mass spectrography. *Geochim. Cosmochim. Acta*, **29**, 1243–1262

Taylor, S.R. (1971) Geochemical applications of spark source mass spectrography. II Photoplate data processing. *Geochim. Cosmochim. Acta*, **35**, 1187–1196

Taylor, S.R. and Gorton, M.P. (1977) Geochemical application of spark source mass spectrography. III Element sensitivity, precision and accuracy. *Geochim. Cosmochim. Acta*, **41**, 1375–1380

Technologist (1959) The minor constituents of coal: their deleterious effects on boiler plant. *S. Afr. Min. Eng. J.*, **70**, Part 2, 1247–1249

Temple, P.J., McLaughlin, D.L., Linzon, S.N. and Wills, R. (1981) Moss bags as monitors of atmospheric deposition. *J. Air Pollut. Contr. Assoc.*, **31**, 668–670

Tennant, W.C. (1967) General spectrographic technique for the determination of volatile trace elements in silicates. *Appl. Spectrosc.*, **21**, 282–285

Ter Meulen, H. (1932) Distribution of molybdenum. *Nature (London)*, **130**, 966

Thiers, R.E. (1957) Contamination in trace element analysis and its control. In *Methods of Biochemical Analysis* (ed. D. Glick), Vol. V, pp. 273–335, New York: Interscience

Thiessen, G. (1945) Composition and origin of the mineral matter in coal. In Lowry (1945), pp. 485–495

Thilo, E. (1934) Results of analysis of two coal ashes. *Z. Anorg. Allg. Chem.*, **218**, 201–209

Thomas, J., Glass, H.D., White, W.A. and Trendel, R.M. (1977) Fluoride content of clay minerals and argillaceous earth materials. *Clays Clay Miner.*, **25**, 278–284

Thomas, J.T. and Gluskoter, H.J. (1974) Determination of fluoride in coal with the fluoride ion-selective electrode. *Anal. Chem.*, **46**, 1321–1323

Thompson, A.P. and Musgrave, J.R. (1953) Germanium, produced as a by-product, has become of primary importance. *Min. Eng. (NY)*, **5**, 42–44

Thompson, G. and Bankston, D.C. (1970) Sample contamination from grinding and sieving determined by emission spectrometry. *Appl. Spectrosc.*, **24**, 210–219

Thompson, M. (1985) The capabilities of inductively coupled plasma atomic-emission spectrometry–some conjectures and refutations. *Analyst (London)*, **110**, 443–449

Thompson, M. and Walsh, J.N. (1983) *A Handbook of Inductively Coupled Plasma Spectrometry*. Glasgow: Blackie, 273 pp

Thorne, L., McCormick, G., Downing, B. and Price, B. (1983) Some aspects of the analysis of coal by X-ray fluorescence spectroscopy. *Fuel*, **62**, 1053–1057

Thürauf, W. and Assenmacher, H. (1963) Photometric determination of zinc in coals and their processed products. *Brennst.-Chem.*, **44**, 21–22

Timofeev, P.P., Bogolyubova, L.I., Miesserova, L.V. and Fedorovskaya, N.P. (1967) Features of boron distribution in Jurassic coals of the Angara-Chulym depression in relation to the paleogeographic conditions of sediment and peat accumulation. *Khim. Tverd. Topl.* (5), 96–102; *CA*, **68**, 61567 (1968)

Tkach, B.I. (1975) On the role of organic matter in concentration of mercury. *Geochem. Int.*, **11**, 973–975

Tölg, G. (1972) Extreme trace analysis of the elements. I Methods and problems of sample treatment, separation and enrichment. *Talanta*, **19**, 1489–1521

Tomza, U. (1987) Trace element content of some Polish hard and brown coals. *J. Radioanal. Nucl. Chem. Lett.*, **119**, 387–396

Tomza, U. and Kaleta, P. (1986) Trace elements in brown coal and its products of combustion. *J. Radioanal. Nucl. Chem. Lett.*, **107**, 1–10

Tramontano, J.M., Scudlark, J.R. and Church, T.M. (1987) A method for the collection, handling, and analysis of trace metals in precipitation. *Environ. Sci. Technol.*, **21**, 749–753

Traves, D.M. and King, D. (eds.) (1975) *Economic Geology of Australia and Papua New Guinea 2. Coal.* Parkville: Australasian Inst. Min. Met., 398 pp

Treibs, A. (1935) Organic mineral substances. V Porphyrins in coals. *Justus Liebigs Ann. Chem.*, **502**, 144–150

Treibs, A. (1936) Chlorophyll and hemin derivatives in organic mineral substances (a review). *Angew. Chem.*, **49**, 682–686

Tripathi, P.S.M. (1979) Neutron activation analysis in coal research. *Erdoel Kohle, Erdgas, Petrochem. Brennast.-Chem.*, **32**, 256–262

Triplehorn, D. and Bohor, B. (1984) Volcanic ash layers in coal: origin, distribution, composition and significance. *Am. Chem. Soc., Div. Fuel Chem., Prepr.*, **29**(4), 48–55

Trudinger, P.A. and Swaine, D.J. (eds.) (1979) *Biogeochemical Cycling of Mineral-Forming Elements.* Amsterdam: Elsevier, 612 pp

Trudinger, P.A., Swaine, D.J. and Skyring, G.W. (1979) Biogeochemical cycling of elements–general considerations. In Trudinger and Swaine (1979), pp. 1–27

Turner, K.E. (1981) Direct determination of trace elements in coal and coal products. Part 2 Computer program package for the calculation of trace element concentrations in coal determined by X-ray fluorescence spectrometry. *Rep. NERDDP*, No. EG81/20, 122 pp

Turner, R.C., Radley, J.M. and Mayneord, W.V. (1958) α-ray activities of humans and their environment. *Nature (London)*, **181**, 518–521

Uchikawa, H., Furuta, R. and Mihara, Y. (1982) Determination of trace mercury in solid fuel by atomic absorption spectrophotometry. *Onoda Kenkyu Hokoku*, **34**, 19–25; *CA*, **100**, 88350 (1984)

Ujhelyi, C. (1955) Determination of extremely small quantities of uranium in coal. *Magyar Kem. Foly.*, **61**, 437–442; *CA*, **52**, 7658 (1958)

Underwood, E.J. (1977) *Trace Elements in Human and Animal Nutrition.* New York: Academic Press, 545 pp

Unfried, E. and Boeck, H. (1982) Natural radioactivity of fossil fuels used in Austria and of their combustion wastes. *Rep. AIAU*-82301, 11 pp

Urbanek, E. (1961) Simultaneous determination of germanium and gallium in coal and fly ash from electricity works. *Paliva*, **41**, 88–91

Uriano, G.A. and Gravatt, C.C. (1977) The role of reference materials and reference methods in chemical analysis. *Crit. Rev. Anal. Chem.*, **6**, 361–411

USGS (1896) *17th Ann. Rep.*, 1895–1896, Part III, pp. 282–283

USGS (1970) Mercury in the environment. *US Geol. Surv. Prof. Pap.*, No. 713, 67 pp

Uzunov, I. (1976) Vanadium in the coal basins in Bulgaria. *Geol. Balc.*, **6**(2), 35–61; *CA*, **85**, 180447 (1976)

Uzunov, I. (1980) Geochemical nature of petrographic components of coal and mechanisms of vanadium concentration in them. *Geol. Balc.*, **10**, 57–74; *CA*, **93**, 222879 (1980)

Uzunov, I. and Karadzhova, B. (1968) Distribution of rare and trace elements in a productive horizon of the Burgas coal basin. *Izv. Geol. Inst. Bulg. Akad. Nauk, Ser. Geokhim., Mineral Petrogr.*, **17**, 21–31; *CA*, **70**, 98822 (1969)

Valeska, F. and Havlova, A. (1959) Quantitative spectrographic determination of titanium and manganese in coal. *Sb. Pr. Horn. Ustavu*, **1**, 182–190

Valeska, F., Malan, O. and Kessler, M.F. (1967) Bonding of some microelements in coal. *Vysledky Banskeho Vysk.* (5), 212–218; *CA*, **72**, 81259 (1970)

Valiev, Y.Y., Pachadzhanov, D.N. and Adamchuk, I.P. (1977) Geochemistry of boron in Jurassic coals of the Gissar Range. *Dokl. Akad. Nauk Tadzh. SSR*, **20**(1), 57–60; *CA*, **87**, 70689 (1977)

Valkovic, V. (1981) *Trace Elements in Coal: A Bibliography.* Zagreb: Inst. 'Ruder Boskovic', 251 pp

Valkovic, V. (1983) *Trace Elements in Coal.* Boca Raton: CRC Press, Vol. I, 210 pp., Vol. II, 281 pp

Valkovic, V., Orlic, I., Makjanic, J., Rendic, D., Miklavzic, U. and Budnar, M. (1984) Comparison of different modes of excitation in X-ray emission spectroscopy in the detection of trace elements in coal and coal ash. *Nucl. Instrum. Methods Phys. Res. Sect. B*, **232**, 127–131

Van der Flier, E. and Fyfe, W.S. (1985) Uranium–thorium systematics of two Canadian coals. *Int. J. Coal Geol.*, **4**, 335–353
Van der Flier-Keller, E. and Fyfe, W.S. (1987a) Geochemistry of two Cretaceous coal-bearing sequences: James Bay lowlands, northern Ontario, and Peace River basin, northeast British Columbia. *Can. J. Earth Sci.*, **24**, 1038–1052
Van der Flier-Keller, E. and Fyfe, W.S. (1987b) Relationships between inorganic constituents and organic matter in a northern Ontario lignite. *Am. Chem. Soc., Div. Fuel Chem., Prepr.*, **32**(1), 41–48
Van der Flier-Keller, E. and Fyfe, W.S. (1988) Relationships between inorganic constituents and organic matter in a northern Ontario lignite. *Fuel*, **67**, 1048–1052
Van der Kraan, A.M., Gerkema, E., and van Loef, J.J. (1983) Investigation into the applicability of Mössbauer spectrometry for a quantitative determination of pyrite in coal and as a control possibility in coal cleaning processes. *Rep. IRI*-132-83-02, 51 pp; *CA*, **100**, 106172 (1984)
Van der Sloot, H.A. (1981) Cluster analysis of coal data: a useful tool? *Proc. Int. Conf. Coal Sci.*, pp. 762–767
Van der Sloot, H.A., Hoede, D., Klinkers, J.L. and Das, H.A. (1982a) The determination of arsenic, selenium and antimony in rocks, sediments, fly ash and slag. *J. Radioanal. Chem.*, **71**, 463–478
Van der Sloot, H.A., Wijkstra, J., van Delen, A., Das, H.A., Slanina, J. Dekkers, J.J. and Wals, G.D. (1982b) Leaching of trace elements from coal solid waste. *Rep. ECN*-120, Verlag Glückauf GmbH, Essen, 90 pp
Van Krevelen, D.W. (1961) *Coal Topology–Chemistry–Physics–Constitution*. Amsterdam: Elsevier, 514 pp
Van Krevelen, D.W. (1963) Geochemistry of coal. In *Organic Geochemistry* (ed. I.A. Breger), pp. 183–247. Oxford: Pergamon
Van Loon, J.C. (1986) Environmental analysis related to the mining and processing of geological materials. In *Butler* (1986), pp. 223–241
Van Voris, P., Page, T.L., Rickard, W.H., Droppo, J.G. and Vaughan, B.E. (1985) Environmental implications of trace element releases from Canadian coal-fired generating stations. *Rep. Can. Elec. Assoc.*, Vol. 1, Phase II, 192 pp
Varekamp, J.C. and Buseck, P.R. (1986) Global mercury flux from volcanic and geothermal sources. *Appl. Geochem.*, **1**, 65–73
Varekamp, J.C., Thomas, E., Germani, M. and Buseck, P.R. (1986) Particle geochemistry of volcanic plumes of Etna and Mount St. Helens. *J. Geophys. Res.*, **91**, No. B12, 12233–12248
Varga, E., Bella, M. and Benocs, S.K. (1972) Comparative survey of the trends of trace elements concentration in Hungarian coal fields. *Publ. Hung. Min. Res. Inst.*, No. 15, 221–236
Vasileva, I.V. and Vekhov, V.A. (1972) Determination of beryllium in coals and ashes. *Met. Koksokhim.*, No. 32, 100–101; *CA*, **79**, 94536 (1973)
Vasilevskaya, A.E. and Shcherbakov, V.P. (1963) Forms of mercury compounds in Donets Basin coals. *Dopov. Akad. Nauk Ukr. RSR* (11), 1494–1496; *CA*, **60**, 10433 (1964)
Vasilevskaya, A.E., Shcherbakov, V.P. and Karakozova, E.V. (1964) Determination of mercury in coals. *Zh. Anal. Khim.*, **19**, 1200–1203; *CA*, **62**, 2636 (1965)
Vasilevskaya, A.E., Shcherbakov, V.P. and Klimenchuk, V.I. (1962) Determination of mercury in coal by dithizone. *Zavod. Lab.*, **28**, 415; *CA*, **58**, 1269 (1963)
Velikovsky, I. (1956) *Earth in Upheaval*. London: Gollancz, 263 pp.; Francis (1961)
Vine, J.D. (1956) Uranium-bearing coal in the United States. *US Geol. Surv. Prof. Pap.*, No. 300, 405–41
Vistelius, A.B. (1947) New verification of the observations of Goldschmidt on the role of germanium in mineral coal. *Dokl. Akad. Nauk SSSR*, **58**, 1455–1457; *CA*, **46**, 7304 (1952)
Volborth, A. (ed.) (1987) *Coal Science and Chemistry*. Amsterdam: Elsevier, 478 pp
Volkov, K.Y. (1958) The regularities of distribution of germanium in coals of the Moscow region deposits. *Mater. Geol. Polezn. Iskop. Tsentr. Rainov. Evr. Chasti SSSR, Moscow, Sb.*, No. 1, 228–242; *CA*, **54**, 24180 (1960)
Volodarskii, I.K., Zharov, Y.N. and Ratynskii, V.M. (1976) Nature of gallium distribution in coals and their combustion products. *Khim. Tverd. Topl.* (6), 18–21; *CA*, **87**, 87576 (1977)
Von Borstel, D. and Halbach, P. (1982). Binding and oxidation states of uranium in recent

phytogenic sediments, peats and lignites of selected areas as influenced by physicochemical parameters. *Ber. Bunsenges. Phys. Chem.*, **86**, 1066–1069; *CA*, **98**, 57294 (1983)

Von Fellenberg, T. (1923) Investigations of the occurrence of iodine in nature. *Biochem. Z.*, **139**, 371–451

Von Lehmden, D.J., Jungers, R.H. and Lee, R.E. (1974) Determination of trace elements in coal, fly ash, fuel oil, and gasoline–a preliminary comparison of selected analytical techniques. *Anal. Chem.*, **46**, 239–245

Vorobev, A.L. (1940) Vanadium and nickel in coals of the Upper Silurian of the Alai and Turkestan mountains. *C.R. Acad. Sci. URSS*, **28**, 250–251; *CA*, **35**, 3060 (1941)

Vorres, K.S. (ed.) (1986) *Mineral Matter and Ash in Coal*. Washington: ACS, 537 pp

Vorres, K.S. (1987) The preparation and distribution of Argonne Premium coal samples. *Am. Chem. Soc., Div. Fuel Chem., Prepr.*, **32**(4), 221–225

Voskresenskaya, N.T. (1960) Calculation of uranium in coal. II Study of the reaction of humic acid with uranyl salts. *Izvest. Akad. Nauk Kirg. SSR Ser. Estestv. Tekh. Nauk*, **2**, 57–64; *CA*, **55**, 8814 (1961)

Voskresenskaya, N.T. (1968) Thallium in coal. *Geochem. Int.*, **5**, 158–168

Voskresenskaya, N.T., Timofeyeva, N.V. and Topkhana, M. (1962) Thallium in some minerals and ores of sedimentary origin. *Geochemistry (USSR)* (8), 851–857

Vyalov, V.I. and Stepanosov, A.R. (1986) An attempt to use X-ray spectral fluorescence analysis for investigating the coals of the Donets Basin. *Solid Fuel Chem.*, **20**(6), 10–13

Wache, R. (1931) Investigations of the occurrence of iodine in waters and coal and some experiments on the use of iodine as a fertilizer. *Mitt. Lab. Preuss. Geol. Landesanst.* (13), 43–52; *CA*, **26**, 4125 (1932)

Wagner, P. and Greiner, N.R. (1981) Second annual report on radioactive emissions from coal production and utilization. *Prog. Rep.*, No. LA-8825-PR, 14 pp

Wagner, P., Dreesen, D.R., Wewerka, E.M. and Greiner, N.R. (1980) Radioactive emissions from coal production and utilization. *Prog. Rep.*, No. LA-8618-PR, 28 pp

Wait, C.E. (1896) The occurrence of titanium. *J. Chem. Soc.*, **18**, 402

Wald, F. (1953) Iodine in the raw materials and in the main by-products of the iron industry. *Hutn. Listy*, **8**, 81–83; *CA*, **49**, 3754 (1955)

Wallace, A. and Berry, W.L. (1979) Trace elements in the environment–effects and potential toxicity of those associated with coal. In *Ecology and Coal Resource Development* (ed. M.K. Wali), pp. 95–114, New York: Pergamon

Walling, J.F., Evans, G., Hinners, T.A., Lambert, J.P., Long, S.J. and Wilshire, F.W. (1978) Assessment of arsenic losses during ashing: a comparison of two methods applied to atmospheric particulates. *J. Air Pollut. Control Assoc.*, **28**, 1134–1136

Walsh, A. (1955) The application of atomic absorption spectra to chemical analysis. *Spectrochim. Acta*, **7**, 108–117

Walsh, P.R., Duce, R.A. and Fasching, J.L. (1979) Considerations of the enrichment, sources, and flux of arsenic in the troposphere. *J. Geophys. Res.*, **84**(C4), 1719–1726

Wandless, A.M. (1958) British coal seams: a review of their properties with suggestions for research. *Proc. Conf. 'Science in the Use of Coal'*, Sheffield, A1–A12

Wandless, A.M. (1959) The occurrence of sulphur in British coals. *J. Inst. Fuel*, **32**, 258–266

Wang, E.H. (1978) Polarography of nickel in ammonia-based solution of vanadium-containing coal. *Fen Hsi Hua Hsueh*, **6**, 231; *CA*, **92**, 166132 (1980)

Warbrooke, P.R. and Doolan, K.J. (1986) The distribution of elements in some Sydney Basin coals. *Proc. Aust. Coal Sci. Conf.*, 2 Newcastle, Aust. Inst. Energy, pp. 404–409

Ward, C.R. (1974) Isolation of mineral matter from Australian bituminous coals using hydrogen peroxide. *Fuel*, **53**, 220–221

Ward, C.R. (1978) Mineral matter in Australian bituminous coals. *Australas. Inst. Min. Metall. Proc.*, No. 267, 7–25

Ward, C. (1980) Mode of occurrence of trace elements in some Australian coals. *Coal Geol.*, **2**, 77–98

Ward, C.R. (1985) *Coal Geology and Coal Technology*. Oxford: Blackwell, 352 pp

Ward, N.I., Kerr, S.A. and Otsuka, T. (1986) Multielement content of coal by neutron activation analysis techniques. *Proc. 7 Int. Conf. Modern Trends Activation Anal.*, pp. 1077–1086

Warf, J.C., Cline, W.D. and Tevebaugh, R.D. (1954) Pyrohydrolysis in the determination of fluoride and other halides. *Anal. Chem.*, **26**, 342–346

Waring, C.L. and Tucker, W.P. (1954) Effect of ashing temperature on the volatility of germanium in low-rank coal samples. *Anal. Chem.*, **26**, 1198–1199

Warne, S. St J. (1979) Differential thermal analysis of coal minerals. In Karr (1979), pp. 447–477

Watling, R.J. and Watling, H.R. (1976) Trace-element loss during ashing of South African coals. *Abstr. Int. Symp. Analytical Chemistry in the Exploration, Mining and Processing of Materials*, Johannesburg, Pergamon Press, Oxford, pp. 160–163

Watling, R.J., Watling, H.R. and Wardale, I.M. (1976) The determination of trace metals in coal by emission spectroscopy. *Abstr. Int. Symp. Analytical Chemistry in the Exploration, Mining and Processing of Materials*, Johannesburg, Pergamon Press, Oxford, pp. 156–159

Watt, J.D. (1968) The physical and chemical behaviour of the mineral matter in coal under the conditions met in combustion plant. Part I The occurrence, origin, identity, distribution and estimation of the mineral species in British coals. *Br. Coal Util. Res. Assoc. Lit. Surv.*, BCURA, Leatherhead, 102 pp

WC (1980) Coal mining methods in Australia. *World Coal*, **6**(9), 36–39

Weaver, C. (1967) The determination of zinc in coal. *Fuel*, **46**, 407–414

Weaver, J.N. (1978) Neutron activation analysis of trace elements in coal, fly ash, and fuel oils. In Karr (1978a), pp. 377–401

Webber, H.F.P. and Taylor, L. (1952) Fluorine (and arsenic) content of malt. *J. Inst. Brew. London*, **58**, 197–198

Weclewska, M. (1960) Polarographic determination of trace metals in coal ash. *Pr. Gl. Inst. Gorn. Komun. Ser. B* (242), 1–12; *CA*, **55**, 8813 (1961)

Weclewska, M. and Popanda, G. (1958) Polarographic determination of germanium in coal ash. *Chem. Anal. (Warsaw)*, **3**, 889–892; *CA*, **53**, 5636 (1959)

Wedepohl, K.H. (ed.) (1969) *Handbook of Geochemistry*. Berlin: Springer-Verlag, Vol. I, 442 pp

Wedge, W.K., Bhatia, D.M.S. and Rueff, A.W. (1976) Chemical analyses of selected Missouri coals and some statistical implications. *Mo. Dep. Nat. Resour. Div. Geol. Land Surv. Rep. Invest.*, No. 60, 33 pp

Weeks, M.E. (1945) *Discovery of the Elements*. Easton, PA: Journal of Chemical Education, 578 pp

Weller, M., Völkl, J. and Wert, C. (1983) Macromolecular structure of coal. *Proc. Int. Conf. Coal Sci.*, pp. 283–286

Wender, I., Heredy, L.A., Neuworth, M.B. and Dryden, I.G.C. (1981) Chemical reactions and the constitution of coal. In Elliott (1981), pp. 425–521

Wendt, R.H. and Fassel, V.A. (1965) Induction-coupled plasma spectrometric excitation source. *Anal. Chem.*, **37**, 920–922

Wershaw, R.L. and Mikita, M.A. (eds.) (1987) *NMR of Humic Substances and Coal: Techniques, Problems and Solutions*. Chelsea: Lewis Publishers, 236 pp

Wert, C.A. and Weller, M. (1982) Polymeric character of coal. *J. Appl. Phys.*, **53**, 6505–6512

Wert, C.A., Hsieh, Kuang-Chien, Tseng, Bae-Heng and Ge, Yun-Pei (1987) Applications of transmission electron microscopy to coal chemistry. *Fuel*, **66**, 914–920

West, T.S. (1986) Foreword. *J. Anal. At. Spectrom.*, **1**, 1

West, T.S. and Williams, X.K. (1969) Atomic absorption and fluorescence spectroscopy with a carbon filament reservoir. I Construction and operation of atom reservoir. *Anal. Chim. Acta*, **45**, 27–41

Westgate, L.M. and Anderson, T.F. (1984) Isotopic evidence for the origin of sulfur in the Herrin (No. 6) coal member of Illinois. *Int. J. Coal Geol.*, **4**, 1–20

Wheeler, B.D. and Jacobus, N.C. (1980) Elemental analysis of whole coal using energy dispersive X-ray fluorescence. *Pap. Am. Ceram. Soc. Mtg.*, 15 pp

Wheeler, B.D. (1983) Utilizing artificial standards in chemical analysis of coal. *J. Coal Qual.*, **2**(2), 30–33

Whelan, J.F., Cobb, J.C. and Rye, R.O. (1988) Stable isotope geochemistry of sphalerite and other mineral matter in coal beds of the Illinois and Forest City Basins. *Econ. Geol.*, **83**, 990–1007

White, C.M. and Lee, M.L. (1980) Identification and geochemical significance of some aromatic components of coal. *Geochim. Cosmochim. Acta*, **44**, 1825–1832

White, D.M., Edwards, L.O. and Du Bose, D.A. (1983) Trace elements in Texas lignite. *Rep. TENRAC/EDF-094*, 97 pp

Widawska-Kusmierska, J. (1975) Dependence of the gallium content in coal ashes and rocks accompanying coal seams on their chemical composition. *Bull. Acad. Pol. Sci., Ser. Sci. Terre*, **23**(2), 97–106; *CA*, **85**, 35711 (1976)

Widawska-Kusmierska, J. (1981) The occurrence of trace elements in Polish hard coals. *Przegl. Gorn.*, **37**, 455–459

Wilke-Dörfurt, E. and Römersperger, H. (1930) The iodine content of coal. *Z. Anorg. Alg. Chem.*, **186**, 159–164; *CA*, **25**, 188 (1931)

Wilkinson, J.R., Ebdon, L. and Jackson, K.W. (1982) Determination of volatile trace metals in coal by analytical atomic spectroscopy. *Anal. Proc.*, **19**, 305–307

Willis, J.B. (1975) Atomic absorption spectrometric analysis by direct introduction of powders into the flame. *Anal. Chem.*, **47**, 1752–1758

Willis, J.P. (1981) Elemental characterization of South African coal and fly ash. *Proc. Int. Conf. Coal Sci.*, Verlag Glückauf GmbH, Essen, pp. 745–750

Willis, J.P. (1983) Trace-element studies on South African coals and fly ash. *Spec. Publ. Geol. Soc. S. Afr.*, **7**, 129–135

Willis, J.P. (1986) Applications of X-ray fluorescence spectrometry and the electron microprobe in the exploration, mining and processing of materials. In Butler (1986), pp. 45–56

Willis, J.P. (1988) XRFS and PIXE: are they complementary or competitive techniques? A critical comparison. *Nucl. Instrum. Methods Phys. Res.*, **B35**, 378–387

Willis, J.P. and Hart, R.J. (1985) Trace element analysis of coals by INAA and XRFS. *J. Trace Microprobe Tech.*, **3**, 109–127

Willis, J.P. and Hart, R.J. (1986) Geochemical characterization of South African coals. *Abstr. Int. Earth Sci. Congr.* Johannesburg, pp. 703–706

Wilson, J.H. (1923) An occurrence of carnotite near Denver. *Eng. Min. J.*, **116**, 239–240

Wilson, M.A. (1987) *N.M.R. Techniques and Applications in Geochemistry and Soil Chemistry*. Oxford: Pergamon Press, 353 pp

Wilson, M.A. and Pugmire, R.J. (1984) New solid state NMR techniques in coal analysis. *TrAC Trends Anal. Chem.*, **3**(6), 144–147

Wilson, M.A., Verheyen, T.V., Vassallo, A.M., Hill, R.S. and Perry, G.J. (1987) Selective loss of carbohydrates from plant remains during coalification. *Org. Geochem.*, **11**, 265–271

Wilson, M.A., Pugmire, R.J., Karas, J., Alemany, L.B., Woolfendon, W.R., Grant, D.M. and Given, P.H. (1984) Carbon distribution in coals and coal macerals by cross polarization magic angle spinning carbon-13 nuclear magnetic resonance spectrometry. *Anal. Chem.*, **56**, 933–943

Wimberley, J.W. (1975) The determination of total mercury at the part per billion level in soils, ores, and organic materials. *Anal. Chim. Acta*, **76**, 337–343

Winter, R., Leibnitz, E. and Otto, F. (1963) Occurrence of boron as a trace element in brown coals, tar distillates and electrode coke. *Freiberg. Forschungsh. A*, **299**, 55–57; *CA*, **60**, 7837 (1954)

Wiser, W.H. (1978) Some new evidence pertaining to the chemistry and mechanisms of coal liquefaction. *Am. Chem. Soc., Div. Fuel Chem., Prepr.*, **23**(4), 116

Wolfrum, E.A. (1984) Correlations between petrographic properties, chemical structure, and technological behaviour of Rhenish brown coal. In Schobert (1984), pp. 15–37

Wong, J., Maylotte, D.H., Lytle, F.W., Greegor, R.B. and St Peters, R.L. (1983) EXAFS and XANES investigations of trace vanadium and titanium in coal. *Springer Ser. Chem. Phys.* (27), 206–209

Wong, J., Maylotte, D.H., St Peters, R.L., Lytle, F.W. and Greegor, R.B. (1982) High-resolution X-ray spectroscopy of vanadium in coal. *Process Mineral., Proc. Symp.*, 2nd, pp. 335–353; *CA*, **98**, 201021 (1983) Metall. Soc. AIME, Warrendale, PA

Woo, I.H., Nishiyama, H., Hashimoto, Y. and Lee, Y.K. (1985) Determination of selenium in coal using graphite furnace atomic absorption spectrometry after chemical separation. *Bunseki Kagaku*, **34**, 595–599; *CA*, **104**, 53296 (1986)

Woo, I.H., Watanabe, K., Hashimoto, Y. and Lee, Y.K. (1987) Determination of selenium and tellurium in coal by graphite furnace atomic absorption spectrometry after coprecipitation with arsenic. *Anal. Sci.*, **3**, 49–52

Wood, G.H., Kehn, T.M., Carter, M.D. and Culbertson, W.C. (1983) Coal resource classification system of the U.S. Geological Survey. *US Geol. Surv. Circ.*, No. 891, 65 pp

Xu, D.-R., Qian, Z.-G., Qiu, D.-R. and Li, Q.-Y. (1980) Extraction spectrophotometric determination of gallium in anthracite. *Chem. chiang Ta Hsueh Hsueh Pao* (1), 38–47; *CA*, **94**, 106094 (1981)

Yamashige, T., Yamamoto, M., Shigetomi, Y. and Yamamoto, Y. (1984) Comparison of acid digestion procedures with and without hydrofluoric acid for the determination of heavy metals in coal fly ash by atomic absorption spectrometry. *Bunseki Kagaku*, **33**, 221–223; *CA*, **101**, 9762 (1984)

Yang, G., Wang, S., Zhou, R. and Sun, S. (1983) Endemic selenium intoxication of humans in China. *Am. J. Clin. Nutr.*, **37**, 872–881

Yang, G., Zhou, R., Sun, S., Wang, S. and Li, S. (1982) Endemic selenium intoxication of man in China and the selenium levels of human body and environment. *Yingyang Xuebao*, **4**(2), 81–89

Yeakel, J.D. and Finkelman, R.B. (1984) Petrography and chemistry of two North Dakota lignites. *Geol. Soc. Amer. Abstr. Programs*, **16**(6), No. 33043

Yohe, G.R. (1958) Oxidation of coal. *Ill. State Geol. Surv., Rep. Invest.*, No. 207, 51 pp

Youh, C.C. (1978) A study of the mineral matter and trace elements in Miocene coals from the Keelung-Taipei region, Taiwan. *K'uang Yeh*, **22**, 107–111; *CA*, **90**, 124251 (1979)

Yudovich, Y.E. (1978) *Geochemistry of Coals (Inorganic Components)*. Leningrad: Nauka, 262 pp

Yudovich, Y.E., Ketris, M.P. and Merts, A.V. (1985) *Impurity Elements in Coals*. Leningrad: Nauka, 239 pp

Yudovich, Y.E., Korycheva, A.A., Obruchnikov, A.S. and Stepanov, Y.V. (1972) Mean trace-element contents in coals. *Geochem. Int.*, **9**, 712–720

Yurovskii, A.Z. (1960) *Sulfur in Coals*. Moscow: Izdatel. Akad. Nauk SSSR, 295 pp

Zahradnik, L., Tyroler, J. and Vondrakova, Z. (1960) The geochemistry of germanium in the Plzen coal basin. *Vest. Ustred. Ustavu Geol.*, **25**, 459–468; *CA*, **55**, 12184 (1961)

Zaidel, A.N., Prokofev, V.K., Raiskii, S.M., Slavnyi, V.A. and Shreider, E.Y. (1970) *Tables of Spectral Lines*. New York: IFI/Plenum, 782 pp

Zarcinas, B.A. and Cartwright, B. (1987) Acid dissolution of soils and rocks for the determination of boron by inductively coupled plasma atomic emission spectrometry. *Analyst (London)*, **112**, 1107–1111

Zhou, L., Chao, T.T. and Meier, A.L. (1984) Determination of indium in geological materials by electrothermal-atomization atomic absorption spectrometry with a tungsten-impregnated graphite furnace. *Anal. Chim. Acta*, **161**, 369–373

Zhou, Y. and Ren, Y. (1981) Gallium distribution in coal of Late Permian coal fields, southwestern China, and its geochemical characteristics in the oxidized zone of coal seams. *Int. J. Coal Geol.*, **1**, 235–260

Zief, M. and Mitchell, J.W. (1976) *Contamination Control in Trace Element Analysis*. New York: Wiley, 262 pp

Zieve, R. and Peterson, P.J. (1984) Volatilization of selenium from plants and soils. *Sci. Total Environ.*, **32**, 197–202

Zilbermintz, V.A. (1935) Occurrence of vanadium in fossil coals. *C.R. Acad. Sci. URSS*, **3**, 117–120; *CA*, **30**, 1335 (1936)

Zilbermintz, V.A. and Rusanov, A.K. (1936) On the distribution of beryllium in coal. *Dokl. Akad. Nauk SSSR Ser. A* (1), 27–31

Zimmer, K. and Haypay-Galocsy, E. (1970) Rapid spectrographic analysis of certain coal ash components. *Magy. Kem. Foly.*, **76**, 83–86; *CA*, **72**, 113549 (1970)

Zimmermeyer, G. (1978) Radiation release from coal-fired and nuclear power stations. *Glueckauf*, **114**, 295–296

Zingaro, R.A. and Cooper, W.C. (eds.) (1974) *Selenium*. New York: Van Nostrand Reinhold, 835 pp

Zitko, V. (1975) Toxicity and pollution potential of thallium. *Sci. Total Environ.*, **4**, 185–192

Zodrow, E.L. (1986) Coal-stratigraphic geochemistry: trends in coal samples from Sydney Coalfield, Upper Carboniferous, Nova Scotia. *CIM Bull.*, **79** (893), 83–85

Zodrow, E.L. and Zentilli, M. (1979) Uranium content of rocks, coal and associated minerals from the Sydney coalfield, Cape Breton Island, Nova Scotia (Canada). *Geol. Surv. Can., Pap.*, No. 79–1C, 31–36

Zubovic, P. (1960) Minor element content of coal from Illinois Beds 5 and 6 and their correlatives in Indiana and Western Kentucky. *US Geol. Surv. Open File Rep.*, No. 537, 79 pp

Zubovic, P. (1966a) Minor element distribution in coal samples of the Interior coal province. *Adv. Chem. Ser.*, No. 55, 232–247

Zubovic, P. (1966b) Physicochemical properties of certain minor elements as controlling factors in their distribution in coal. *Adv. Chem. Ser.*, No. 55, 221–231

Zubovic, P. (1976) Geochemistry of trace elements in coal. *Proc. Symp. Envl. Aspects Fuel*

Conversion. US Environ. Protection Agency, Washington, DC, II EPA-600/2-76-149, pp. 47–63

Zubovic, P., Hatch, J.R. and Medlin, J.H. (1979) Assessment of the chemical composition of coal resources. *UN Symp. World Coal Prospects Rep.*, No. TCD/NRET/AC.12/EP/15, 24 pp

Zubovic, P., Stadnichenko, T. and Sheffey, N.B. (1960a) Relation of the minor element content of coal to possible source rocks. *US Geol. Surv. Prof. Pap.*, No. 400-B, B82–B84

Zubovic, P., Stadnichenko, T. and Sheffey, N.B. (1960b) Comparative abundance of the minor elements in coals from different parts of the United States. *US Geol. Surv. Prof. Pap.*, No. 400-B, B87–B88

Zubovic, P., Stadnichenko, T. and Sheffey, N.B. (1960c) The association of some minor elements with organic and inorganic phases of coal. *US Geol. Surv. Prof. Pap.*, No. 400-B, B84–B87

Zubovic, P., Stadnichenko, T. and Sheffey, N.B. (1961a) Geochemistry of minor elements in coals of the Northern Great Plains coal province. *US Geol. Surv. Bull.*, No. 1117-A, 58 pp

Zubovic, P., Stadnichenko, T. and Sheffey, N.B. (1961b) Chemical basis of minor-element associations in coal and other carbonaceous sediments. *US Geol. Surv. Prof. Pap.*, No. 424-D, D345–D348

Zubovic, P., Stadnichenko, T. and Sheffey, N.B. (1964) Distribution of minor elements in coal beds of the Eastern Interior region. *US Geol. Surv. Bull.*, No. 1117-B, 41 pp

Zubovic, P., Stadnichenko, T. and Sheffey, N.B. (1966) Distribution of minor elements in coals of the Appalachian region. *US Geol. Surv. Bull.*, No. 1117-C, 3 pp

Zubovic, P., Oman, C.L., Bragg, L.J., Coleman, S.L., Rega, N.H., Lemaster, M.E., Rose, H.J. and Golightly, D.W. (1980) Chemical analysis of 659 coal samples from the eastern United States. *US Geol. Surv. Open-file Rep.*, No. 80-2003, 513 pp

Zubovic, P., Oman, C.L., Coleman, S.L., Bragg, L.J., Kerr, P.T., Kozey, K.M., Simon, F.O., Rowe, J.J., Medlin, J.H. and Walker, F.E. (1979) Chemical analysis of 617 coal samples from the Eastern United States. *US Geol. Surv. Open-file Rep.*, No. 79-665, 460 pp